MULTI-SCALE (TIME AND MASS) DYNAMICS OF SPACE OBJECTS

IAU SYMPOSIUM 364

COVER ILLUSTRATION:

Multi-scale dynamics of space objects

IAU SYMPOSIUM PROCEEDINGS SERIES

INTERNATIONAL ASTRONOMICAL UNION

UNION ASTRONOMIQUE INTERNATIONALE

MULTI-SCALE (TIME AND MASS) DYNAMICS OF SPACE OBJECTS

PROCEEDINGS OF THE 364th SYMPOSIUM OF THE INTERNATIONAL ASTRONOMICAL UNION HYBRID MEETING, IASI, ROMANIA 18–22 OCTOBER, 2021

Edited by

ALESSANDRA CELLETTI (co-chair)
University of Rome Tor Vergata, Italy

CĂTĂLIN GALEŞ (co-chair)
University Al. I. Cuza Iaşi, Romania

CRISTIAN BEAUGÉ
Observatory of Cordoba, Argentina

and

ANNE LEMAÎTRE
University of Namur, Belgium

Shaftesbury Road, Cambridge CB2 8EA, United Kingdom

One Liberty Plaza, 20th Floor, New York, NY 10006, USA

477 Williamstown Road, Port Melbourne, VIC 3207, Australia

314–321, 3rd Floor, Plot 3, Splendor Forum, Jasola District Centre, New Delhi – 110025, India

103 Penang Road, #05–06/07, Visioncrest Commercial, Singapore 238467

Cambridge University Press is part of Cambridge University Press & Assessment, a department of the University of Cambridge.

We share the University's mission to contribute to society through the pursuit of education, learning and research at the highest international levels of excellence.

www.cambridge.org
Information on this title: www.cambridge.org/9781108490764

First published 2022

A catalogue record for this publication is available from the British Library

ISBN 978-1-108-49076-4 Hardback

Table of Contents

Preface

With the advent of powerful telescopes, instruments and computation facilities, as well as the results from space missions ventured towards the edge of the Solar system, we are witnessing a new era of extraordinary discoveries, that is pushing the frontier of science toward new horizons. Different or refined theories, methods and techniques are needed to deal with the enormous amount of highly accurate observational data on the celestial bodies. The emergence of new open problems, such as the formation, habitability and long-term evolution of planetary systems, the complex dynamical behavior of minor bodies in the Solar system, the increased traffic in Earth's orbit, the exploration and exploitation of space objects, stimulates the birth of new lines of investigation, the search for novel scientific methods and techniques, as well as the development of technologies.

The range of phenomena that manifest at all different time and length scales and the wide range of sizes of space objects, from minor bodies in the Solar system to exoplanets, from dust particles to Jupiter-size bodies, require the development of dynamics modelling and analysis tools that can handle these different scales. The understanding of the dynamics of space objects of various sizes, both natural and artificial, is a key to the advancement of various branches of science, such as celestial mechanics, astrodynamics, planetary sciences, applied mathematics and dynamical systems, with considerable benefits to society and economy.

These topics motivated the organization of the **IAU Symposium 364, Multi-scale (time and mass) dynamics of space objects**, held online from Iasi (Romania) during the period October 18–22, 2021. Although the pandemic situation did not allow to gather together in Iasi, the Symposium represented a unique opportunity to share ideas and projects. This book is a collection of contributions given by distinguished scientists at **the IAU Symposium 364**. The methods in dynamics modeling of space objects have already reached a state of maturity, and their implementation provided a large number of results of particular importance both in theory and in applications. The contributions in this volume deal with a variety of important topics covering the recent advances in the multi-scale dynamics of natural and artificial space objects from various perspectives, among which:

a) dynamics modelling of space objects at different time and length scales (multi-scale): dust particles, asteroids and comets, planets and exoplanets, satellites and space debris;
b) theories and tools to analyze the long-term evolution of space objects: perturbation methods, numerical, semi-analytical and analytical techniques, computer-algebraic techniques, planetary ephemerides, special manipulators and computational environments, dynamical systems methods;
c) multi-scale stability analysis of celestial bodies: resonances, mechanisms of onset of chaos, chaos indicators, equilibrium points, invariant manifolds, local and global analysis;

The Symposium was attended by an overall number of 199 participants from different institutions all over the world. The Symposium was made possible thanks to the support of the International Astronomical Union, with the endorsement of Division A Fundamental Astronomy, A4-Inter-Division A-F Commission Celestial Mechanics and Dynamical Astronomy. The Symposium was organized thanks to the collaboration of the University of Rome Tor Vergata (Italy), the University Alexandru Ioan Cuza of Iaşi (Romania) and the Romanian National Committee for Astronomy.

We take the opportunity to thank all members of the Scientific Organizing Committee (SOC) of the Symposium and all members of the Local Organizing Committee. We acknowledge the Department of Mathematics of the University of Al. I. Cuza, Iasi, Romania, for hosting the Symposium and, in particular, we warmly thank the dean of the Faculty, Prof. Răzvan Liţcanu, and the vice-dean, Prof. Marius Apetrii.

Alessandra Celletti and Cătălin Galeş (co-chairs)
Cristian Beaugé and Anne Lemaitre (co-editors)

Editors

Alessandra Celletti (co-chair)
University of Rome Tor Vergata, Italy

Cătălin Galeş (co-chair)
University Al. I. Cuza Iaşi, Romania

Cristian Beaugé
Observatory of Cordoba, Argentina

Anne Lemaître
University of Namur, Belgium

Scientific Organising Committee

Local Organising Committee

List of Participants

1.	AGGARWAL Rajiv	Deshbandhu College, University of Delhi, India
2.	ALESSI Elisa Maria	Consiglio Nazionale delle Ricerche, Italy
3.	ALVES Raphael	University of Sao Paulo, Brazil
4.	ANGHEL Simon	Astronomical Institute of the Romanian Academy / Faculty of Physics, University of Bucharest / IMCCE, Observatoire de Paris, Romania
5.	APETRII Marius	UAIC, Romania
6.	ARUSOAIE Andreea	Faculty of Computer Science, Alexandru Ioan Cuza University of Iasi, Romania
7.	AZANFIREI Gabriela- Ana	Faculty of Mathematics, Al. I. Cuza University of Iasi, Romania
8.	BALYAEV Ivan	Saint Petersburg State University, Russia
9.	BARBOSA Gerson	UNESP, Brazil
10.	BAU′ Giulio	University of Pisa, Italy
11	BEAUGÉ Cristian	Instituto de Astronomía Teórica y Experimental
12.	BERNARDI Fabrizio	SpaceDyS, Italy
13.	BERTOLUCCI Alessia	SpaceDyS, Italy
14.	BIRLAN Mirel	Astronomical Institute of the Romanian Academy & IMCCE, Paris Observatory, Romania
15.	BOACA Ioana-lucia	Astronomical Institute of the Romanian Academy, Romania
16.	BOLDEA Afrodita Liliana	National Institut for Physics and Nuclear Engineering, Bucharest, University of Craiova, Romania
17.	BORDERES MOTTA Gabriel	Universidad Carlos III de Madrid, Spain
18.	BOUÉ Gwenaël	IMCCE, France
19.	BRAGA CAMARGO Barbara Celi	UNESP, Brazil
20.	CĂLIMAN Alexandru	Alexandru Ioan Cuza University of Iasi, Romania
21.	CALLEGARI JR. Nelson	São Paulo State University (Unesp), Institute of Geosciences and Exact Sciences (IGCE), Brazil
22.	CARDOSO DOS SANTOS Josué	ITA - Aeronautics Institute of Technology (Brazil) and Technion - Israel Institute of Technology (Israel), Brazil
23.	CARLOS EDUARDO Eligio	Department of Physics, UNESP Rio Claro., Brazil
24.	CARRUBA Valerio	UNESP, Brazil
25.	CASTRO GUIMARÃES Millena	UNESP, Brazil
26.	CAVALLARI Irene	Universita' di Pisa, Italy
27.	CECCATTO Demétrio Tadeu	Universidade Estadual Paulista, Brazil
28.	CELLETTI Alessandra	University of Rome Tor Vergata, Italy
29.	CHARALAMBOUS Carolina	UNamur, Belgium
30.	CHAUDHARY Harindri	Deshbandhu College, University of Delhi, India
31.	CHAUHAN Shipra	Department of Mathematics, University of Delhi, India
32.	CHUVASHOV Ivan	Institute of Astronomy, Russian Academy of Sciences, The Russian Federation
33.	CINELLI Marco	Tor Vergata - University of Rome, Italy
34.	CORREIA Alexandre	University of Coimbra, Portugal
35.	COUTURIER Jérémy	IMCCE, Observatoire de Paris, France
36.	COYETTE Alexis	University of Namur, Belgium
37.	DA SILVA SOARES Paulo Victor	Ana Maria da Silva, Brazil

38.	DANESI Veronica	University of Rome Tor Vergata, Italy
39.	DAQUIN Jerome	University of Namur, Belgium
40.	DE BLASI Irene	University of Turin, Italy
41.	DELL'ELCE Lamberto	Inria, France
42.	DERMOTT Stanley	University of Florida, USA
43.	DI CINTIO Pierfrancesco	Enrico Fermi Researche Centre (CREF) and INFN, Italy
44.	DI RUZZA Sara	Università di Padova, Italy
45.	DOLGAKOV Ivan	Institute of Applied Astronomy of the Russian Academy of Sciences, Russia
46.	DUBEIBE Fredy	Universidad de los Llanos, Colombia
47.	EFIMOV Sergey	Moscow Institute of Physics and Technology, Russia
48.	EFTHYMIOPOULOS Christos	Dipartimento di Matematica, Universita degli Studi di Padova, Italy
49.	EMEL'YANENKO Vacheslav	Institute of Astronomy, Moscow, Russia
50.	ESMER Ekrem Murat	Ankara University, Turkey
51.	FENUCCI Marco	University of Belgrade, Serbia
52.	FERNINI Ilias	Sharjah Academy for Astronomy, Space Sciences, and Technology, UAE
53.	FERRAZ-MELLO Sylvio	Universidade de São Paulo, Brazil
54.	FERREIRA Lucas S.	Grupo de Dinâmica Orbital & Planetologia - São Paulo State University - UNESP - Brazil, Brasil, Brazil
55.	FIENGA Agnes	Observatoire de la Côte d'Azur, France
56.	FOLTRAN Bruno	UNESP, Brazil
57.	FUNATO Yoko	Univsersity of Tokyo, Graduate Division of International and Interdisciplinary Studies, Japan
58.	GALES Catalin	Al. I. Cuza University of Iasi, Romania
59.	GALLARDO Tabare	Facultad de Ciencias, Udelar, Uruguay
60.	GASLAC GALLARDO Daniel Martin	Sao Paulo State University UNESP, Brazil
61.	GEVORGYAN Yeva	University of São Paulo, Brazil
62.	GIMENO Joan	University of Rome Tor Vergata, Italy
63.	GIULIATTI WINTER Silvia	UNESP, Brazil
64.	GIUPPONE Cristian	Iate - Conicet, Argentina
65.	GKOLIAS Ioannis	Aristotle University of Thessaloniki, Greece
66.	GOMES Luiz	UNESP, Brazil
67.	GOMES Sérgio	University of Coimbra, Portugal
68.	GRASSI Clara	University of Pisa, Italy
69.	GRONCHI Giovanni Federico	University of Pisa, Italy
70.	GUERRA Francesca	SpaceDyS, Italy
71.	GULIYEV Rustam	Shamakhy Astrophysical Observatory, Azerbaijan
72.	GUZZO Massimiliano	University of Padova, Italy
73.	HAGHIGHIPOUR Nader	Institute for Astronomy, University of Hawaii, USA
74.	HAMILTON Douglas	University of Maryland, USA
75.	HERASIMENKA Alesia	Université Côte d'Azur, CNRS, Inria, LJAD, France
76.	HESTROFFER Daniel	Paris observatory, France
77.	HILTON James	U.S. Naval Observatory, USA
78.	HOANG Hoai Nam	IMCCE, observartory of Paris, France
79.	HOWELL Kathleen	Purdue University, USA
80.	IBRAIMOVA Aigerim	Fesenkov Astrophysical Institute, Kazakhstan

81.	IPATOV Sergei	Vernadsky Institute of Geochemistry and Analytical Chemistry of Russian Academy of Sciences, Moscow, Russia
82.	JAFARI NADOUSHAN Mahdi	K N Toosi University of Technology, Iran
83.	JHA Devanshu	MVJCE, India
84.	JUNQUEIRA Camila	UNESP, Brazil
85.	KARAMPOTSIOU Efsevia	University of Rome Tor Vergata, Aristotle University of Thessaloniki, Greece
86.	KARTHICK Chrisphin	Indian Institue Of Astrophysics (Iia), India
87.	KARYDIS Dionysios	Aristotle University of Thessaloniki, Greece
88.	KAUR Dr Bhavneet	University of Delhi, India
89.	KNEŽEVIĆ Zoran	Serbian Academy of Sciences and Arts, Serbia
90.	KOKUBO Eiichiro	National Astronomical Observatory of Japan, Japan
91.	KOTOULAS Thomas	Department of Physics, A.U.Th., Greece
92.	KUMAR Bhanu	Georgia Institute of Technology, USA
93.	KUMAR Dinesh	Department of Mathematics, University of Delhi, India
94.	KUMAR Sumit	University of Delhi, New Delhi-110007, India
95.	KUZNETSOV Eduard	Ural Federal University, Russian Federation
96.	LARI Giacomo	University of Pisa
97.	LASKAR Jacques	Paris Observatory, France
98.	LATTARI Victor	São Paulo State University - UNESP, Brazil
99.	LECLERE Nicolas	University of Liege, Belgium
100.	LEGNARO Edoardo	Academy of Athens, Italy
101.	LEMAITRE Anne	University of Namur, Belgium
102.	LEVKINA Polina	The Institute of Astronomy of the Russian Academy of Sciences, Russian Federation
103.	LHOTKA Christoph	Department of Astrophysics, University of Vienna, Austria
104.	LIBERT Anne-sophie	naXys, University of Namur, Belgium
105.	LIN Houyuan	Purple Mountain Observatory, China
106.	LITCANU Razvan	University Al. I. Cuza of Iasi, Romania
107.	LOCATELLI Ugo	Dipartimento di Matematica, Università degli Studi di Roma "Tor Vergata", Italy
108.	LOIBNEGGER Birgit	University of Vienna, Department of Astrophysics, Türkenschanzstraße 17, 1180 Vienna, Austria
109.	MACHADO Raí	São Paulo State University, Brazil
110.	MADEIRA Gustavo	São Paulo State University, Brazil
111.	MAKO Zoltan	Sapientia Hungarian University of Transylvania, Romania
112.	MALHOTRA Renu	The University of Arizona, USA
113.	MANCHENKO Liliia	V.N. Karazin Kharkiv National University, Department of Theoratical Physics named by academician I. M. Lifshits, Ukraine
114.	MARO′ Stefano	University of Pisa, Italy
115.	MARTIN Andreza	São Paulo State University, Brazil
116.	MASTROIANNI Rita	University of Padova, department of Mathematics, Italy
117.	MEENA Om Prakash	University of Delhi, India
118.	MILIĆ ŽITNIK Ivana	Astronomical Observatory Belgrade, Assistant Research Professor, Serbia
119.	MINGLIBAYEV Mukhtar	Fesenkov Astrophysical Institute, Almaty
120.	MISQUERO Mauricio	University of Rome Tor Vergata, Italy

121.	MITTAL Amit	University of Delhi, India
122.	MOGAVERO Federico	Institut de mécanique céleste et calcul des éphémérides, France
123.	MORAIS Helena	UNESP (São Paulo State University), Brazil
124.	MORBIDELLI Alessandro	CNRS/OCA, France
125.	MORINJ Bruno	Unesp/undergraduate, Brazil
126.	MOURA Tamires	UNESP, Brazil
127.	MOURÃO Daniela	UNESP - São Paulo State University, Brazil
128.	MOURSI Ahmed	National Research Institute of Astronomy and Geophysics, Egypt
129.	NDUNGE Mbonteh Roland	Cameroon Astronomy and Space Research Organization, Cameroon
130.	NICOLÁS Begoña	University of Barcelona, Spain
131.	NUNES Daniel	Grupo de Dinâmica Orbital & Planetologia - São Paulo State University - UNESP - Brazil, Brazil
132.	OLIVEIRA Patrick	National Observatory, Brazil
133.	PAGANELLI Flora	NRAO, USA
134.	PAOLI Roberto	UAIC, Romania
135.	PAVLOV Dmitry	St. Petersburg Electrotechnical University (LETI), Russian Federation
136.	PEÑARROYA Pelayo	Deimos Space S.L.U., Spain
137.	PERMINOV Alexander	Ural Federal University, Russia
138.	PETIT Antoine	Lund University, Sweden
139.	PICHIERRI Gabriele	MPIA, Germany
140.	PILAT-LOHINGER Elke	Department of Astrophysics, University of Vienna, Austria
141.	PINHEIRO Tiago	São Paulo State University, UNESP, Brazil
142.	PIRES Pryscilla	Rio de Janeiro State University, Brazil
143.	PLÁVALOVá Eva	Mathematical Institute Slovak Academy of Sciences, Slovakia
144.	POMET Jean-baptiste	INRIA Sophia Antipolis, France
145.	PONS Juan	UdelaR, Uruguay
146.	POPESCU Marcel	Astronomical Institute of the Romanian Academy, Romania
147.	POUSSE Alexandre	IMATI-CNR, Italy
148.	RIOFRIO Louise	International Lunar Observatory, USA
149.	ROBUTEL Philippe	IMCCE/Observatoire de Paris/PSL, France
150.	RODRÍGUEZ DEL RÍO Óscar	Universitat Politècnica de Catalunya & Università di Pisa, Italy
151.	RODRIGUEZ Adrian	Universidade Federal do Rio de Janeiro, Brazil
152.	ROIG Fernando	Observatorio Nacional, Brazil
153.	ROISIN Arnaud	University of Namur, naXys, Belgium
154.	ROSAEV Alexey	Research and Educational Center "Nonlinear Dynamics", Yaroslavl State University, Russia
155.	ROSENGREN Aaron Jay	University of California San Diego, USA
156.	ROSSI Alessandro	IFAC-CNR, Italy
157.	ROSSI Mattia	Department of Mathematics - Università degli Studi di Padova, Italy
158.	RUIZ DOS SANTOS Lucas	UNIFEI - Brazil, Brazil
159.	SACHAN Prachi	Department of Mathematics, University of Delhi, India
160.	SAILLENFEST Melaine	IMCCE, Paris Observatory, France
161.	SCHEERES Daniel	University of Colorado Boulder, USA

162.	SFAIR Rafael	UNESP, Brazil
163.	SHOAIB Muhammad	Smart and Scientific Solutions, Pakistan
164.	SIDORENKO Vladislav	Keldysh Institute of Applied Mathematics, Moscow, Russia, Russian Federation
165.	SINGH Rishabh	Narayana Etechno School, India
166.	SLYUSAREV Ivan	V.N. Karazin Kharkiv National University, Ukraine
167.	SOMMER Maximilian	Institute of Space Systems, University of Stuttgart, Germany
168.	STEVES Bonnie	Glasgow Caledonian University, Scotland, UK
169.	SURAJ Md. Sanam	University of Delhi, India
170.	SWEATMAN Winston	Massey University, New Zealand
171.	SZÜCS-CSILLIK Iharka-magdolna	Romanian Academy. Astronomical Institute of Cluj-Napoca., Romania
172.	TAN Pan	School of Astronomy and Space Science, Nanjing University, China
173.	TARNOPOLSKI Mariusz	Jagiellonian University, Poland
174.	TCHAPTCHET TCHAPTCHET William Christian	Astronomy Club of Cameroon / University of Dschang, Cameroon
175.	TEIXEIRA GUIMARÃES Gabriel	IAG-USP, Brazil
176.	TODOROVIĆ Nataša	Astronomical Observatory in Belgrade, Serbia
177.	TRUONG LE Gia Bao	International University - Vietnam National University, Vietnam
178.	TSIGANIS Kleomenis	Aristotle University if Thessaloniki, Greece
179.	VAILLANT Timothée	CFisUC, Universidade de Coimbra, Portugal
180.	VALENTE Ema	University of Coimbra, Portugal
181.	VALSECCHI Giovanni	IAPS-INAF, Italy
182.	VALVANO Giulia	Student from UNESP, Brazil
183.	VARTOLOMEI Tudor	University of Rome Tor Vergata, Italy
184.	VASILE Massimiliano	University of Strathclyde, UK
185.	VASILEVA Mariia	UrFU, Russia
186.	VASYLENKO Maksym	Main Astronomical Observatory of NAS of Ukraine, Ukraine
187.	VAVILOVA Iryna	Main Astronomical Observatory of the NAS of Ukraine, Ukraine
188.	VOLPI Mara	University of Rome Tor Vergata, Italy
189.	VOYATZIS George	Aristotle University of Thessaloniki, Greece
190.	WILLIET NYUYWIYNI Dinka	Astronomy club of Cameroon (program Officer), Cameroon
191.	WINTER Othon	São Paulo State University - UNESP, Brazil
192.	XI Xiaojin	National Time Service Center, Chinese Academy of Sciences, China
193.	YESILIRMAK Burcak	Akdeniz University - Space Science and Technologies Department, Turkey
194.	YOSHIDA Haruo	National Astronomical Observatory of Japan, Japan
195.	YOUSUF Saleem	Central University of Rajasthan, India
196.	YSEBOODT Marie	Royal Observatory of Belgium, Belgium
197.	ZHUMABEK Torebek	Al-Farabi Kazakh National University, Faculty of Mechanics and Mathematics, Kazakhstan
198.	ZIMMERMANN Max	University of Vienna, Departement of Astrophysics, Austria
199.	ZOPPETTI Federico	Observatorio Astronómico de Córdoba, Argentina

Multi-scale (time and mass) dynamics of space objects
Proceedings IAU Symposium No. 364, 2022
A. Celletti, C. Galeş, C. Beaugé, A. Lemaitre, eds.
doi:10.1017/S174392132100140X

Dynamical constraints on the evolution of the inner asteroid belt and the sources of meteorites

Stanley F. Dermott[1], Dan Li[2] and Apostolos A. Christou[3]

[1]Department of Astronomy, University of Florida, Gainesville, FL 32611, US
email: sdermott@ufl.edu

[2]NSF's National Optical-Infrared Astronomy Research Laboratory, Tucson, AZ 85719, US
email: dan.li@noirlab.edu

[3]Armagh Observatory and Planetarium, College Hill, Armagh, BT61 9DG
email: Apostolos.Christou@armagh.ac.uk

Abstract. We have shown that in the inner belt the loss of asteroids from the ν_6 secular resonance and the 3:1 Jovian mean motion resonance accounts for the observation that the mean size of the asteroids increases with increasing orbital inclination. We have used that observation to constrain the Yarkovsky loss timescale and to show that the family asteroids are embedded in a background population of old ghost families. We argue that all the asteroids in the inner belt originated from a small number of asteroids and that the initial mass of the belt was similar to that of the present belt. We also show that the observed size frequency distribution of the Vesta asteroid family was determined by the action of Yarkovsky forces, and that the age of this family is comparable to the age of the solar system.

Keywords. asteroids

---

1. Introduction

Small fragments of many asteroids exist in our meteorite collections and while these fragments provide invaluable information on the origin and evolution of the remnants of the primitive building blocks that formed the rocky planets, some important dynamical questions remain unanswered. Ideally, we would like to link specific meteorites or meteorite classes to known asteroids. In one case at least, given the strong links between 4 Vesta and the HED meteorites, that goal has been achieved (McSween *et al.* 2013). We also have small samples of material from the near-Earth asteroid (NEA) Itokawa and soon we expect to have samples from the NEAs Ryugu and Bennu. However, these small NEAs are rubble-pile asteroids that originate from the collisional disruption of much larger main-belt asteroids. One aim of this paper is to discuss some of the dynamical constraints on the likely number of precursor asteroids in the inner main belt (IMB) that are the root sources of a large fraction of the NEAs and meteorites.

We assume that the asteroids accreted in two separate reservoirs of carbonaceous (CC) and non-carbonaceous (NC) material, interior and exterior to their current locations, and were then scattered by planetary perturbations into the present belt (Walsh *et al.* 2011; Kruijer *et al.* 2017). We also assume that after all planetary migration and the scattering that resulted from that migration ceased, further evolution of the dynamically excited belt was driven by: (1) the collisional and (2) the rotational destruction of asteroids (Dohnanyi 1969; Jacobson *et al.* 2014); (3) chaotic orbital evolution (Wisdom 1985;

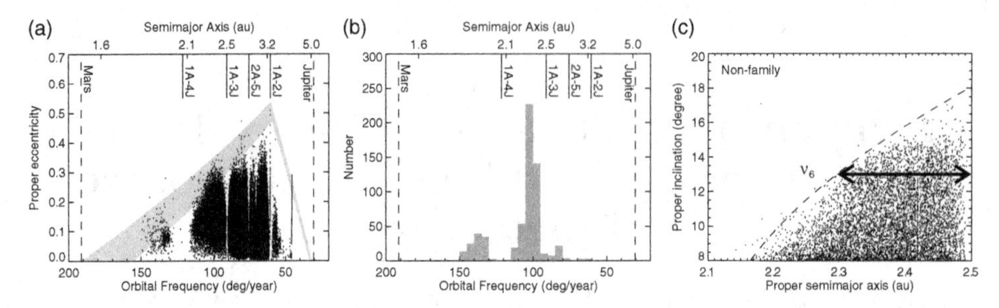

Figure 1. Panel (a): Scatter plot of the proper eccentricity e and the semimajor axis, a of the asteroids in the IMB with absolute magnitude $H < 15$. The shaded zone on the left is the Mars-crossing zone. Asteroids in that zone can, over time, cross the orbit of Mars. Panel (b): Histogram of the semimajor axes of the asteroids in the Mars-crossing zone. Panel (c): non-family asteroids in the IMB with $H < 16.5$ and high proper inclinations (Dermott *et al.* 2021).

Farinella *et al.* 1994; Morbidelli & Nesvorný 1999; Minton & Malhotra 2010); and (4) Yarkovsky-driven transport of small asteroids to the escape hatches located at orbital resonances (Migliorini *et al.* 1998; Farinella & Vokrouhlický 1999; Vokrouhlický & Farinella 2000). Insight into this dynamical evolution and estimates of the loss timescales are gained from an analysis of: (1) the observed variations with asteroid size of the mean orbital inclinations and eccentricities of the non-family asteroids; and (2) the size-frequency distributions of the small asteroids in the major families (Dermott *et al.* 2018; Dermott *et al.* 2021); (3) the cosmic-ray exposure ages of meteorites (Eugster *et al.* 2006); (4) the spin directions of near-Earth asteroids (Greenberg *et al.* 2020); and (5) the distribution of family asteroids in $a - 1/D$ space, where D is the asteroid diameter.

2. Asteroid size - orbital element correlations

The orbital eccentricities of main-belt asteroids are largely capped by the Mars-crossing zone (Fig. 1a) indicating that Mars has scattered some asteroids into the inner solar system. Most of the asteroids in the crossing-zone are in the inner main belt (IMB) (Fig. 1b), suggesting that the IMB is a major source of near-Earth asteroids (NEAs) and meteorites, a conclusion that is supported by the results of numerical investigations of the likely escape routes (Gladman *et al.* 1997; Granvik *et al.* 2017, 2018). Using the Hierarchical Clustering Method (HCM) developed by Zappala *et al.* (1990), Nesvorný (2015) has classified about half of the asteroids in the IMB with absolute magnitudes, $H < 16.5$ as family asteroids. But this fraction is an underestimate because some of the remaining asteroids are halo asteroids (Nesvorný *et al.* 2015), that is, they are also family asteroids, but because of the unavoidable limitations of the HCM it is not possible to classify these asteroids unambiguously. The remaining asteroids that are neither family nor halo asteroids are currently classified as non-family and an understanding of the evolution of the asteroid belt is not complete without an understanding of the nature and origin of these unclassified asteroids.

The family and halo asteroids are, by definition, tightly clustered in proper orbital element space. However, the family asteroids in that space are embedded in a background population of asteroids that could be members of old ghost families with dispersed orbital elements. To explore this background population, we need to find windows in orbital element space that are not obscured by the asteroids in the major families. Fortunately, one very large window exists in the IMB where all the asteroids in the major families and their halos have proper orbital inclinations, $I < 9$ *deg* (Dermott *et al.* 2018). In Fig. 1c, we see that the remaining non-family asteroids in the IMB with $I > 9$ *deg* are bound in $a - I$ space by the ν_6 secular resonance and the 3:1 Jovian mean motion resonance. These

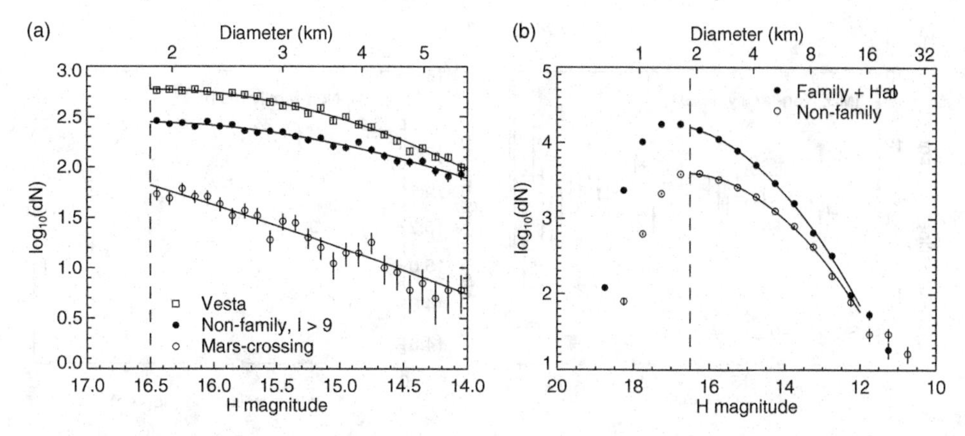

Figure 2. Panel (a): Quadratic polynomial fits to the SFDs of the asteroids in the Vesta family, for non-family asteroids with high inclinations ($I > 9$ deg), and for asteroids in the Mars-crossing zone. Panel (b): Comparison of the SFD of the non-family asteroids in the IMB with an estimate of the SFD of the family asteroids when combined with the asteroids in their halos.

resonances are two of the major escape hatches for asteroids in the IMB (Gladman *et al.* 1997; Granvik *et al.* 2017, 2018), but a third escape route is provided by a dense web of high-order Martian and Jovian resonances (Morbidelli & Nesvorný 1999; Milani *et al.* 2014) and we have argued that there is observational evidence that these high-order resonances also provide a significant loss mechanism (Dermott *et al.* 2021).

The size-frequency distribution (SFD) of the high inclination non-family asteroids shown in Fig. 2a shows a lack of small asteroids that is consistent with these asteroids being members of old ghost families that have lost small asteroids through collisional and rotational disruptions and the action of Yarkovsky forces. By assuming that the number density in $a - I$ space of the high-inclination, non-family asteroids shown in Fig. 1c applies to the IMB as a whole, we have shown that the fraction of asteroids in the IMB with $H < 16.5$ that are members of the major families or their halos is 76% and that the remaining 24% of the asteroids in the IMB are members of old ghost families (Dermott *et al.* 2021). If we further assume that the SFD of the high-inclination, non-family asteroids shown in Fig. 2a applies to all the non-family asteroids in the IMB and that the SFD of the halo asteroids (as a whole) is the same as that of the family members, then we can compare the SFD of the ghost family members with that of the family members and their halos. Accepting these simplifying assumptions, Fig. 2b shows that the smaller asteroids in the IMB with $H \sim 16$ are predominantly members of the major families. However, for asteroids with $H \lesssim 12$ and diameters, $D \gtrsim 16$ km this is not the case. This has several implications. Firstly, the probability of an asteroid that is currently classified as a family member being a family interloper increases with increasing asteroid size. Secondly, the fractions of the asteroids that are currently classified as S-type or C-type, etc., could change with asteroid size (see DeMeo & Carry 2014). Thirdly, our estimate of the number of asteroids that are the root sources of the NEAs and meteorites that originate from the IMB depends on the typical size of the asteroids whose disruption resulted in the injection of NEAs and meteorites into the inner solar system. Here, because the cosmic ray exposure ages of meteorites (Eugster *et al.* 2006) are much less than the ages of the asteroid families, we assume that the NEAs and meteorites do not originate directly from the initial disruptions of the root precursor asteroids, that is, from the events that formed the families, but from secondary disruptions of the family members. If the secondary asteroids were totally disrupted and typically had diameters

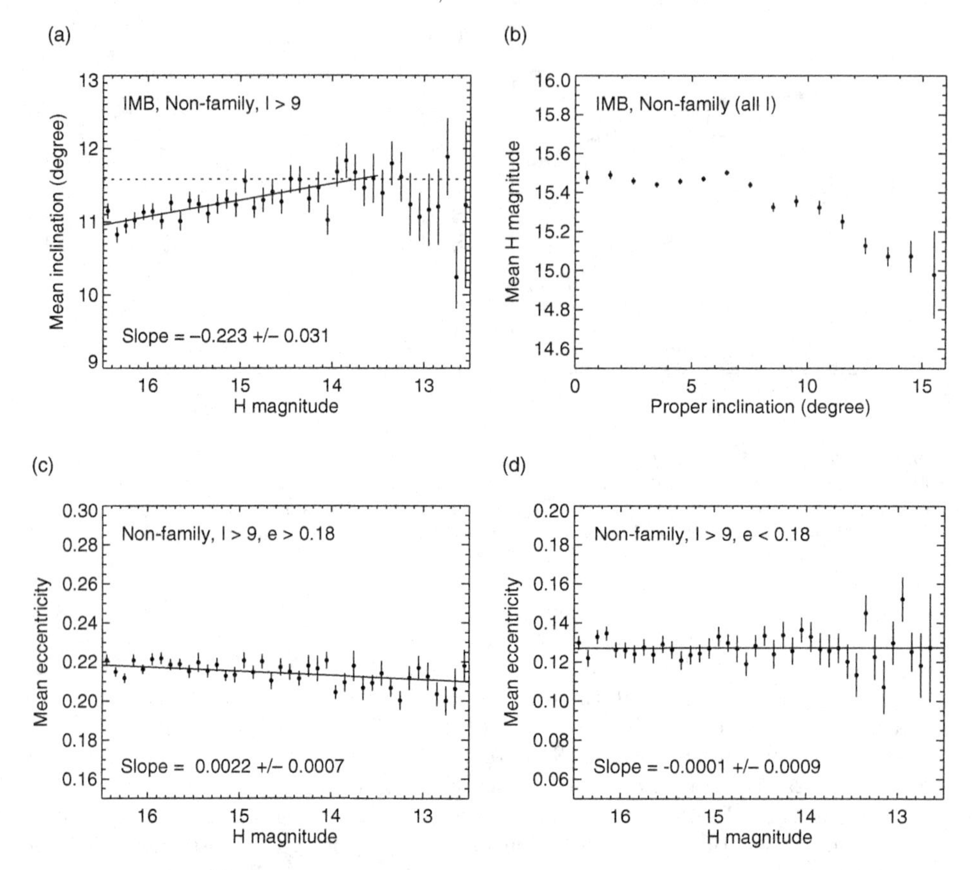

Figure 3. Panel (a): Variation with absolute magnitude, H of the mean proper inclination of the high inclination ($I > 9$ deg), non-family asteroids. The data is shown binned in H, but the slope has been determined from the individual points in the range $16.5 > H > 13.5$. Panel (b): Variation with proper inclination, I of the mean absolute magnitude, H of the all the non-family asteroids in the IMB. Panel (c): Variation with absolute magnitude, H of the mean proper eccentricity of the high inclination ($I > 9$ deg), non-family asteroids with $e > 0.18$. Panel (d) A similar plot to Panel (c) for those asteroids with $e < 0.18$ (Dermott *et al.* 2021).

~ 1 km, as suggested by Jenniskens (2020), then these asteroids were most likely members of the 5 or 6 major families currently dominating the IMB in that size range (Fig. 2b). This estimate of the number of precursors is small and could be reduced even further by considering the proximity of the major families to the most likely escape hatches. However, if the meteorite sources were larger, then we must also consider the asteroids in the ghost families as possible precursors, and this increases our estimated number of root precursors to $\lesssim 20$ (Dermott *et al.* 2021).

Asteroids are lost from the IMB through at least four mechanisms: collisional and rotational destruction, chaotic orbital evolution, and Yarkovsky-driven transport of small asteroids to the resonant escape hatches. The timescales of these loss mechanisms are uncertain and there is a need for observational constraints. Of particular interest are the observed correlations between the mean asteroid sizes and their proper orbital elements. In Fig. 3a, we see that the mean proper inclination of the high-inclination ($I > 9$ deg), non-family asteroids in the IMB increases with increasing asteroid size. In contrast, in

Fig. 3c we see that the mean proper eccentricity of the high-eccentricity asteroids (that also have high inclination), increases with decreasing asteroid size. The size-inclination correlation can be accounted for by the action of Yarkovsky forces driving small asteroids to the two bounding resonances (Dermott *et al.* 2021). The length of the escape route (Fig. 1c) decreases with increasing inclination and this leads, inevitably, to a correlation between the sizes and inclinations of the remaining asteroids. The distribution of the asteroids in $a - I$ space shown in Fig. 1c appears to be approximately uniform. If we assume that the initial distribution was also uniform, that is, not dependent on a or I, then the observed size-inclination correlation is determined by the Yarkovsky timescale alone and this timescale can be determined without knowing the initial SFD. We write

$$\frac{1}{a}\frac{da}{dt} = \pm\left(\frac{1}{T_{\mathrm{Y}}}\right)\left(\frac{1\,km}{D}\right)^{\alpha}, \tag{1}$$

where T_{Y} is the Yarkovsky timescale and the coefficient α is determined by the size dependence of the Yarkovsky force. The other loss mechanisms that do not depend on the orbital inclination include the net effect of catastrophic destruction and creation, and rotational disruption. These loss mechanisms are size dependent and should be modeled separately, but we are able to show that, in the size range that we model, Yarkovsky loss is the dominant loss mechanism and therefore it is expedient to reduce the number of variables in our models by writing

$$\frac{1}{N(D)}\frac{dN(D)}{dt} = -\left(\frac{1}{T_{\mathrm{L}}}\right)\left(\frac{1\,km}{D}\right)^{\beta}, \tag{2}$$

where T_{L} is the timescale of the combined inclination-independent loss mechanisms and $N(D)$ is the number of asteroids of diameter D.

Some of our model results, obtained using both loss mechanisms, are shown in Fig. 4. By adjusting the values of the five parameters b, α, T_{Y}, β, T_{L}, we can account for both the observed size-inclination correlation (Figs. 4c and 4f) and the observed SFD (Fig. 4b). These results show that for asteroids with absolute magnitudes in the range $13.5 < H < 16.5$, Yarkovsky transport of asteroids to the resonant escape hatches is the dominant loss mechanism (Fig. 4a). This conclusion is supported by the model results shown in Figs. 4d and 4e in which we use the Yarkovsky loss mechanism alone. For asteroids with $H < 16.5$ the inclination-independent asteroid loss mechanism has only a small effect on the fit for the SFD, and no effect on the observed asteroid size-inclination correlation. There is a large difference to the SFD fit for those asteroids with $H > 16.5$, but, at present, these very small asteroids are observationally incomplete. When the IMB completeness level has been extended from $H = 16.5$ to, say, $H = 18$, we will be able to constrain the loss timescales for the collisional and rotational disruption of the asteroids.

Using both loss mechanisms, we calculate that if these mechanisms have operated without change over the age of the solar system, then the Yarkovsky loss timescale, T_{Y} needed to account for the size-inclination correlation is 13.4 Gyr. This timescale is unacceptably longer than the result, $T_{\mathrm{Y}} \approx 4$ Gyr for asteroids with $a = 2.4$ au derived from the value that Greenberg *et al.* (2020) obtained from an analysis of the orbital evolution of 247 small NEAs. If the Yarkovsky timescale was as short as 4 Gyr, then many more asteroids would have been lost from the IMB and the size-inclination correlation would have been much stronger. We have argued that the most likely explanation for this large discrepancy is that the asteroids in the IMB are not as old the solar system but are collision products and members of old ghost families. However, this explanation for the observed size-inclination correlation needs to be explored further. Previously, we

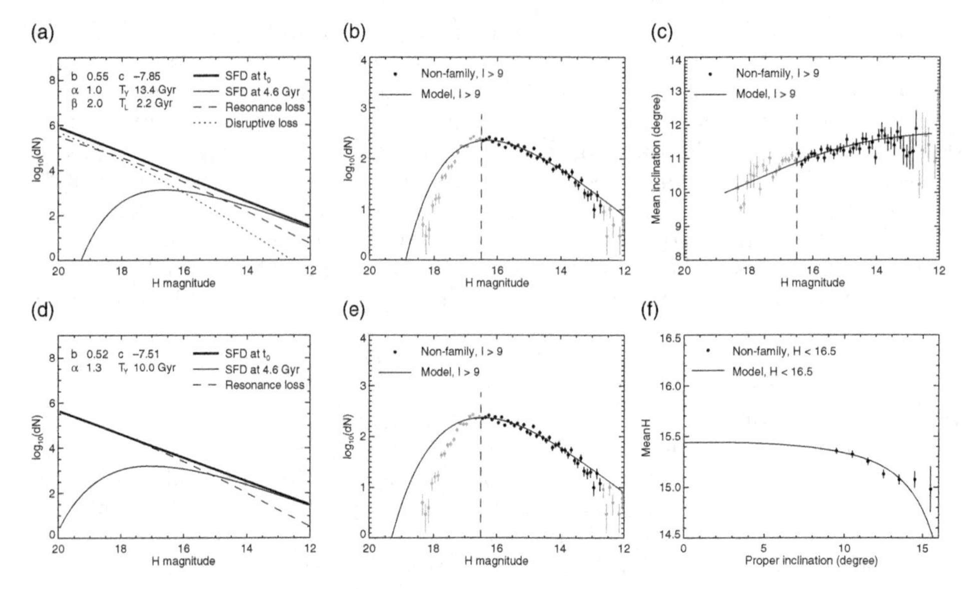

Figure 4. Models for the depletion of all the high-inclination, non-family asteroids in the IMB due to a Yarkovsky force that changes the semimajor axes on a timescale T_{Y}, and all other loss mechanisms that do not depend on the proper inclination, I and result in the loss of asteroids on a timescale T_{L}. α and β describe the dependence of these two timescales on the asteroid diameter.

argued that the asteroid size-orbital element correlations of the non-family asteroids in the IMB are evidence for the existence of ghost families (Dermott *et al.* 2018), because if we had, say, two families and the members in one family had a common inclination that was different from the common inclination of the members in the second family, and if these two families had different SFDs, then merging these two families could result in a ghost family with correlated inclinations and sizes. The difference between these two ideas is that the Yarkovsky loss model results, inevitably, in a size-inclination correlation of predictable sign and magnitude, whereas with the second idea a correlation of unpredictable sign and magnitude is only a possibility. If the number of merged families increases, then the possibility of a significant correlation due to the second mechanism alone decreases. However, what these two ideas have in common is that they both argue for the existence of ghost families, and we now need to examine other evidence for the existence of these families.

Figure 5 is a scatter plot of the high-inclination, IMB asteroids in $e - I$ space. Using the WISE albedos (Masiero *et al.* 2014), these asteroids have been separated into CC and NC groups. About 12% of the asteroids are members of small families (Figs. 5a and 5b). The remaining 88% (Figs. 5c and 5d) are non-family asteroids. Inspection of this figure shows that some of the CC non-family asteroids could be halo asteroids originating from the Klio and Chaldaea families, but some other apparent clumps could be large ghost families. The SFDs of the non-family CC and NC asteroids are shown in Fig. 6. These SFDs are significantly different which could indicate families of different ages. However, this is not a reliable conclusion because only about half of the IMB asteroids with $H < 16.5$ have WISE albedos and therefore the data set is incomplete and not bias free. A more reliable indication of the existence of ghost families is the observation that the CC and NC asteroids have markedly different mean eccentricities and inclinations (Dermott *et al.* 2021).

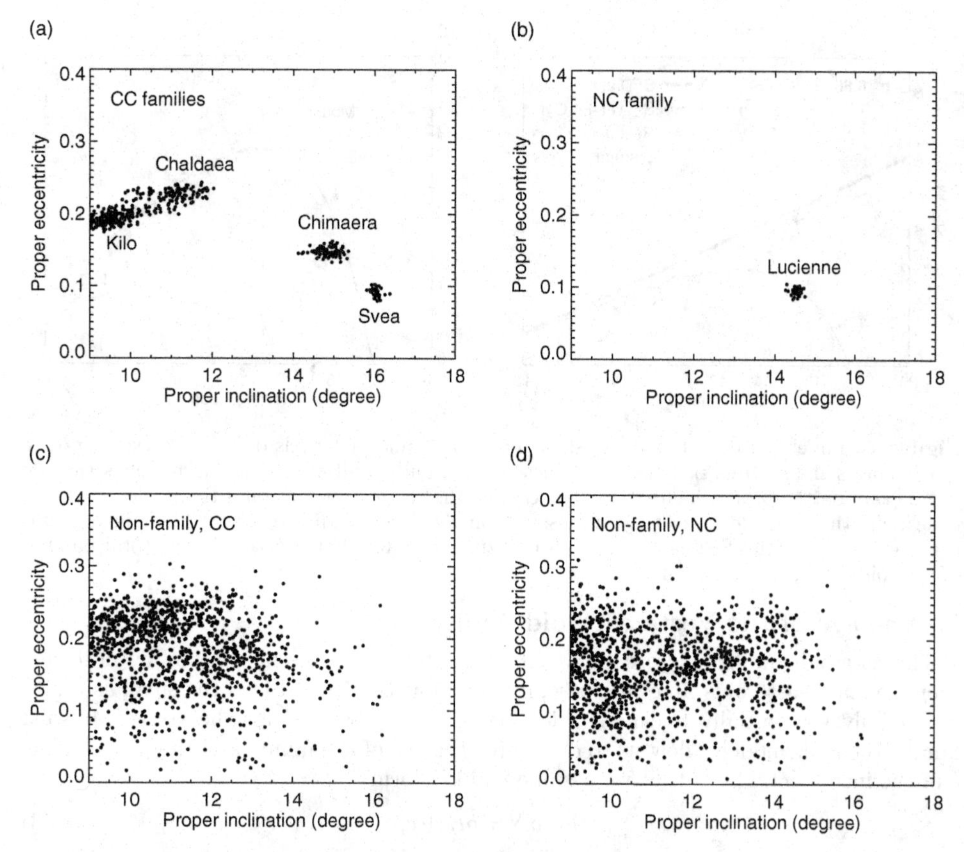

Figure 5. The asteroids in the IMB with high inclinations have been divided into two groups of CC and NC asteroids according to their WISE albedos, A (Masiero *et al.* 2014). CC have $A < 0.13$ and NC have $A > 0.13$. The panels show e and I scatter plots of all the asteroids with $H > 16.5$ divided into CC and NC family and non-family groups.

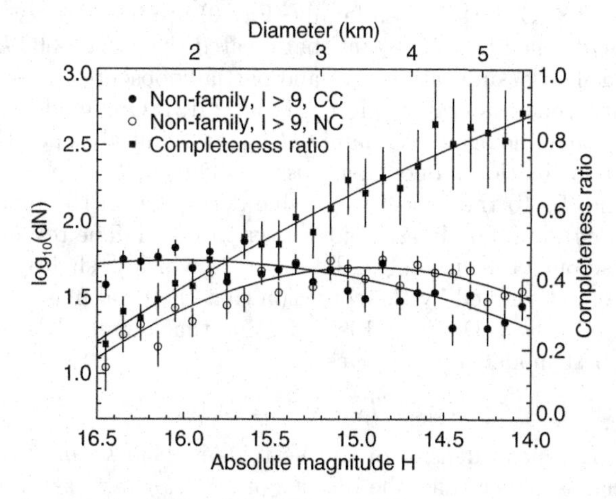

Figure 6. Comparison of the SFDs of the CC and NC non-family asteroids in the IMB with $I > 9$ *deg* and $H < 16.5$. The completeness ratio is the fraction of the asteroids with WISE albedos.

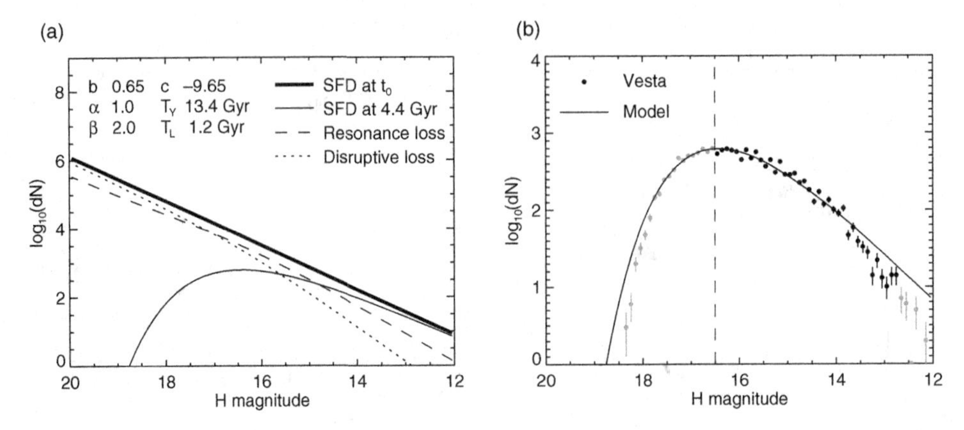

Figure 7. Panel (a): model for the loss of the Vesta family asteroids due to a Yarkovsky force that changes the semimajor axis on a timescale $T_{\rm Y}$, and to all other loss mechanisms that do not depend on the proper inclination that change the semimajor axis on a timescale $T_{\rm L}$. α and β describe the dependence of these timescales on the asteroid diameter, D. The constant b is the slope of the initial SFD and c is a normalizing constant that determines the total number of asteroids in the distribution.

3. The age of the Vesta asteroid family

The asteroid families were created either by collisional disruptions or by cratering events on large asteroids. In either case, if we assume that the SFD of the small asteroids in a family was initially linear on a log-log scale, as observed for the young Massalia family, then our models allow us to determine the age of the family and the initial slope, b of the incremental SFD that we assume had the form

$$\log dN = bH + c, \tag{3}$$

where dN is the number of asteroids in a box of width dH and c is a constant for asteroids with the same albedo. Because the asteroids in a family have, by definition, approximately the same inclination, and because the slope, b of the initial SFD is an unknown parameter, we only have one observational constraint on our model and that is the current SFD. Model results for the SFD of the Vesta family (Fig. 2) are particularly interesting. This family was probably formed by the impact that created the giant $\sim$500 km diameter and $\sim$20 km deep Rheasilvia basin that overlies and partially obscures the smaller ($\sim$400 km diameter) and older Veneneia crater (Marchi *et al.* 2012). Our model results are shown in Fig. 7. We allow that the age of the family, $t_{\rm Vesta}$, the initial slope of the SFD, b, and the timescale of the inclination-independent loss mechanism, $T_{\rm L}$, are free parameters, but set $\alpha = 1$, $\beta = 2$ and $T_{\rm Y}$ to the best-fitting value derived from the model for the high-inclination ($I > 9$ deg), non-family asteroids shown in Fig. 4. The results are only partly satisfactory because our best-fit solution has $b = 0.65$ which is slightly greater than the upper limit, $b = 0.6$, determined by the condition that the total mass in the distribution cannot be infinite (Durda & Dermott 1997). Noting that our models only derive ratios of timescales, our best model (Fig. 7) gives

$$t_{\rm Vesta}/T_{\rm Y} = 0.33 \pm 0.015. \tag{4}$$

If we assume that the mean density of the Vestoids is 2850 $kg\ m^{-3}$ (Jenniskens *et al.* 2021) - a value that is larger than the density of 2000 $kg\ m^{-3}$ assumed in our earlier paper (Dermott *et al.* 2021), then using the Greenberg *et al.* (2020) NEA observations and their estimate that the thermal efficiency, $\xi = 0.12$, we estimate that $T_{\rm Y} = 6$ Gyr and $t_{\rm Vesta} = 2.0 \pm 0.1$ Gyr. This age is greater than the age $\sim$1.3 Gyr obtained from Vesta

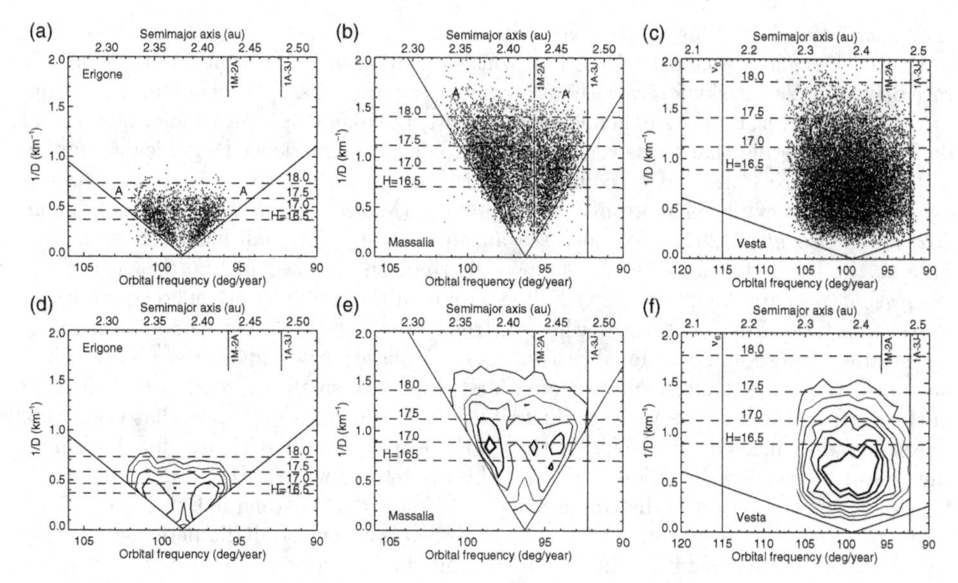

Figure 8. V-shaped distributions in $1/D - a$ space of the asteroids in the Erigone, Massalia and Vesta families with no limit on H. Diameters, D have been calculated assuming albedos of 0.06, 0.22 and 0.35 for, respectively, Erigone, Massalia and Vesta. The V-shaped lines for Massalia correspond to the fit determined by Milani *et al.* (2014). The lower panels are the corresponding number density plots with linear spacing of the contour plots.

surface crater counts (Marchi *et al.* 2012). However, the difference between these two age estimates could be even larger. Using the NEA orbital evolution timescale implies that the orbital evolution rates are constant, and we consider that to be unlikely. The NEA observations do not allow for asteroid disruptions or for changes in the asteroid spin directions and therefore the NEA observations are not applicable to the small asteroids in a family that is old. Thus, our estimate of the age of the Vesta family is likely to be a significant underestimate. We note that for both the high-inclination, non-family asteroids and the Vesta family, $t_{\rm evol}/T_{\rm Y} = 1/3$ (where $t_{\rm evol}$ is the average time of orbital evolution), consistent with our argument that both groups of asteroids have had similar evolutionary histories and therefore that both have ages comparable with the age of the solar system.

We support this conclusion by considering the dynamical evolution of small asteroids due to mechanisms other than Yarkovsky driven orbital evolution. The observations and models shown in Figs. 2, 4 and 7 reveal that the incremental SFDs of all the family and non-family asteroids are close to peaks at $H = 16.5$. For asteroids with $H < 16.5$, Yarkovsky loss is the dominant loss mechanism. The asteroids with $H > 16.5$ may not be observationally complete, but our models suggest that the number of small asteroids drops off sharply with increasing H. Small asteroids may be created by the disruption of larger asteroids, but this must be at a rate less than the loss rate. The distributions in $a - 1/D$ space of the asteroids in the Erigone, Massalia and Vesta families are shown in Fig. 8. The three sets of V-shaped lines correspond to ages of 1.6 10^8 *Gyr* (Erigone), 1.6 10^8 *Gyr* (Massalia) and 1.3 10^9 *Gyr* (Vesta). These ages were calculated using Yarkovsky timescales calculated from the NEA observations (Greenberg *et al.* 2020) for asteroids with $a = 2.4$ au, assuming asteroid densities of 1570 $kg\ m^{-3}$ (Erigone), 3000 $kg\ m^{-3}$ (Massalia) and 2850 $kg\ m^{-3}$ (Vesta, Jenniskens *et al.* (2021)). These three families were formed by cratering events (Spoto *et al.* 2015). However, their distributions in $a - 1/D$ space show significant differences. In Fig. 8a, we see that in the young Erigone family, the distribution of the larger asteroids with inverse diameters, $1/D \lesssim$, 0.5 has two well-separated lobes and a central depletion, consistent with orbital evolution at a constant rate driven by Yarkovsky forces without significant asteroid disruptions or changes in spin direction. However, for the smaller asteroids with $1/D \gtrsim 0.5$, we see an absence of a central depletion suggesting that these very small asteroids have experienced significant changes in their spin directions and/or collisional disruptions with the result that their orbital evolution may have been more of a random walk (Marzari *et al.* 2020; Dermott *et al.* 2021; Dell'Oro *et al.* 2021). We also see an absence of very small asteroids with high $1/D$ at the two extremes of the V-shaped distribution (marked in Fig. 8a with an A) suggesting, perhaps, that the very small asteroids in the family experienced disruptions preventing them from evolving to those extreme locations. However, this conclusion is not secure because any collisional disruption of the larger asteroids in the family would have produced an enhancement of the number density of small asteroids with high $1/D$ in the central region of the V-shaped distribution. In Fig. 8b, we observe that the young Massalia family has a distribution like that of the Erigone family, although for this family the loss of the central depletion occurs for those asteroids with $1/D \gtrsim 1.2$. Significantly, in Fig. 8c, we see that the distribution for the Vesta family is markedly different from both young families. For Vesta, the central depletion is absent at all diameters suggesting that most of the small asteroids in the Vesta family have experienced collisional evolution and/or significant changes in their spin directions, supporting our argument, based on the observed SFD, that the Vesta family could have an age comparable with the age of the solar system.

The plots in Fig. 8 for the Erigone and Massalia families contain information on the dynamical evolution of the families in addition to their ages. This may allow us to determine the timescale, $T_{\rm L}$ of the combined inclination-independent loss mechanisms separately from our estimate of the Yarkovsky loss timescale, but this information may only be useful when the observational completeness limit has been extended to, say, $H \sim 18$. The Erigone and Massalia families have approximately the same mean semimajor axis. Therefore, the degree of completeness of each family for a given value of H should be closely similar, although we note that the Erigone family has a mean eccentricity, $e = 0.191$ which is slightly greater than the mean value, $e = 0.142$, of the Massalia family, making the Erigone family asteroids slightly easier to discover. Current estimates of the observational completeness limit at $a = 2.4$ au are $H = 16.5$ (Dermott *et al.* 2018) and $H = 17.6$ (Hendler & Malhotra 2020). For families with ages greater than the timescale for the loss of asteroids of diameter $D_{\rm L}$ due to mechanisms other than Yarkovsky loss, we should observe a drop off in the number density of asteroids in $a - 1/D$ space of those asteroids with diameters $D < D_{\rm L}$, consistent with the observed shape of the SFD. For Erigone, which is a C-type asteroid with an albedo, A 0.06, there is a marked drop off in the number of asteroids with $H \gtrsim 17.0$, whereas for Massalia, which is an S-type asteroid with A 0.22, the drop off occurs for those asteroids with $H \gtrsim 18.0$. These values of H correspond to diameters $\sim$2.2 km for Erigone and $\sim$0.7 km for Massalia. While recognizing that for $H \gtrsim 17.0$ the distributions may be observationally incomplete, it is worth discussing the information that will be available when the completeness limit is extended to $H \sim 18.0$.

The Erigone and Massalia families both appear to have an age of $\sim$1.6 10^8 Myr, although this estimate assumes that the Yarkovsky timescale can be deduced from the NEA observations with allowance for the differences in the mean densities, but with no allowance for possible differences in their thermal efficiencies, even though these asteroids are of different types. We also note that the mean value of the thermal efficiencies of the NEAs is an average value that does not distinguish between NEAs of different types. The number of asteroids with $H < 16.5$ in the Massalia family is 1450, whereas the corresponding number in the Erigone family is 777. This difference of a factor of 2 is not enough to account for the observation that the number of asteroids in the Massalia family with $H > 17.5$ is 1859, whereas for the Erigone family the number is 166. If this large difference is not due solely to observational selection, then it suggests that C-type asteroids of a given diameter are easier to disrupt than S-type asteroids. We note that Morbidelli *et al.* (1997) consider that there is observational evidence that C-type NEAs may be easier to disrupt than S-type NEAs.

Small asteroids are lost through Yarkovsky forces, rotational disruption, and catastrophic collisions. With respect to the latter mechanism, we note that because of their near identical locations in the IMB, that only differ because of the small differences in their mean eccentricities and inclinations, the two families are impacted by a common population of asteroids. If the asteroids are rubble-piles, then the critical mass, m of the asteroid bullet needed to overcome the gravitational forces binding an asteroid of mass M is given by

$$f\frac{1}{2}mV^2 = \frac{12GM^2}{5D}, \tag{5}$$

where V is the impact velocity and f is the fraction of the kinetic energy available to disrupt the larger asteroid. This implies that the critical diameter for asteroids of a given age varies as

$$D_{\mathrm{L}} \propto \left(\frac{f}{\rho^2}\right)^{1/5}, \tag{6}$$

where ρ is the bulk density. Thus, D_{L} increases with decreasing density and increasing f, but to account for C-type and S-type asteroids in families of approximately the same age having values of D_{L} that differ by a factor as large as 2, we need f/ρ^2 to differ by a factor of 2^5. If we ignore the differences in the values D_{L} and assume that for both families the number drop-off occurs for $D_{\mathrm{L}} \sim 1\ km$, then the collisional lifetime of these $1\ km$ asteroids is $\sim 1.6\ 10^8\ yr$, the estimated age of the families. Jacobson *et al.* (2014) argue that collisional disruption is not the dominant loss mechanism because YORP forces increase the spin rates and drive asteroids to destruction on a timescale given by

$$T_{\mathrm{YORP}} = 10\,Myr\left(\frac{D}{1\,km}\right)^2. \tag{7}$$

This is a factor $\sim$10 less than our very rough estimate of the non-Yarkovsky loss timescale and a timescale as small as $10^7\ yr$ appears to conflict with the observations. However, these conclusions are not firm, partly because of possible observational incompleteness, but also because it is not known how the Yarkovsky loss timescale, that determines the age of the families, varies with asteroid type.

4. Martian mean motion resonances

Greenberg *et al.* (2020) determined from observations of the orbital evolution of 247 NEAs that the ratio of the retrograde to prograde spins of these objects is as high as $2.7^{+0.3}_{-0.7}$ and argued, following Nugent *et al.* (2012) and Fanocchia *et al.* (2013), that this ratio places bounds on the fraction of the NEAs that are scattered into the inner solar system through the ν_6 secular resonance. If this resonance and the Jovian 3:1 mean motion resonance were the only escape hatches from the IMB, and if asteroids with both retrograde and prograde spins escape through the 3:1 resonance, but only asteroids with retrograde spins and shrinking semimajor axes escape through the ν_6 resonance, then their conclusion that the ν_6 resonance is the dominant escape route is valid. However, in the IMB we observe that the mean proper eccentricity of the non-family asteroids increases with decreasing asteroid size (Fig. 3c). In addition, the proper

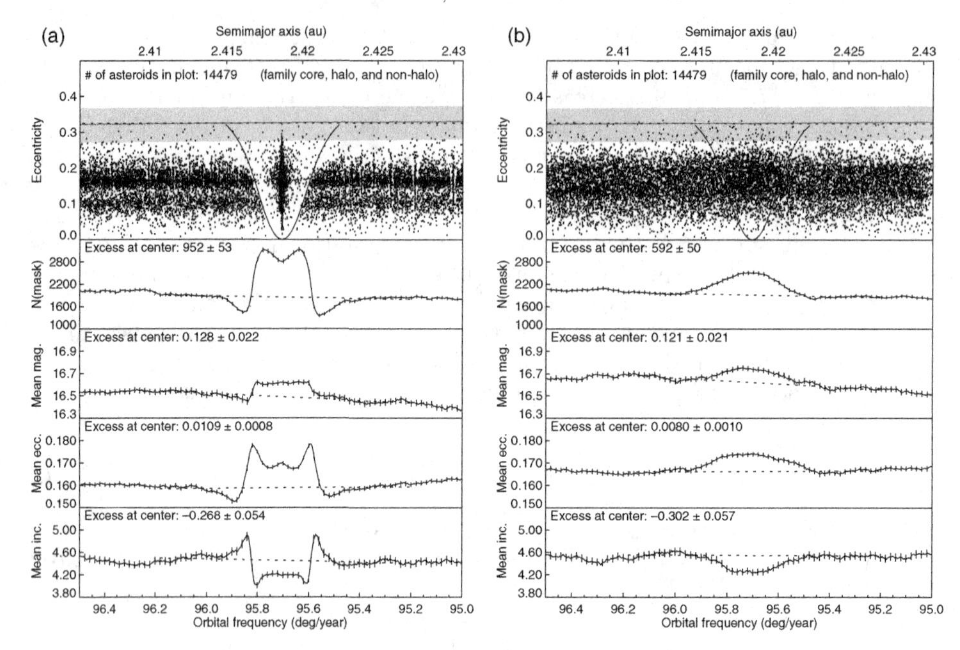

Figure 9. Panel (a): Scatter plot of the proper eccentricities and semimajor axes of the asteroids near the Martian 1:2 mean motion resonance (no limit on H). The mask shown in the top panel is based on the maximum libration width of the strongest term in the disturbing function of the resonance. By sliding this mask through the asteroid population, we compare the mean values of the number density, the mean absolute magnitude, the mean proper eccentricity, and the mean proper inclination of those asteroids trapped in the resonance with those of their near neighbors. Panel (b): A similar plot based on sliding the same mask through the osculating orbital elements.

inclinations of the asteroids in the Mars-crossing zone have distributions in $a - I$ space like those of the asteroids in the major families and, for asteroids with the observed distribution of semimajor axes, the inclinations are mostly too low for the asteroids to have entered the Mars-crossing zone through the ν_6 resonance. Following the work of Morbidelli & Nesvorný (1999), we have argued that these two observations suggest that the escape of asteroids through a dense web of high-order mean motion resonances may be a significant loss mechanism. Yarkovsky forces will feed the asteroids into these high-order resonances, and it may be that the asteroids with retrograde spins experience increases in their eccentricities and escape at a greater rate than those with prograde spins. However, obtaining numerical results to support this suggestion, that consider the passage of asteroids through many high-order resonances, and include both point-mass gravitational forces, that do not depend on the sense of asteroid rotation, and size-dependent Yarkovsky forces, is a task that has not yet been undertaken.

Here, we discuss the simpler problem of the dynamics of the Martian 1:2 mean motion resonance. Gallardo (2007); Gallardo *et al.* (2011) was the first to show that, in contrast to the prominent Kirkwood gaps in the main belt, this Martian resonance has a large excess of asteroids and the asteroids in the resonance have different mean sizes, mean eccentricities, and mean inclinations from those of their non-resonant near-neighbors. These observed differences can be quantified precisely, and with further work they could shed light on the role of the Martian mean motion resonances in the dynamical evolution of the IMB. The top panel in Fig. 9a is a scatter plot of the proper eccentricities and semimajor axes of the asteroids close to the Martian 1:2 mean motion resonance (no limit on H). Note that this plot contains 14,479 asteroids and has a total width of only $\sim$0.02 *au*. The mask shown in this panel is based on the maximum libration width of the strongest term in the disturbing function of the resonance, calculated using the full disturbing function with the simplifying assumption that the orbits are coplanar.

The approximate width, Δa of the 1:2 first-order resonance that defines the libration mask is given by

$$\frac{\Delta a}{a} = 0.00341\sqrt{e}. \tag{8}$$

By sliding the libration mask through the asteroid population, we can compare the mean values of the number density, the mean absolute magnitude, the mean proper eccentricity, and the mean proper inclination of the asteroids trapped in the resonance, defined here as lying within the libration mask, with those of their near-neighbors. We observe that the number of asteroids residing in the resonance is clearly excessive, and that those asteroids have higher mean proper eccentricities, are smaller, and have lower mean proper inclinations, than their near-neighbors. The plots in Fig. 9a are based on the synthetic proper orbital elements obtained by Knežević & Milani (2000) by numerically integrating the orbits of the planets and asteroids over millions of years and extracting the forced and proper elements by filtering. Their integrations do not include Yarkovsky forces, even though, for very small asteroids, the effects of these forces may not be negligible. One could question the use of proper, or long-term average, orbital elements when any analysis of the dynamics of resonance involves the osculating elements. For those reasons, in Fig. 9b we have included plots based on the osculating elements. The changes that we observe with the osculating elements are like those obtained with the proper elements shown in Fig. 9a, although there are differences in the magnitudes of those changes. In future work, involving the numerical integration of the orbits of the asteroids as they evolve through the resonance under the action of Yarkovsky forces, the observations for the osculating orbits should prove to be the most useful because they can be compared directly with the results of the numerical integrations without the use of any filters.

The dynamics of first-order resonance involving satellites with small eccentricities and well-separated libration frequencies and widths, such that only one resonant argument needs to be retained in the analysis, has had many successful applications. But the Martian 1:2 resonance for asteroids with moderate eccentricities is a multiplet consisting of many significant overlapping resonances and those methods cannot be universally applied. For example, simple first-order theory predicts that capture into resonance can occur only if asteroids are evolving on converging orbits, that is, towards the resonance and towards Mars (Murray & Dermott 1999), and that the probability of capture into resonance decreases markedly with increasing eccentricity. But Gallardo *et al.* (2011) has shown through numerical integrations that capture into resonance can occur for asteroids with large eccentricities, evolving on both converging and diverging orbits. We observe that the libration mask, based on a single resonant argument, does give a good account of the distribution of the proper eccentricities for $e \leq 0.2$ (Fig. 9a), but breaks down for higher eccentricities. This could be partly because gravitational interactions between the Martian 1:2 resonance and the 3:1 Jovian resonance result in two series of 3-body resonances with asteroid frequencies, n_A given by

$$n_{\rm A} - 3n_{\rm J} = k(n_{\rm M} - 2n_{\rm A}), \tag{9}$$

where the integer $k = \pm(1, 2, 3, ...)$ and the mean motions of the asteroid, Mars and Jupiter are denoted, respectively, by $n_{\rm A}$, $n_{\rm M}$ and $n_{\rm J}$. These 3-body resonances have mean motions, $n_{\rm A}$ either greater or less than $n_{\rm M}/2$, depending on the sign of k. As $|k|$

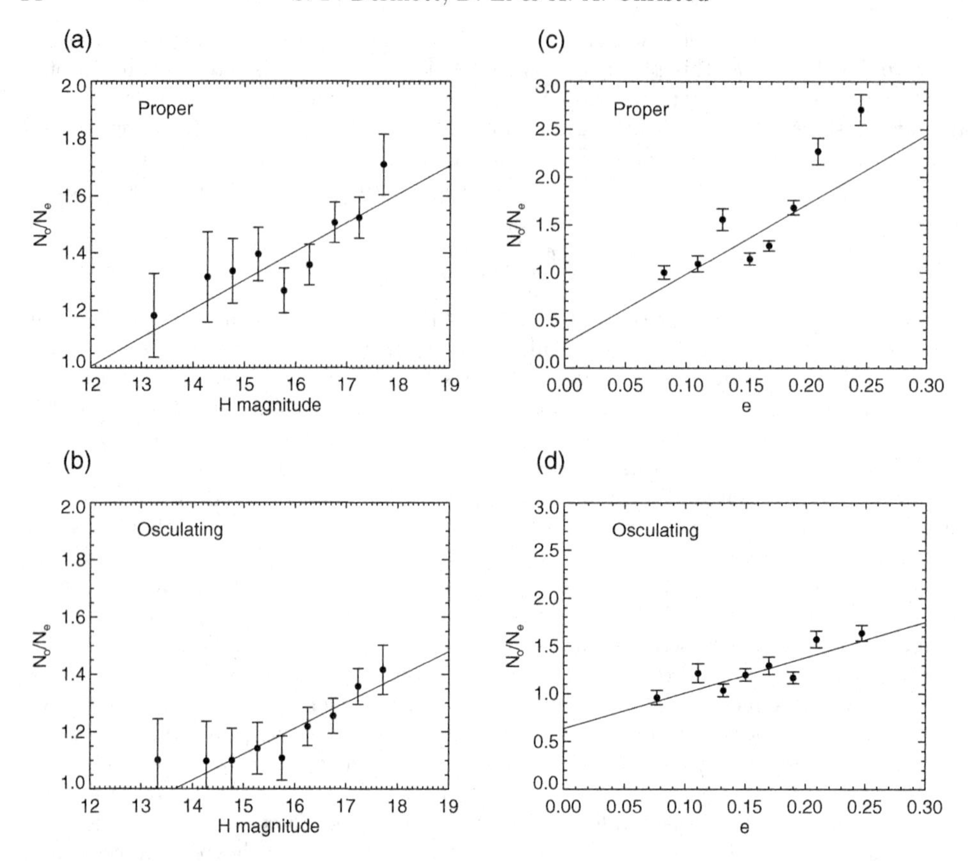

Figure 10. Plots of the variation of the excess number of asteroids trapped in the 1:2 Martian resonance with, on the left, absolute magnitude and, on the right, orbital eccentricity. The upper and lower plots show, respectively, the proper and the osculating elements.

increases, the separation of neighboring resonances, Δn_{A} decreases as

$$\Delta n_{\mathrm{A}} = (n_{\mathrm{M}}/2 - n_{\mathrm{A}})/k \tag{10}$$

and both series converge on the Martian 1:2 resonance. These 3-body resonances account for the columns of asteroids with $e \sim 0.2$ shown in Fig. 9a. As $|k|$ increases, the 3-body multiplets become close enough to overlap, resulting in a highly chaotic zone that extends well beyond the bounds of the libration mask. This could account for the observed lack of asteroids with $e \gtrsim 0.2$ immediately outside of those bounds (Fig. 9a).

The number of asteroids shown in Fig. 9 is 14,479 (no limit on H). This number is so high that we can usefully divide the population into several subgroups. In Fig. 10 we show, using both the proper and the osculating orbital elements, how the number excess in the libration mask varies with absolute magnitude and eccentricity. If we consider the passage of the asteroids through the resonance as a flow problem, then the ratio of the observed to the expected number of asteroids in the mask, $N_{\mathrm{o}}/N_{\mathrm{e}}$, is determined by the ratio of the average time, Δt_{mask} that the asteroids reside within the mask to the time needed by the asteroids to traverse the width of the mask in the absence of the resonance, as determined by the Yarkovsky timescale. This argument does not require the asteroids to have a common direction of flow. From eqns. (1) and (8), we obtain

$$\Delta t_{\mathrm{mask}} = 0.00341 (N_{\mathrm{o}}/N_{\mathrm{e}}) T_{\mathrm{y}} (D/1\,km)\sqrt{e}. \tag{11}$$

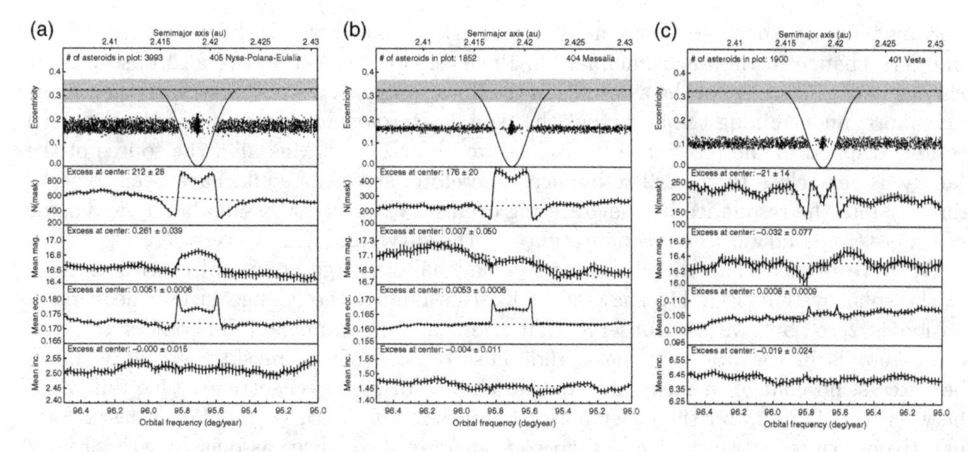

Figure 11. These plots are similar to those shown in Figure 9 and use the proper elements of the asteroids in (a) the Nysa-Polana-Eulalia family complex, (b) the Massalia family, and (c) the Vesta family (no limit on H).

The mean osculating e of the asteroids in Fig. 10b is 0.166. Assuming an average albedo, A of 0.13, $H = 16$ corresponds to a diameter, D of 2 km. The observed number excess for $H = 16$ shown in Fig. 10b is 1.2. Assuming that $T_{\mathrm{y}} = 4$ Gyr, we calculate that $\Delta t_{\mathrm{mask}} = 13$ Myr. This is probably less than both the collisional lifetime and the timescale for changes in the spin directions, suggesting that the changes due to the resonance that we observe in Fig. 10 are caused solely by the actions of gravitational forces and Yarkovsky forces. Thus, it should be possible, through numerical experiments, to determine the magnitudes of the Yarkovsky forces from the observations. However, we need to discuss some problems with this suggestion.

In Fig. 9, we observe that the resonance has an excess number of asteroids and that the asteroids are, on average, smaller than their neighbors. This suggests that Yarkovsky forces transport small asteroids to the resonance, and this is supported by the observations shown in Fig. 10 that the various changes are size dependent. We also observe an increase in the mean eccentricity. Point-mass gravitational forces acting alone on asteroids trapped in the Martian 1:2 resonance increase the dispersions of the eccentricities and inclinations, but they do not change their mean values (Christou *et al.* 2020). However, if the spin directions of the asteroids in the resonance are not random, then Yarkovsky forces will act to change the mean eccentricities, but the changes in the mean inclinations will be very much less than any changes in the mean eccentricities, and this conflicts with the observations. However, the large changes in the mean inclination observed in Fig. 9 can be explained by the existence of discrete families in the IMB. Consider two families of comparable size that have distinctly different mean inclinations and eccentricities. In Fig. 10 we observe that the number excess varies with eccentricity. Therefore, if the families have different mean eccentricities, then we might expect these families to have different number excesses and it follows that the mean inclination of the combined families outside of the resonance will differ from that of the asteroids in the resonance, even though the change in the mean inclination of each family is zero.

We support this argument by analyzing the distributions of the asteroids in the separate families. When we separate the asteroids shown in Fig. 9 into families, we observe that in each case the change in the mean inclination is zero. Fig. 11a is for the Nysa-Polana-Eulalia complex, which consists of at least two families, but these families have a common mean eccentricity and this may explain why the inclination change is zero. Fig. 11b is for the Massalia family, a young family that appears to be devoid of any secondary families.

For this family, we observe a large number excess and an increase in the mean eccentricity, but zero change in the mean inclination. However, for this family we also observe zero change in the mean size of the asteroids in the libration mask as compared with their near-neighbors, undermining the argument that small asteroids are preferentially transported to and trapped in the resonance. In Fig. 8, we observe that Massalia, the source of the family, is very close to the 1:2 resonance. Therefore, asteroids could have been injected directly into the resonance by the cratering event that formed Massalia family and some of the asteroids now in the resonance may be the survivors of that event.

We have argued that the Vesta family may have an age comparable with the age of the solar system. Fig. 11c shows that for this family, the change in the mean inclination is zero, but we also observe that there are no significant differences between the asteroids in the libration mask and those outside of the mask: there is no number excess, no change in mean size and no change in mean eccentricity. One difference between this family and the other major families in the IMB, is that Vesta has a comparatively low eccentricity. There are two first-order resonances associated with the 1:2 resonance: one resonance involves the eccentricity of the asteroid, while the other involves the eccentricity of Mars. Our analysis of the distribution of the proper elements of the asteroids in the resonance shows that for $e_{\rm A} \gtrsim 0.1$ the resonances predominantly involve the eccentricity, $e_{\rm A}$ of the asteroid, whereas for $e_{\rm A} \lesssim 0.1$ the resonances predominantly involve the eccentricity, $e_{\rm M}$ of Mars. We conclude that while an analysis of the dynamics of the mean motion resonances in the IMB could yield information on the Yarkovsky timescales, there are important differences between the interactions of the separate family asteroids with these resonances and each family and each resonance needs a separate analysis.

5. Discussion

Our analysis of the distributions of the orbital elements and sizes of the high-inclination non-family asteroids in the inner main belt suggests that these asteroids belong to old ghost families with dispersed orbital elements. The observations that support this conclusion are: (1) the lack of small asteroids; (2) the mean size of the asteroids increases with increasing orbital inclination; and (3) the CC and NC asteroids in the population, as defined by their WISE albedos, have significantly different proper eccentricity and proper inclination distributions. This analysis supports the work of Delbó *et al.* (2019) who, using independent arguments, reached some of the same conclusions. They searched for correlations between the semimajor axes and the inverse sizes of asteroids and found evidence for two previously unknown families in the inner main belt among the moderate-albedo X-complex asteroids. Delbó *et al.* (2017) also detected a small, old family of large dark asteroids in the inner main belt.

We have shown that the action of Yarkovsky forces driving small asteroids to the ν_6 secular resonance and the 3:1 Jovian mean motion resonance can account for the observation that the mean size of these asteroids increases with increasing orbital inclination. However, if we assume that the Yarkovsky forces acted on the asteroids over the age of the solar system, then the Yarkovsky timescale obtained from the observed size-inclination correlation is much longer than that expected from observations of the orbital evolution of NEAs. We conclude that either the spin directions of the asteroids were not constant over the age of the solar system, or that these asteroids originate from the disruption of larger asteroids and their ages are therefore less than the age of the solar system, or that both explanations are valid. However, we note that the CC and NC asteroids in the inner belt, as defined by their WISE albedos, have significantly different proper eccentricity and proper inclination distributions and this suggests that all the asteroids in the inner belt originate from a small number of large asteroids. This result has implications for the number of root sources of meteorites and NEAs.

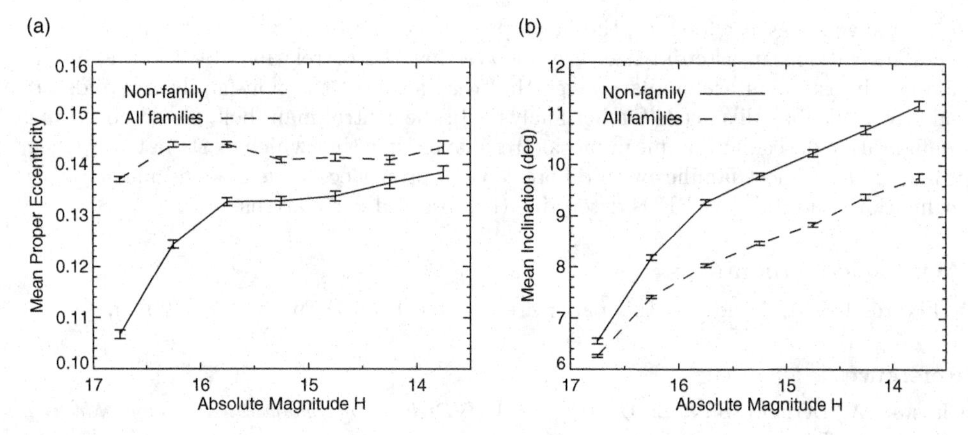

Figure 12. Variation with absolute magnitude, H of the mean proper eccentricity and the mean proper inclination of the family and non-family asteroids in the central main belt with $2.50 < a < 2.82$ *au*.

For asteroids in the inner belt with absolute magnitudes in the range $13.5 < H < 16.5$, we have shown that Yarkovsky transport of asteroids to the resonant escape hatches is the dominant asteroid loss mechanism and that other, inclination-independent, loss mechanisms have only a small effect on the observed SFD, and no effect on the observed asteroid size-inclination correlation. This result implies that the initial mass of the inner belt was like that of the present inner belt. However, the current observational completeness of the IMB is less than $H = 18$, and until that higher completeness limit is achieved it may be difficult to constrain the collisional and rotational disruption timescales. We have discussed the distribution of asteroids in $a - 1/D$ space and suggested that these distributions could give information on the timescale for changes in the spin directions and on the collisional disruption timescales. We note that the same information could be obtained from an analysis of the distributions of family asteroids in proper orbital element space. For example, Pavela *et al.* (2014) have analysed the evolution of the Karma family asteroids in $a - I$ space and shown that some regions of this space appear to lack small asteroids. They suggested that this could be due to observational incompleteness, but another explanation for this interesting result is that some small asteroids are absent because they have been disrupted. If that is the case, then knowing the age of the family, we should be able to constrain the asteroid loss timescale.

If we allow that the asteroids in the Vesta family, like those in the putative inner belt ghost families, have experienced collisional disruptions and/or changes in their spin directions, then the observed size frequency distribution of the Vesta family can be accounted for by the action of Yarkovsky forces, but only if, in addition, the age of this family is comparable to the age of the solar system. This result conflicts with an estimate of the age of the family that has been derived from surface crater counts, a result obtained by assuming that the population of impactors in the solar system has remained constant. Thus, our result calls into question the reliability of that dating method. To derive the age of the Vesta family, we have assumed that the initial SFD was linear on a log-log scale, an assumption that is supported by the observed SFD of the young Massalia family. From our analysis, we deduce that the initial slope of the Vesta family SFD was similar to that of the Massalia family, but we have not given a reason why families created by cratering events should necessarily have similar initial SFDs.

In this paper, we have only discussed the observed correlations of the orbital elements and asteroid sizes in the IMB, but these correlations are not confined to the IMB. In Fig. 12, we show that for the asteroids in the central main belt, with semimajor axes in the range $2.50 < a < 2.82$ *au*, the mean proper inclination of the non-family asteroids increases with mean asteroid size. We observe that the correlations for the family and non-family asteroids are similar, suggesting that some of the non-family asteroids are halo asteroids, but this needs further analysis. In the central main belt, we do not have a configuration of secular and mean motion resonances like that which exists in the IMB, so it may be that ghost families with different mean proper eccentricities and mean proper inclinations and different SFDs determine the observed correlations.

Acknowledgements

Figures 1, 3, 4, 7 and 8 in this paper are adapted from Dermott *et al.* (2021).

References

Christou, A., Dermott, S., & Li, D. 2020, in: *AAS/Division of Dynamical Astronomy Meeting*, 52, 100.05.
Delbó, M., Walsh, K., Bolin, B., Avdellidou, C., & Morbidelli, A. 2017, *Science*, 357, 1026.
Delbó, M., Avdellidou, C., & Morbidelli, A. 2019, *A&A*, 624, A69.
Dell'Oro, A., Boccenti, J., Spoto, F., Paolicchi, P., & Knežević, Z. 2021, *MNRAS*, 506, 4302.
DeMeo, F.E., & Carry, B. 2014, *Nature*, 505, 629.
Dermott, S.F., Christou, A.A., Li, D., Kehoe, T. J.J., & Robinson, J.M. 2018, *Nature Astronomy*, 2, 549.
Dermott, S.F., Li, D., Christou, A.A., Kehoe, T. J.J., Murray, C.D., & Robinson, J.M. 2021, *MNRAS*, 505, 1917.
Dohnanyi, J.S. 1969, *Journal of Geophysical Research*, 74, 2531.
Durda, D.D., & Dermott, S.F. 1997, *Icarus*, 130, 140.
Eugster, O., Herzog, G.F., Marti, K., & Caffee, M.W. 2006, in: Lauretta, Dante S. & McSween, H.Y. (eds.), *Meteorites and the Early Solar System II*, 829
Farinella, P., Froeschlé, C., Froeschlé, C., Gonczi, R., Hahn, G., Morbidelli, A., & Valsecchi, G.B. 1994, *Nature*, 371, 314.
Farinella, P. & Vokrouhlický, D. 1999, *Science*, 283, 1507.
Farnocchia, D., Chesley, S.R., Chodas, P.W., Micheli, M., Tholen, D.J., Milani, A., Elliott, G.T., & Bernardi, F. 2013, *Icarus*, 224, 192.
Gallardo, T. 2007, *Icarus*, 190, 280.
Gallardo, T., Venturini, J., Roig, F., & Gil-Hutton, R. 2011, *Icarus*, 214, 632.
Gladman, B.J., Migliorini, F., Morbidelli, A., Zappala, V., Michel, P., Cellino, A., Froeschle, C., Levison, H.F., Bailey, M., & Duncan, M. 1997, *Science*, 277, 197.
Granvik, M., Morbidelli, A., Vokrouhlický, D., Bottke, W.F., Nesvorný, D., & Jedicke, R. 2017, *A&A*, 598, A52.
Granvik, M., Morbidelli, A., Jedicke, R., Bolin, B., Bottke, W.F., Beshore, E., Vokrouhlický, D., Nesvorný, D., & Michel, P. 2018, *Icarus*, 312, 181.
Greenberg, A.H., Margot, J.L., Verma, A.K., Taylor, P.A., & Hodge, S.E. 2020, *AJ*, 159, 92.
Hendler, N.P. & Malhotra, R. 2020, *The Planetary Science Journal*, 1, 75.
Jacobson, S.A., Marzari, F., Rossi, A., Scheeres, D.J., & Davis, D.R. 2014, *MNRAS* (Letters), 439, L95.
Jenniskens, P. 2018, in: *AAS/Division of Dynamical Astronomy Meeting*, 49, 102.04.
Jenniskens, P. 2020, in: *IAU General Assembly*, 30, 9.
Jenniskens, P., Gabadirwe, M., Yin, Q-Z., et al. 2021, *Meteoritics & Planetary Science*, 56, 844.
Knežević, Z., & Milani, A. 2000, *Cel. Mech. Dyn. Astr.*, 78, 17
Kruijer, T.S., Burkhardt, C., Budde, G., & Kleine, T. 2017, *PNAS*, 114, 6712.
Marchi, S., McSween, H.Y., O'Brien, D.P., Schenk, P., De Sanctis, M.C., Gaskell, R., Jaumann, R., Mottola, S., Preusker, F., Raymond, C.A., Roatsch, T., & Russell, C.T. 2012, *Science*, 336, 690.
Marzari, F., Rossi, A., Golubov, O., & Scheeres, D.J. 2020, *AJ*, 160, 128.
Masiero, J.R., Grav, T., Mainzer, A.K., Nugent, C.R., Bauer, J.M., Stevenson, R. & Sonnett, S. 2014, *ApJ*, 791, 121.

McSween, H.Y., Mittlefehldt, D.W., Russell, C.T., & Raymond, C.A. 2013, *Meteoritics & Planetary Science*, 48, 2073.

Migliorini, F., Michel, P., Morbidelli, A., Nesvorný, D., & Zappala, V. 1998, *Science*, 281, 2022.

Milani, A., Cellino, A., Knežević, Z., Novaković, B., Spoto, F., & Paolicchi, P. 2014, *Icarus*, 234, 46.

Minton, D.A. & Malhotra, R. 2010, *Icarus*, 207, 744.

Morbidelli, A. & Nesvorný, D. 1999, *Icarus*, 139, 295.

Morbidelli, A., Delbo, M., Granvik, M., Bottke, W.F., Jedicke, R., Bolin, B., Michel, P., & Vokrouhlicky, D. 1997, *Icarus*, 340, 113631.

Murray, C.D. & Dermott, S.F. 1999, *Solar System Dynamics*, Cambridge University Press, Cambridge.

Nesvorný, D. 2015, HCM Asteroid Families V3.0. *NASA Planetary Data System.*

Nesvorný, D., Brož, M., & Carruba, V 2015, in: Michel, P., DeMeo, F.E., & Bottke, W.F. (eds.), *Asteroids IV*, 297.

Nugent, C.R., Margot, J.L., Chesley, S.R., & Vokrouhlický, D. 2012, *AJ*, 144, 60.

Pavela, D., Novaković, B., Carruba, V., & Radović, V, 2021, *MNRAS*, 501, 356.

Spoto, F., Milani, A., & Knežević, Z. 2015, *Icarus*, 257, 275.

Vokrouhlický, D. & Farinella, P. 2000, *Nature*, 407, 606.

Walsh, K.J., Morbidelli, A., Raymond, S.N., O'Brien, D.P., & Mandell, A.M. 2011, *Nature*, 475, 206.

Wisdom, J. 1985, *Nature*, 315, 731.

Zappala, V., Cellino, A., Farinella, P., & Knežević, Z. 1990, *AJ*, 100, 2030.

Multi-scale (time and mass) dynamics of space objects
Proceedings IAU Symposium No. 364, 2022
A. Celletti, C. Galeş, C. Beaugé, A. Lemaître, eds.
doi:10.1017/S1743921322000059

On Tides and Exoplanets

S. Ferraz-Mello

Instituto de Astronomia, Geofísica e Ciências Atmosféricas, Universidade de São Paulo, Brasil
email: sylvio@usp.br

Abstract. This paper reviews the basic equations used in the study of the tidal variations of the rotational and orbital elements of a system formed by one star and one close-in planet as given by the creep tide theory and Darwin's constant time lag (CTL) theory. At the end, it reviews and discusses the determinations of the relaxation factors (and time lags) in the case of host stars and hot Jupiters based on actual observations of orbital decay, stellar rotation and age, etc. It also includes a recollection of the basic facts concerning the variations of the rotation of host stars due to the leakage of angular momentum associated with stellar winds.

1. Introduction

Our current knowledge of dynamical tides is mainly based on Darwin's theory of 1880. All studies done before were based on the static or stationary ellipsoidal models of Jacobi, Maclaurin and Roche and were focused on the relative equilibrium of the forces acting on the considered bodies.

Darwin was the first to take into account that one body's response to tidal forces is not instantaneous, but suffers a delay that depends on the viscosity of the body.

His first approach using hydrodynamical equations (Darwin 1879) showed that fluid bodies respond to tidal torques with a lag that, in case of viscous bodies with small viscosity, is proportional to the frequency of the torque. In his 1880 theory, Darwin (1880) used this result as an insight to assume that the Earth's tides, due to the stresses generated by the lunar attraction, have an 'ad hoc' lag proportional to the tide frequency and calculated the secular changes in the orbital elements of the Moon. This approach is followed even today. It is the basis of the so-called CTL (constant time lag) theories.

In the extended reformulation of Darwin's general theory by Kaula (1964), the 'ad hoc' lags were kept arbitrary, thus allowing for variants of Darwin's theory, as the CPL (constant phase lag) theory (MacDonald 1964; Jackson *et al.* 2008). It also opened the way for considering more complex laws regulating the lags and allowed the construction of theories able to describe dissipation in rocky planets (Efroimsky and Lainey 2007; Gevorgyan *et al.* 2020).

An alternative model valid for both gaseous and stiff bodies, the creep tide theory, was proposed by Ferraz-Mello (2012, 2013). It is based on an approximate solution of the Navier-Stokes equation and is equivalent to Darwin's theory in the case of viscous bodies. In this theory, the actual surface of the body creeps towards the instantaneous equilibrium surface with a speed proportional, at each point, to their separation. This dynamics may be written as $dZ/dt = -\gamma(Z - Z_0)$ (Newtonian creep), where $Z - Z_0$ is the height of one point above the equilibrium ellipsoid Z_0, normal to the ellipsoid, and γ is a relaxation factor inversely proportional to the viscosity of the body at the surface. The equilibrium surface $Z_0(\phi, \lambda)$ is a triaxial ellipsoid (ϕ, λ are the spherical coordinates on the surface of the ellipsoid). The integration of this differential equation allows us to obtain the dynamical figure of equilibrium $Z(\phi, \lambda)$ and to define the boundaries of the integrals giving the torque due to the tidal forces as well as the disturbing forces acting on the system.

2. Tidal evolution

Tidal evolution affects both the orbital elements of the pair and the rotation of the two bodies in interaction. The basic planar equations of this process showing how each of the two bodies contributes to the variations of the semi-major axis and eccentricity of the system are (Folonier *et al.* 2018, Online suppl.; Ferraz-Mello 2019):

$$[\langle \dot{a} \rangle]_i = \frac{3k_{2i} n m_j R_i^5}{m_i a^4} \left((1-5e^2) \frac{\gamma_i \nu_i}{\gamma_i^2 + \nu_i^2} - \frac{3e^2}{4} \frac{\gamma_i n}{\gamma_i^2 + n^2} \right. \tag{2.1}$$

$$\left. + \frac{e^2}{8} \frac{\gamma_i(\nu_i + n)}{\gamma_i^2 + (\nu_i + n)^2} + \frac{147e^2}{8} \frac{\gamma_i(\nu_i - n)}{\gamma_i^2 + (\nu_i - n)^2} \right) + \mathcal{O}(e^4).$$

and

$$[\langle \dot{e} \rangle]_i = -\frac{3k_{2i} n e m_j R_i^5}{4 m_i a^5} \left(\frac{\gamma_i \nu_i}{\gamma_i^2 + \nu_i^2} + \frac{3}{2} \frac{\gamma_i n}{\gamma_i^2 + n^2} \right. \tag{2.2}$$

$$\left. + \frac{1}{4} \frac{\gamma_i(\nu_i + n)}{\gamma_i^2 + (\nu_i + n)^2} - \frac{49}{4} \frac{\gamma(\nu_i - n)}{\gamma_i^2 + (\nu_i - n)^2} \right) + \mathcal{O}(e^3).$$

where the subscript i refers to the body deformed by the tidal stress and j refers to the body whose gravitational attraction is creating the stress (they can be the star and the planet or vice versa); a, e, n are the semi-major axis, eccentricity and mean-motion of the system; m, R, γ are the masses, equatorial radii and relaxation factors of the considered bodies, and

$$k_{2i} = \frac{15 C_i}{4 m_i R_i^2}$$

(C_i is the moment of inertia). The quantities

$$\nu_i = 2\Omega_i - 2n$$

are the semi-diurnal frequencies. The rotation velocities of the bodies, Ω_i are affected by the tidal evolution and ruled by the equations

$$\langle \dot{\Omega}_i \rangle = -\frac{3k_{2i} G m_j^2 R_i^5}{2 C_i a^6} \left((1-5e^2) \frac{\gamma_i \nu_i}{\gamma_i^2 + \nu_i^2} + \frac{e^2}{4} \frac{\gamma_i(\nu_i + n)}{\gamma_i^2 + (\nu_i + n)^2} \right. \tag{2.3}$$

$$\left. + \frac{49e^2}{4} \frac{\gamma_i(\nu_i - n)}{\gamma_i^2 + (\nu_i - n)^2} \right) + \mathcal{O}(e^4).$$

These equations do not appear so simple as the corresponding equations found in the literature. The reason is that they are general. The parameters of extra-solar systems span various orders of magnitude and it is not possible to encompass all cases with simplified formulas. If more accurate equations are needed, one may use Mignard's (1980) or Hut's (1981) versions of Darwin's CTL theory with rational functions of the eccentricities, instead of truncated series expansions or, yet, the parametric version of the creep tide theory (Folonier *et al.* 2018; Ferraz-Mello *et al.* 2020) where the instantaneous polar oblateness, equatorial prolateness and lag are given by closed differential equations which may be integrated numerically together with the equations for the variation of the elements.

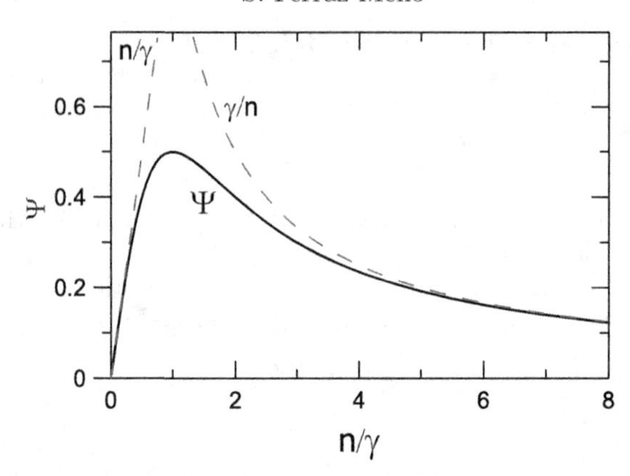

Figure 1. The function $\Psi = (\gamma/n + n/\gamma)^{-1}$

2.1. *Synchronisation*

The right-hand side of eqn. (2.3) is dominated by the term independent of the eccentricity. Given the negative sign of the coefficient in front of the brackets, the dominating term will have a sign contrary to the sign of the semi-diurnal frequency ν_i. Then the tidal torque accelerates or brakes the rotation of the body so as to make it almost synchronous (exactly synchronous if $e = 0$). The contribution of the terms dependent on the eccentricity, however, alters this rule. In order to see the actual location of the stationary rotation, let us simplify eqn. 2.3 by assuming that we may neglect terms of the order of ν_i/n and let us solve the equation $\langle\dot{\Omega}_i\rangle = 0$. There results

$$\nu_{i\,(\mathrm{stat})} = 12e^2\gamma_i\Psi_i + \mathcal{O}(e^4). \tag{2.4}$$

where

$$\Psi_i = \frac{\gamma_i n}{\gamma_i^2 + n^2}. \tag{2.5}$$

Since Ψ_i is positive and always less than 0.5 (see fig. 1), we see that the stationary value of ν_i is always positive. This means that $\Omega_{i\,(\mathrm{stat})} > n$. Then, the final state of the rotation of one planet is not exactly synchronous, but supersynchronous, unless the eccentricity is equal to zero.

2.2. *Hot Jupiters and host stars*

More simple equations can be obtained in the case of systems harboring a close-in hot Jupiter (or other gaseous companions). In these cases, both the planet and the star have relaxation factors larger than 1 s^{-1} (see section 6), such that, always, $\gamma_i \gg n$, allowing us to neglect terms of order n/γ. Hence, Eqns. (2.1-2.2) become

$$[\langle\dot{a}\rangle]_i \simeq \frac{6k_{2i}n^2m_jR_i^5}{\gamma_i m_i a^4}\left((1+\frac{27}{2}e^2)\frac{\Omega_i}{n} - (1+23e^2)\right) + \mathcal{O}(e^4); \tag{2.6}$$

$$[\langle\dot{e}\rangle]_i = -\frac{27k_{2i}n^2em_jR_i^5}{\gamma_i m_i a^5}\left(1 - \frac{11}{18}\frac{\Omega_i}{n}\right) + \mathcal{O}(e^3). \tag{2.7}$$

These equations are the same obtained with Darwin's theory when a constant time lag τ_i is assumed (see Hut 1981; Ferraz-Mello *et al.* 2008), and the correspondence formula

$$\tau_i = \frac{1}{\gamma_i} \tag{2.8}$$

(valid when $\gamma_i \ll n$) is adopted.

In this case, the synchronization process leads to a supersynchronous stationary rotation of the body whose angular velocity is the same found in Darwin's theory with the CTL hypothesis:

$$\Omega_i = n + 6ne^2. \tag{2.9}$$

2.3. *Telluric exoplanets*

In the case of exoplanets with a surface composed at least partially of solid silicate material, like the Earth's crust, γ is much smaller than in the case of gaseous planets. In the Solar System, the relaxation factor of terrestrial planets and ice satellites is $\gamma \ll 10^{-6}\mathrm{s}^{-1}$ (see Ferraz-Mello 2019, Table 4).

In addition, as the synchronization of close-in telluric planets occurs quickly, we may introduce the additional hypothesis that the system reached the stationary state and fix a value for ν_i in accordance with eqn. (2.4). In this case, eqns. (2.1) and (2.2) become

$$[\langle \dot{a} \rangle]_i \simeq -\frac{21 k_{2i} n e^2 m_j R_i^5}{m_i a^4} \Psi_i \tag{2.10}$$

$$[\langle \dot{e} \rangle]_i \simeq -\frac{21 k_{2i} n e m_j R_i^5}{2 m_i a^5} \Psi_i. \tag{2.11}$$

It is worth noting that these equations are also valid for hot Jupiters trapped in stationary rotation, in which case, $\Psi_i \simeq n/\gamma_i$.

3. Inclination and obliquity

In the previous section, we collected the main perturbations in the orbital motion and in the planetary rotation, in the coplanar case, that is, when the rotation axis of the body is perpendicular to the orbital plane of the system. In the inclined case, one correction to the given formulas must be added. According to Darwin's CTL theory, to the formulas given above, we have to add (Hut 1981; Ferraz-Mello *et al.* 2008)

$$[\langle \delta\dot{a} \rangle]_i \simeq -\frac{3 k_{2i} n m_j R_i^5 \Omega_i}{\gamma_i m_i a^4} \sin^2 I_i, \tag{3.1}$$

$$\langle \delta\dot{\Omega}_i \rangle \simeq \frac{3 k_{2i} G m_j^2 R_i^5 (\Omega_i - n)}{2 C_i \gamma_i a^6} \sin^2 I_i. \tag{3.2}$$

where I_i is the inclination of the equator of the tidally deformed body (i) with respect to the orbital plane. When this body is one planet moving around one star, the I_i is often called obliquity. At the order considered, $[\langle \delta\dot{e} \rangle]_i \simeq 0$.

We note that the parameter τ_i of the CTL theories has been substituted by the inverse of the relaxation factor γ_i following what was established in Eqn. 2.8. It is also worth stressing the fact that CTL theories are only applicable to gaseous bodies (in which case, $\gamma_i \ll n$).

3.1. *Tidal evolution of obliquity*

In the coplanar case, if we neglect the viscosity, the force is radial, its torque is equal to zero and, therefore, the rotation of the body is not affected by the static tide. There is no exchange of angular momentum between the rotation of the body and the orbital motion. Besides, the work done by the static tide force is an exact differential and, therefore, the total mechanical energy of the system remains constant in one cycle. There is no dissipation of energy (see discussion in Ferraz-Mello 2019). The perturbations given in the previous section vanish if the viscosity of the body is neglected (that is, if we are in the limit $\gamma \to \infty$).

In the spatial case, however, even in the static (or inviscid) case, the vertex of the ellipsoid representing the equilibrium figure is not directed to the companion (Folonier *et al.* 2022) and the torque no longer vanishes. However, the contributions of the tidal and of the rotational oblateness may be considered separately and, assuming that the orbital plane has a constant precession (forced by a third body in an external orbit, for instance), the Cassini stationary states known in the case of rigid bodies may subsist. In this case, initially, if the star rotation is slower than the orbital motion, the tides raised by the planet on the star act decreasing the orbital semi-major axis. In the fall towards the planet, if the planet is sufficiently far from the star, it may cross the situation that corresponds to the so-called Cassini 2 stationary state in which the fall of the planet towards the star brings as consequence an increase in the obliquity, which, in its turn, accelerates the fall to a point in which the permanence in the stationary state is no longer possible. The whole problem is very complex as it involves the variations of the equatorial and orbital planes due to tides in both bodies, and also external factors forcing an important precession of the orbital plane. A complete theory is not available. However, it may justify the existence of high-obliquity short-period planets (Fabricky *et al.* 2007; Millholland *et al.* 2020).

4. Dissipation

If one planet evolves according to the equations of section 2, the tides raised on the planet by the star change the orbital energy of the system and the rotational energy of the planet with the power

$$[\langle \dot{E}_{\rm orb} \rangle]_p = \frac{3k_{2p} G m_s^2 n R_p^5}{2a^6} \left((1-5e^2)\frac{\gamma_p \nu_p}{\gamma_p^2+\nu_p^2} - \frac{3e^2}{4}\frac{\gamma_p n}{\gamma_p^2+n^2} \right. \tag{4.1}$$

$$\left. + \frac{e^2}{8}\frac{\gamma_p(\nu_p+n)}{\gamma_p^2+(\nu_p+n)^2} + \frac{147e^2}{8}\frac{\gamma_p(\nu_p-n)}{\gamma_p^2+(\nu_p-n)^2} \right) + \mathcal{O}(e^4).$$

and

$$[\langle \dot{E}_{\rm rot} \rangle]_p = -\frac{3k_{2p} G m_s^2 \Omega R_p^5}{2a^6} \left((1-5e^2)\frac{\gamma_p \nu_p}{\gamma_p^2+\nu_p^2} + \frac{e^2}{4}\frac{\gamma_p(\nu_p+n)}{\gamma_p^2+(\nu_p+n)^2} \right. \tag{4.2}$$

$$\left. + \frac{49e^2}{4}\frac{\gamma_p(\nu_p-n)}{\gamma_p^2+(\nu_p-n)^2} \right) + \mathcal{O}(e^4).$$

where the subscripts p, s refer to the planet and the star, respectively (Ferraz-Mello 2015). These are the two only sources of the mechanical energy dissipated inside the planet. Strictly speaking, to have a complete picture, we should also consider the internal mechanical energy of the planet which changes in the process as the tidal forces affect the semi-axes of the ellipsoid representing the planet. However, this is a small quantity oscillating about zero. It is important to say that in the parametric version of the creep

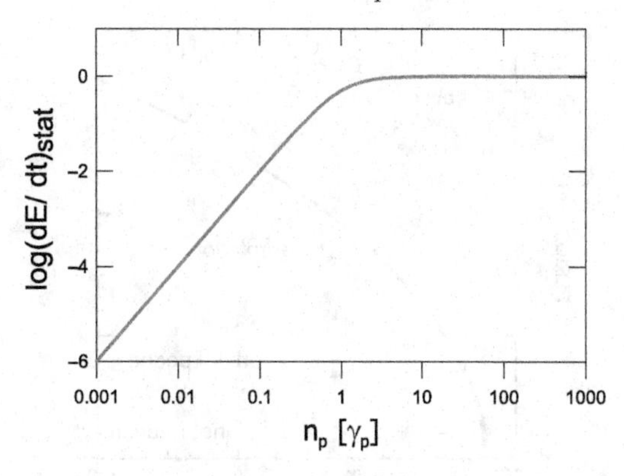

Figure 2. Power of the tidal energy dissipated in a planet in stationary rotation (arbitrary units) as a function of the orbital frequency (in units of γ_p). In the case of gaseous bodies, $n_p[\gamma_p] < 1$ and, in the case of stiff planets, $n_p[\gamma_p] > 1$.

tide theory (Folonier *et al.* 2018), the mechanical energy variation appears as a sum of squares multiplied by a negative coefficient so that, in all circumstances, $[\dot{E}_{\rm tot}]_p \leqslant 0$.

Since the rotation of short-period planets is damped to the stationary state considered in section 2.1 in a few million years, it is convenient to substitute the general formula with a much more simple one:

$$[\langle \dot{E}_{\rm stat} \rangle]_p = -\frac{21 k_{2p} G m_s^2 n R_p^5}{2a^6} e^2 \frac{\gamma_p n}{\gamma_p^2 + n^2} + \mathcal{O}(e^4) \tag{4.3}$$

(Folonier *et al.* 2018). This equation means that when $n \ll \gamma_p$ (e.g. the hot Jupiters), the dissipation power grows with the square of the orbital frequency (i.e. the square of the mean motion) and when $n \gg \gamma$ (e.g. Earths and super-Earths), the dissipation power does not depend on the mean motion (see fig. 2). This figure differs of the often shown Λ-shaped plot of the inverse of the quality factor Q. The reason is that in the latter case, Q^{-1} is proportional to the energy dissipated in one period, and to get the average dissipation per unit time, we have to divide by the period.

In the inclined case, one correction to the given formulas must be added. According with Darwin's CTL theory, valid for gaseous bodies, we have to add, to the formulas given above, the term

$$[\langle \delta\dot{E}_{\rm stat} \rangle]_p = -\frac{3 k_{2p} G m_s^2 n R_p^5}{2a^6} \sin^2 I_p \frac{n}{\gamma_p} + \mathcal{O}(e^4) \tag{4.4}$$

(Ferraz-Mello *et al.* 2008) where I_p is the inclination of the equator of the planet with respect to the orbital plane (obliquity).

5. Star wind braking

The rotational period of the host star plays a determinant role in the tidal orbital evolution of low-eccentricity planets. The leading terms in Eqns. (2.1) and (2.3) are proportional to the semi-diurnal frequency $\nu = 2\Omega - 2n$ and thus have different signs when the orbital period is larger or smaller than the revolution period of the star†. It happens that solar-type stars are prone to lose rotational angular momentum through the

† This fact is at the origin of some incorrect generalizations. One quadratic term in the orbital eccentricity has a large coefficient which is negative if $|\nu_p| < n$. In addition, tides in the almost synchronous planet give also a negative contribution to $\dot{a}$ (see Eqn. 2.10). Hence, except for very

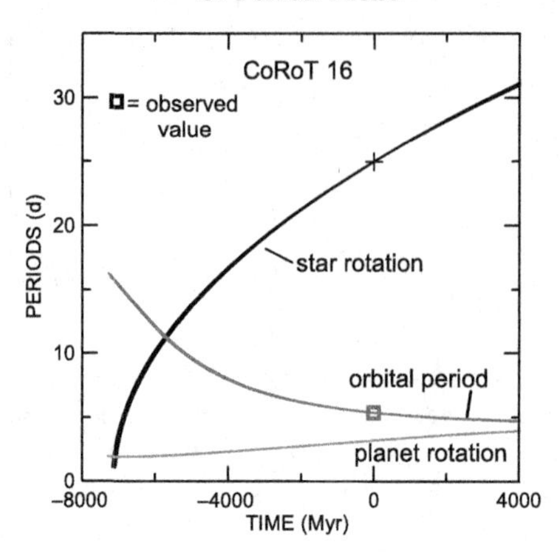

Figure 3. CoRoT-16: Long-term evolution of the stellar (black) and planetary (brown) rotational periods and of the orbital period of the planet (red). CoRoT-16 b is a hot Saturn in an orbit beyond the limit where the tidal interactions with the star influence significantly the stellar rotation. The known isochronal age of the star (6.7 ± 2.8 Gyr) constrains the current stellar rotational period to the neighborhood of 25 days. Taken from Ferraz-Mello (2016).

stellar wind. Initial periods of $1-2$ days in the early phases of the life of an active star may change to tens of days in a few Gyrs. Thus, one close-in planet in a low-eccentricity orbit that, at the beginning of its life was being ejected from the neighborhood of the star because of the tides raised by it on the star, after some time stops receding from the star and starts falling into it. Therefore, the consideration of the star's angular momentum leakage is mandatory in long-term evolutionary studies.

In order to consider the variations of the host star rotation, we may use Bouvier's model for the magnetic braking of low-mass stars ($0.5\mathrm{M}_\odot < m_s < 1.1\mathrm{M}_\odot$) with an outer convective zone:

$$\dot{\Omega}_s = \begin{cases} -B_W \Omega_s^3 & \text{when} \quad \Omega_s \leq \omega_{\text{sat}} \\ -B_W \omega_{\text{sat}}^2 \Omega_s & \text{when} \quad \Omega_s > \omega_{\text{sat}} \end{cases} \tag{5.1}$$

where ω_{sat} is the value at which the angular momentum loss saturates (fixed at $\omega_{\text{sat}} = 3, 8, 14\Omega_\odot$ for 0.5, 0.8, and 1.0 $M_\odot$ stars, respectively) and B_W is a factor depending on the star moment of inertia (C_s), mass (M_s) and radius (R_s) through the relation

$$B_W = 2.7 \times 10^{47} \frac{1}{C_s} \sqrt{\left(\frac{R_s}{R_\odot} \frac{M_\odot}{M_s} \right)} \qquad \text{(cgs units)}. \tag{5.2}$$

(Bouvier *et al.* 1997). In the case of the Sun, the coefficient used in Eqn. (5.1) corresponds to 6.6×10^{30} g cm^2 s^{-2}.

For fully convective low-mass M-stars, more complex models are used in which the coefficient in Eqn. (5.2) is not the same for slow and fast rotators. Typical values in the case of slow rotators range from the same value as for solar-like stars to 1.1×10^{47} g cm^2 s, depending on the adopted saturation value. For fast rotators, the coefficient value is taken at least one order of magnitude smaller (see Irwin *et al.* 2011).

The above form of the law is valid after the star has completed its contraction (the stellar moment of inertia C_s no longer changes significantly) and is fully decoupled from its primeval disk. No significant mass loss needs to be considered.

low eccentricities, the planet will be falling onto the star no matter if the star rotation period is larger or smaller than the orbital period of the planet (see fig. 3).

We may note that the magnetic connection to very close planets may inhibit the stellar wind (Strugarek *et al.* 2014) and thus reduce the braking torque on the star to a minimum of 70 percent of its normal value for $a \simeq 3.5R_s$ (see op.cit. fig. 10) and request a customized model. In the study of the planet around the sub-giant star CoRoT-21, Pätzold *et al.* (2012) have adopted a coefficient with a value smaller than the one adopted for solar-like stars.

F stars with masses larger than 1.3 $M_\odot$ are not expected to be affected by a magnetic braking.

6. The relaxation factor

The parameters γ_i are not ad hoc quantities, but they are not known. In order to allow the application of the given formulas to actual problems, we have to decide which values of γ_i to use. The factors are discussed below for the several classes of objects considered.

6.1. *Stars*

Dissipation values of solar-type stars have been estimated by Hansen (2010, 2012) from the analysis of the survival of short-period exoplanets. His results can be converted into the relaxation factor of the creep tide theory (Ferraz-Mello 2013) through

$$\gamma_s \simeq \frac{2Gk_{2s}|\nu_s|}{3nR_s^5\sigma_s}$$

His mean result $\sigma_s = 8.3 \times 10^{-64}$ g^{-1} cm^{-2} s^{-1}, for a wide range of stars, corresponds to $\gamma_s \sim 5 - 45$ s^{-1} (assuming $|\nu_s/n| \sim 2$). This variation in γ_s is mainly due to the strong dependence on the radius and the larger values correspond to stars having half of the radius of the Sun.

However, the study of transiting hot Jupiters around old stars in significantly non-circular orbits shows that $\gamma_s < 30$ s^{-1} often implies a too large exchange of angular momentum between the orbit and the star rotation inconsistent with eccentricities not damped to zero.

Results in a range slightly higher, $\gamma_s \sim 20 - 80$ s^{-1}, were derived from the comparison of the observed rotational and orbital periods in systems with big close-in companions and known age and rotation (Ferraz-Mello *et al.* 2015). In those cases, the rotational period of the star evolves under the action of the tides raised by the (big) close-in companion and, in the cases of active stars, the wind braking. Their current value is critically determined by the relaxation factor of the star.

One example is given in figure 4 that shows results from simulations of the orbital evolution of the brown dwarf CoRoT-3 b and its host star CoRoT-3, a F3 star. No braking was added. In that example, in order to reproduce the known parameters of the system, we may have $40 < \gamma_s < 80$ s^{-1}. A smaller (resp. larger) γ_s means a faster (resp. slower) evolution and a system younger (resp. older) than given by the age of the star as determined from stellar models.

The limit $\gamma_s > 30$ s^{-1} is also more or less the same obtained by Jackson *et al.* (2011) from the analysis of the distribution of the putative remaining lifetime of hot Jupiters.

A lesser value is obtained from the decay of the planet WASP-12b. In that case, it was possible to fit a tidal model to the decreasing orbital period of the planet (29 ± 2 msec/year) (Yee *et al.* 2019). The result obtained for the quality factor of the star corresponds to $\gamma_s = 18$ s^{-1}.

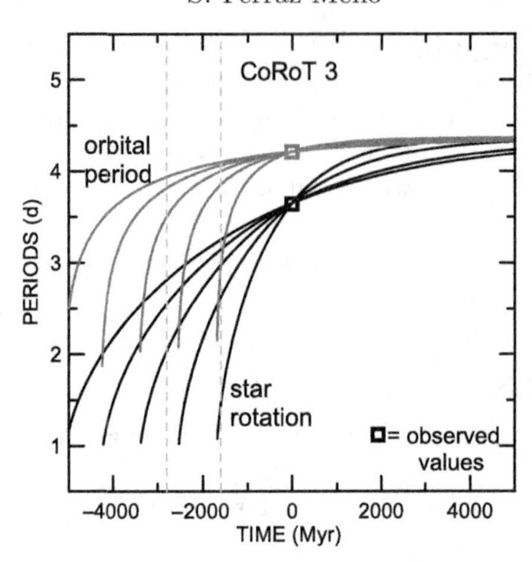

Figure 4. Simulations of the evolution of the rotational period of the host star CoRoT-3 (black) and of the orbital period of its companion, the brown dwarf CoRoT-3 b (red). Adopted stellar tidal relaxation factors (starting from the steepest curves): 40, 60, 80, 100, 120 s^{-1}. The dashed vertical lines show the stellar age range. Taken from Ferraz-Mello (2016).

6.2. *Hot Jupiters*

In his analyses, Hansen (2010) also made a first assessment of the dissipation of hot Jupiters. The comparison of his model to the creep tide theory, in this case, assumes that the planets are trapped in stationary quasi-synchronous rotation. The correspondence formula is

$$\gamma_p \simeq \frac{3Gk_{2p}}{4R_p^5\sigma_p}$$

where σ_p is Hansen's planetary dissipation parameter. Then, using his mean results for WASP-17 b, CoRoT-5 b and Kepler-6 b, transiting planets whose radii have been determined, we obtain values of γ_p in the range $10 - 60$ s^{-1}, the smallest value corresponding to the bloated WASP-17 b of radius $\sim 2R_{\rm Jup}$.

In the case of CoRoT-5 b, a memory of the past evolution is kept by the eccentricity. Backward simulations of this system show a significant rate of circularization and the eccentricity must have been much larger in the past. If the relaxation factor is smaller than a certain limit, the resulting circularization rate would impose eccentricities close to 1 in the past Gyr (Ferraz-Mello 2016), difficult to explain with our current knowledge of the exoplanets orbital evolution .

6.3. *Small planets*

We may transpose to exoplanets the results known for some small planets of our Solar System. We may mention the estimates given for Mercury, $4 - 30 \times 10^{-9}$ s^{-1} (Ferraz-Mello 2015), and for the solid Earth, $1 - 4 \times 10^{-7}$ s^{-1} (conversion of the quality factor adopted by several authors). We may add that the time lag necessary to a CTL theory to reproduce the tidal acceleration of the Moon is ~600 s (Mignard 1979), which is equivalent† to $\gamma \sim 1.2 \times 10^{-5}$ s^{-1}.

† For rocky planets, we cannot use the same correspondence formula $\gamma = 1/\tau$ used for gaseous planets. In this case, we have to use the complete equivalence formula for $Q \simeq 1/\nu\tau$ where the function Ψ is taken without approximation (Ferraz-Mello 2013). Hence, the determination of γ involves a second-degree equation with two solutions. For the whole Earth, besides the value given in the text, we also have the solution $\gamma \sim 1.6 \times 10^{-3}$ s^{-1}, the choice done being just an educated guess.

However, it is not wise just to extend the relaxation factors estimated in the Solar System to most of the known exoplanets. The distances of telluric exoplanets to their host stars is generally too small and the stellar radiation power creates an extreme physical environment for the planet. We cannot exclude values different from the above-indicated ones. Nevertheless, the estimated viscosity of the super-Earth CoRoT-7 b is $\eta > 10^{18}$ Pa s (Léger *et al.* 2011), which corresponds, for this planet, to $\gamma_p < 5 \times 10^{-7}$ s^{-1}.

We may compare these values to the analogous values for Neptune: $3 - 19$ s^{-1}. The large difference in the order of magnitude of the relaxation factors of gaseous and stiff planets may be used to decide, from the study of their putative tidal evolutions, if some potentially habitable small planet is a super-Earth or a mini-Neptune of the same size (Gomes and Ferraz-Mello 2020).

7. Conclusion

We visited the main formulas used in the study of the tidal evolution of one system formed by one star and one close-in planet taking into account the tides raised mutually in the two bodies. Most of these formulas are known, in one way or another, since the work of Darwin. The main ones shown here correspond to the approximation to $\mathcal{O}(e^2)$ of the hydrodynamical approach adopted by Ferraz-Mello (2012, 2013), the creep tide theory. Additionally, are also given, in the inclined problem, the formulas of Darwin's CTL-theory at the same order of approximation.

The state of the tidal theories, in what concerns the inclination, is far from complete. Classical theories consider only the terms proportional to the phase lag. As far as the coplanar case is considered, this is justified by the fact that the static tide (the deformation of the body considered as inviscid) is just a stretching of the body along the direction of the line joining the centers of the two bodies. Because of the symmetry of this configuration, the resulting torque is zero. In the tri-dimensional case with inclined rotation axes, however, this is not so. The bulge of the static tide in each body is no longer directed to its companion. It is pointed to a direction in the plane formed by the rotation axis of the body and the line joining the centers of the two bodies. The symmetry of the planar case is destroyed and the torque due to the static tide is no longer equal to zero (Folonier *et al.* 2022). Therefore, the contribution of the static tide cannot be neglected. In fact, it should not be neglected even in the coplanar case where it does not give long-term contributions to the energy and angular momentum, but affects the angular elements of the orbit which will precess because of the static tide.

We also added a summary of the numerical values of the relaxation factors. Some early estimations of the dissipation in stars and hot Jupiters (Hansen 2010) were used to determine the relaxation factors of those objects. The relaxation factor of the star WASP-12 could also be known from the dissipation determined from the observed decay of the orbit of its planet. In the case of stars with large companions (hot Jupiters or brown dwarfs) and with well-determined physical parameters (mass, radius, age, spin), the tides in the host star affect its period of rotation and models using the given equations allow an estimation of the γ_s (Ferraz-Mello *et al.* 2015). We have, now, a large number of bright and well-known host stars with large companions and we can expect that the reported investigations be soon replicated with larger samples thus improving our knowledge of the tidal parameters of stars. A new analysis allowing a better knowledge of the tidal parameters of the planets would also be welcome.

Acknowledgement

To Maurice Ravel, ever-present companion during this endless quarantine. I thank Prof. C. Beaugé, Dr. H. A. Folonier and G. O. Gomes for all discussions and suggestions.

This investigation is sponsored by CNPq (Proc. 303540/2020-6) and FAPESP (Proc. 2016/13750-6 ref. PLATO mission).

References

Bouvier, J., Forestini, M., Allain, S. 1997, *A&A* 326, 1023–1043
Darwin, G.H. 1879, *Philos. Trans.* 170, 1–35 (repr. Scientific Papers, Cambridge, Vol. II, 1908)
Darwin, G.H. 1880, *Philos. Trans.* 171, 713–891 (repr. Scientific Papers, Cambridge, Vol. II, 1908)
Efroimsky, M., Lainey, V. 2007, *J. Geophys. Res.*112, E12003
Fabricky, D.C., Johnson, E.T., Goodman, J. 2007, *ApJ* 665, 754.
Ferraz-Mello, S., Rodríguez, A., Hussmann, H. 2008, *Cel.Mech.Dyn.Astr.* 101, 171–201. Errata: *Cel.Mech.Dyn.Astr.* 104, 319–320 (2009). (ArXiv: 0712.1156)
Ferraz-Mello, S. 2012, *AAS/DDA meeting* 43, id.8.06 (ArXiv: 1204.3957)
Ferraz-Mello, S., 2013, *Cel.Mech.Dyn.Astr.* 116, 109–140.
Ferraz-Mello, S. 2015, *Cel.Mech.Dyn.Astr.* 122, 359–389. Errata: *Cel.Mech.Dyn.Astr.* 130:78 (2018), pp. 20-21. (arXiv: 1505.05384)
Ferraz-Mello, S., Folonier, H., Tadeu dos Santos, M., Csizmadia, Sz., do Nascimento, J.D., Pätzold, M. 2015, *ApJ* 807, 78
Ferraz-Mello, S. 2016, In A. Baglin and CoRoT team (eds.) *The CoRoT Legacy Book*, EDP Sciences, pp. 169–176.
Ferraz-Mello, S. 2019, In G.Baù et al. (eds.) *Satellite Dynamics and Space Missions*, Springer, pp. 1–50.
Ferraz-Mello, S., Beaugé, C., Folonier, H.A., Gomes, G.O. 2020, *Eur. Phys. J. ST*, 229, 1441–62.
Folonier, H.A., Ferraz-Mello, S., Andrade-Ines, E. 2018, *Cel. Mech. Dyn. Astr.* 130: 78 (arXiv: 1707.09229v2)
Folonier, H.A., Boué, G., Ferraz-Mello, S. 2022, *Cel.Mech.Dyn.Astr.* 134: 1.
Gevorgyan, Y., Boué, G., Ragazzo, C., Ruiz, L.S., Correia, A.C. 2020, *Icarus*, 343, p. 113610.
Gomes, G.O., Ferraz-Mello, S. 2020, *MNRAS* 494, 5082-5090
Hansen, B.M.S. 2010, *ApJ* 723, 285–299
Hansen, B.M.S. 2012, *ApJ* 757: 6
Hut, P. 1981, *A&A* 99, 126–140
Irwin, J., Berta, Z. K., Burke, C., Charbonneau, D., Nutzman, P. et al. 2011, *ApJ* 727: 56
Jackson, B., Greenberg, R., Barnes, R. 2008, *ApJ* 678, 1396.
Jackson, B., Penev, K., Barnes, R. 2011, *BAAS* 43, #402.06
Kaula, W.M. 1964, *Rev. Geophys.* 3 661–685
Léger, A., Grasset, O., Fegley, B., Codron, F., Albarede, A.F. et al. 2011, *Icarus* 213, 1–11
MacDonald, G.F. 1964, *Rev. Geophys.* 2, 467–541
Mignard, F. 1979, *Moon and Planets* 20, 301–315
Mignard, F. 1980, *Moon and Planets* 23, 185–201
Millholland, S.C., Spalding, C. 2020, *ApJ* 905: 71
Pätzold, M., Endl, M., Csizmadia, Sz., Gandolfi, D., Jorda, L. et al. 2012, *A&A* 545: A6.
Strugarek, A., Brun, A.S., Matt, S.P., Réville, V. 2014, *ApJ* 795: 86
Yee, S.W., Winn, J.N., Knutson, H.A., Patra, K.C., Vissapragada, S. et al. 2019, *ApJ* (Letters) 888: L5.

Multi-scale (time and mass) dynamics of space objects
Proceedings IAU Symposium No. 364, 2022
A. Celletti, C. Galeş, C. Beaugé, A. Lemaitre, eds.
doi:10.1017/S1743921321001277

Evolution of INPOP planetary ephemerides and Bepi-Colombo simulations

A. Fienga[1,2], L. Bigot[4], D. Mary[4], P. Deram[1], A. Di Ruscio[3,1], L. Bernus[2,1], M. Gastineau[2] and J. Laskar[2]

[1]GéoAzur, Observatoire Côte d'Azur, Université Côte d'Azur, CNRS, 250 Av. A. Einstein, Valbonne, 06560, France
email: agnes.fienga@oca.eu

[2]IMCCE, Observatoire de Paris, PSL University, CNRS, Sorbonne Université, 77 avenue Denfert-Rochereau, Paris, 75014, France

[3]Dipartimento di Ingegneria Meccanica e Aerospaziale, Sapienza Università di Roma, via Eudossiana 18, 00184 Rome, Italy

[4]Lagrange, Université Côte d'Azur, Observatoire de la Côte d'Azur, CNRS, Lagrange UMR 7293, CS 34229, 06304, Nice Cedex 4, France

Abstract. We give here a detailed description of the latest INPOP planetary ephemerides INPOP20a. We test the sensitivity of the Sun oblateness determination obtained with INPOP to different models for the Sun core rotation. We also present new evaluations of possible GRT violations with the PPN parameters β,γ and $\dot{\mu}/\mu$. With a new method for selecting acceptable alternative ephemerides we provide conservative limits of about 7.16×10^{-5} and 7.49×10^{-5} for $\beta - 1$ and $\gamma - 1$ respectively using the present day planetary data samples. We also present simulations of Bepi-Colombo range tracking data and their impact on planetary ephemeris construction. We show that the use of future BC range observations should improve these estimates, in particular γ. Finally, interesting perspectives for the detection of the Sun core rotation seem to be reachable thanks to the BC mission and its accurate range measurements in the GRT frame.

Keywords. Planetary ephemerides; Space mission tracking; Fundamental physics; solar physics.

---

1. Introduction

The INPOP (Intégrateur Numérique Planétaire de l'Observatoire de Paris) planetary ephemeris has started to be built in 2003. It consists in numerically integrating the Einstein-Imfeld-Hoffman equations of motion and relativistic time-scale definitions (TT,TDB,TCG,TCB) for the eight planets of our solar system, and the Moon (orbit and rotation) (Fienga et al. 2009). In our latest version, INPOP20a presented in this paper, orbits of about 343 Main Belt asteroids and 500 Trans-Neptunian Objects (TNO) are also integrated. An adjustment of 402 parameters, including planetary initial conditions, gravitational mass of the Sun and its oblateness, the Earth Moon mass ratio, 343 Main Belt asteroid masses and one global mass for the 500 TNOs, obtained in using 150,000 observations from spacecraft tracking and ground-based optical observations.

In Solar physics, important questions are still pending. In particular, the solar differential rotation has been determined by helioseismology with great precision using the acoustic modes (p-modes) trapped into the solar cavity (Thompson et al. 2003, for a review). It was shown that rotation depends on both the depth r and the latitude θ in the convection zone ($r > 0.71R_{\odot}$). In the radiative zone, the rotation is near solid (uniform) with a rotation rate of about $\Omega_0/2\pi = 435$ nHz (Komm et al. 2003). The rotation of the solar core in still a matter of debate since it is hardly inferred by p-modes. Indeed, since the sensitivity of the p-modes decreases towards the core due to their decreasing mode amplitude at large sound speed, helioseismic inversions were done with precision, from the surface down to about $r \sim 0.3\,R_{\odot}$. Several instruments both from ground (networks of telescopes around the world) and space missions have tried to infer the solar core rotation using p-modes, but they diverge below $r/R_{\odot} \sim 0.25$ (see Di Mauro 2003, for a comparison of rotation profiles). While space measurements from GOLF on board on the SOHO spacecraft (Roca Cortés et al. 1998; García et al. 2004) and ground-based observations from the networks IRIS (Lazrek et al. 1996) and GONG (Gavryuseva et al. 1998) all show an increase of the rotation in the core, the Bison network result is in favor of a decrease (Chaplin et al. 1999; Gough 2015). It is worth mentioning that tentative detections of gravity mode (g-modes) using the GOLF data have also led to a very fast core rotation roughly 3-4 times the value of the radiative zone (García et al. 2007; Fossat et al. 2017; Fossat and Schmider 2018). These claimed detections are still questioned by the community and must be taken with care (see Schunker et al. 2018; Scherrer and Gough 2019; Appourchaux and Corbard 2019, for discussions about the recent analysis of Fossat et al. 2017). A redetermination of the solar rotation using decades of seismic data up to the most recent ones is under investigation (Christensen-Dalsgaard 2021) since the exact rotation profile in the core is important to understand the past history of our Sun, its angular momentum transport, and the role of the magnetic field on solar structure. In view of these uncertainties of the rotation in the core, we have tested the possibility of using INPOP ephemerides to disentangle these scenarios : faster, slower, or solid rotation in the solar core, through its impact on the Sun gravitational oblateness, $J_2^{\odot}$.

For this work, we use the latest update of the INPOP planetary ephemerides, INPOP20a, for obtaining for the Post Parametrized Newtonian (PPN) parameters β and γ as we did with INPOP15a (Fienga et al. 2015) but with a new criterion for ephemeris selection, a more accurate ephemeris and a more complete dynamical modeling. We also give constraints on the Sun oblateness related to different hypotheses of the Sun core rotation, together with new limits for the Sun gravitational mass loss, based on planetary orbits. We also consider the future evolution of INPOP in simulating Mercury-Earth range observations from the Bepi-Colombo mission and in deducing subsequent improvements for the ephemerides and General Relativity Theory (GRT) tests.

Sect. 2 introduces the INPOP20a planetary ephemeris, describing the modifications brought in the dynamical modeling and the update of the planetary data sets used for its adjustment. New determinations for the Sun oblateness including the Lense-Thirring effect, different scenario for the rotation of the Sun core, and the mass of the Kuiper belt are given in Sect 2.1, 2.2 and 2.3 respectively. Comparisons with our former ephemeris, INPOP19a, are also presented in terms of postfit residuals in Sect. 2.4. In Sect. 2.5 we explain how we simulate Earth-Mercury range measurements for the Bepi-Colombo MORE experiment. In Sect. 3.1 and 3.2, is described the approach associating Monte Carlo sampling and least squares adjustment with WRSS filtering for obtaining new limits for PPN parameters and secular variations of Sun gravitational mass, $\dot{\mu}/\mu$. Finally we give in Sect. 4.1 the results obtained with INPOP20a for the PPN parameters β, γ and $\dot{\mu}/\mu$ where Sect 4.2 provides the one obtained in including the Bepi-Colombo simulations.

2. INPOP20a planetary ephemerides

The INPOP20a planetary ephemerides was built with the same data sample as INPOP19a (Fienga et al. 2019) but with the addition of 5 Jupiter positions deduced from the Juno perijove PJ19 to PJ23, leading to a coverage of more than 4 years with

Table 1. INPOP20a data samples. Column 1 gives the observed planet and information on the type of observations, and Column 2 indicates the number of observations. Columns 3 and 4 give the time interval and the a priori uncertainties provided by space agencies or the navigation teams, respectively. Finally, the WRMS for INPOP19a and INPOP20a are given in the last two columns.

Planet / Type	#	Period	A priori uncertainty	WRMS INPOP19a	WRMS INPOP20a
Mercury					
Direct range [m]	462	1971.29 : 1997.60	900	0.95	0.95
Messenger range [m]	1096	2011.23 : 2014.26	5	0.82	0.82
Mariner range [m]	2	1974.24 : 1976.21	100	0.37	0.42
Venus					
VLBI [mas]	68	1990.70 : 2013.14	2.0	1.13	1.15
Direct range [m]	489	1965.96 : 1990.07	1400	0.98	0.98
Vex range [m]	24783	2006.32 : 2011.45	7.0	0.93	0.94
Mars					
VLBI [mas]	194	1989.13 : 2013.86	0.3	1.26	1.26
Mex range [m]	30669	2005.17 : 2017.37	2.0	0.98	1.0175
		2005.17 : 2016.37	2.0	0.97	1.02
MGS range [m]	2459	1999.31 : 2006.70	2.0	0.93	0.945
MRO/MO range [m]	20985	2002.14 : 2014.00	1.2	1.07	1.016
Viking range [m]	1258	1976.55 : 1982.87	50.0	1.0	1.0
Jupiter					
VLBI [mas]	24	1996.54 : 1997.94	11	1.01	0.998
Optical RA/Dec [arcsec]	6416	1924.34 : 2008.49	0.3	1.0/1.0	1.02/1.01
Flyby RA/Dec [mas]	5	1974.92 : 2001.00	4.0/12.0	0.94/1.0	0.95/1.01
Flyby range [m]	5	1974.92 : 2001.00	2000	0.98	1.24
Juno range [m]	14	2016.65 : 2019.84	14	1.35	1.02
	9	2016.65 : 2018.68	14	1.35	1.01
Saturn					
Optical RA/Dec [arcsec]	7826	1924.22 : 2008.34	0.3	0.96/0.87	0.96/0.88
Cassini					
VLBA RA/Dec [mas]	10	2004.69 : 2009.31	0.6/0.3	0.97/0.99	0.945/0.973
JPL range [m]	165	2004.41 : 2014.38	25.0	0.99	1.033
Grand Finale range [m]	9	2017.35 : 2017.55	1.0	1.71	0.8
Navigation [m]	572	2006.01 : 2009.83	6.0	0.71	0.85
TGF range [m]	42	2006.01 : 2016.61	15.0	1.13	1.30
Uranus					
Optical RA/Dec [arcsec]	12893	1924.62 : 2011.74	0.2/0.3	1.09 / 0.82	1.09 / 0.82
Flyby RA/Dec [mas]	1	1986.07 : 1986.07	50/50	0.12 / 0.42	0.133 / 0.40
Flyby range [m]	1	1986.07 : 1986.07	50	0.92	0.92
Neptune					
Optical RA/Dec [arcsec]	5254	1924.04 : 2007.88	0.25/0.3	1.008 / 0.97	1.008 / 0.97
Flyby RA/Dec [mas]	1	1989.65 : 1989.65	15.0	0.11 / 0.15	0.11 / 0.15
Flyby range [m]	1	1989.65 : 1989.65	2	1.14	3.6805

an accuracy of about 14 meters. Two important modifications have also been brought to the dynamical modeling and are presented in the following.

2.1. *Lense-Thirring effect*

When comparing the estimations of the Sun oblateness †, $J_2^{\odot}$, obtained with planetary ephemerides to values obtained by helioseismology (Antia et al. 2008; Pijpers 1998), it is important to keep in mind that an additional contribution must be included in order to compare consistent estimates: the effect of the Sun rotation on the space-time metric (Lense and Thirring 1918). This effect known as the Lense-Thirring effect has been evaluated to contribute to about 10% (Hees 2015) of the dynamical acceleration induced by the shape of the Sun in General relativity (GRT). With the accuracy of the Bepi-Colombo mission, it is important to include this effect in the INPOP equations of

† this definition corresponds to the gravity field second degree term, -C20

motion. The acceleration induced by the Lense-Thirring effect generated by a central body (at the first post-Newtonian approximation) is given by

$$\vec{a}_{LT} = \frac{(\gamma+1)G}{c^2 r^3} S \left[3\frac{\vec{k}.\vec{r}}{r^2}(\vec{r}\wedge\vec{v}) - (\vec{k}\wedge\vec{v}) \right] \tag{2.1}$$

where G is the gravitational constant, c the speed of light, $\vec{S}$ is the Sun angular momentum such as $\vec{S} = S\vec{k}$ where $\vec{k}$ is the direction of the Sun rotation pole defined according to the IAU right ascension and declination (Archinal et al. 2018), $\vec{r}$ and $\vec{v}$ are the position and velocity vectors of the planet relative to the central body (here the Sun) and γ is the PPN parameter for the light deflection. Depending the model adopted for the rotation of the Sun core (see Sect. 2.2), one can estimate different values for the amplitude of the Sun angular momentum S, implemented in INPOP and presented in Table 2. For each value of the Sun angular momentum, an INPOP adjustment is done and $J_2^{\odot}$ is estimated. The $J_2^{\odot}$ obtained with INPOP20a in considering the Sun angular momentum from helioseismological measurements (Pijpers 1998) is given in the first line of Table 2. This value, $(2.21 \pm 0.01)\times 10^{-7}$, is very close from the values deduced from SOHO $(2.22 \pm 0.009)\times 10^{-7}$ and GONG $(2.18 \pm 0.005)\times 10^{-7}$ (Antia et al. 2008). It is also in good agreement with the previous analysis of the same data made by (Pijpers 1998) giving as an average estimate between GONG and SOHO, $(2.18 \pm 0.06) \times 10^{-7}$. In (Park et al. 2017) estimations for both S and $J_2^{\odot}$ (presented in Table 2) were obtained in considering Messenger tracking data. It is important to stress that there is an important correlation (80 %) between S and $J_2^{\odot}$ when both estimated in a global planetary fit. Because of this high correlation, the Sun angular momentum S is not fitted in the INPOP adjustment instead we use the value from (Pijpers 1998). The same choice has been made by (Genova et al. 2018) who focus on using Messenger data for constraining Mercury and Earth orbits. Their obtained value of $J_2^{\odot}$ is also given in Table 2 and is consistent with our estimate as well as with the one of (Park et al. 2017) and (Pijpers 1998) but not with (Antia et al. 2008). Finally, with the planetary ephemerides determinations, the PPN parameter β and the Sun oblateness $J_2^{\odot}$ are usually strongly correlated. A simultaneous estimation of these two quantities is usually very complex or leads to underestimated uncertainties (Fienga et al. 2015; Genova et al. 2018). For this reason and as it is now possible to directly related INPOP $J_2^{\odot}$ with helioseismological values, we chose to constraint the fitted values of $J_2^{\odot}$ to remain in between the interval of $(2.18 \pm 0.06) \times 10^{-7}$ corresponding to (Pijpers 1998) results. We see in Table 2 that in the case of GRT (with β and γ equal to one), this interval is easily respected. The case where GRT is violated is discussed in Sec. 3.

2.2. *Sun core rotation*

We use the INPOP planetary ephemerides to constrain solar core rotation. In our global solution of the planetary ephemeris, solar rotation is present through the gravitational $J_2^{\odot}$ and the Lense-Thirring effects. While the first effect is coming as a solution of the fit of these ephemerides, the latter needs the knowledge of the solar angular momentum calculated using a calibrated solar model:

$$S = \frac{1}{2}\int_0^{R_\odot} r^2 dm \int_{-1}^{1} (1-\cos^2\theta)\,\Omega(r,\theta)\,d\cos\theta, \tag{2.2}$$

where $dm = 4\pi\rho r^2 dr$ is the mass fraction. The rotation profile is splitted in two parts : $\Omega(r,\theta) = \Omega_{\text{core}}(r) + \Omega_{\text{helio}}(r,\theta)$. We adopt the solution $\Omega_{\text{helio}}(r,\theta)$ proposed in Roxburgh (2001) inferred from helioseismology.

For Ω_{core}, we assume only a radial dependence. The profile of the rotation is assumed to have the parametric form $\Omega_{\text{core}}/2\pi = \mathrm{K}\exp(-\mathrm{r}^2/\mathrm{r}_\mathrm{c}^2)$, where K is a constant to adjust

Table 2. Sun Angular Momentum and oblateness. Is given in Column 3, the values of Sun $J_2^{\odot}$ obtained after fit using the values of the amplitude of the angular momentum given in Column 2. Different models of rotation (identified in Column 1) are used for estimating S. In the first line, are given the results obtained for INPOP20a. See Sect 2.2 for the significance of the different rotation hypothesis.

Type of rotation	**S** $\times 10^{48}$ **g.cm**$^{-2}$**.s**$^{-1}$	$J_2^{\odot} \times 10^7$
INPOP20a with Pitjers 1998	1.90	2.218 ± 0.03
(Park et al. 2017)	1.96 ± 0.7	2.280 ± 0.06
(Genova et al. 2018)	1.90	2.2710 ± 0.003
Slow rotation	1.896	2.208 ± 0.03
uniform rotation at 435 nHz	1.926	2.210 ± 0.03
Fast rotation	1.976	2.213 ± 0.03
Very fast rotation	1.998	2.214 ± 0.03
Pijpers (1998)	1.90 ± 1.5	2.180 ± 0.06
Antia et al. (2008)	1.90 ± 1.5	2.2057 ± 0.007

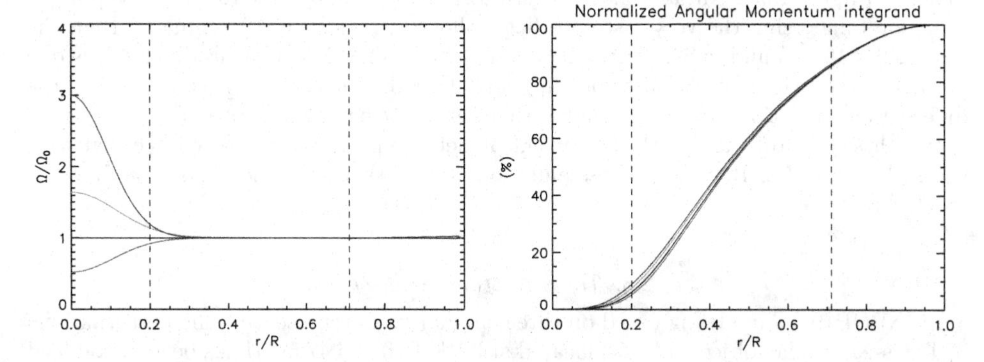

Figure 1. (Left Panel) The four rotation profiles as functions of depth used to computed the angular momentum of the Sun normalized by $\Omega_0 = 435\,\text{nHz}$: very fast rotation (blue), fast rotation (green), slow rotation (red), and uniform rotation (black). The dashed lines represent the approximate boundaries of the solar core at $r \sim 0.2\,R_{\odot}$ and the position of the tachocline, i.e. the transition between the radiative and the convection zones. (Right panel) The corresponding angular momentum integrands in Eq. 2.2 as functions of depth.

and $r_c \sim 0.15\,R_{\odot}$ characterizes the extension of the core. These coefficients are adjusted to reproduce the rotation rates inferred from helioseismology and shown in Fig. 16 of Di Mauro (2003). The result of the fit is shown in Fig. 1 together with the case of a very fast rotation rate, as proposed in Fossat et al. For the present analysis, we consider 4 cases : very fast (GOLF, Fossat et al. (2017)), fast (GOLF, Roca Cortés et al. (1998)), slow (Bison, Chaplin et al. (1999)), and uniform. The values of angular momenta and corresponding $J_2^{\odot}$ are shown in Table 2. The different core rotations change the total angular momentum by about 5% at most between the two extreme cases (very fast and slow core rotations). The reason of such small impact of the core rotation is due to the r^2 dependence in Eq. 2.2 as seen in Fig. 1. Most of the contribution of the angular momentum (80%) is coming from the radiative zone and the core has a small contribution. We have also tested the impact of differential rotation of the convection zone, as inferred by helioseismology, i.e. the radial and latitudinal dependencies, compared to the case of uniform solid rotation. The difference is very small, i.e. 0.02%. Using these different angular momenta to account for the Lense-Thirring effect in our global ephemerides fit, we extract the corresponding gravitational $J_2^{\odot}$, as shown in Tab. 2. Our values found by

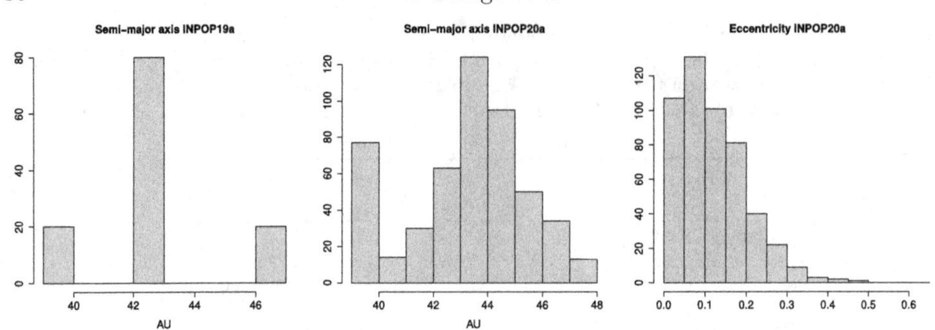

Figure 2. Semi-major axis and eccentricity distributions: Comparison between the INPOP19a TNO modeling and one selection of 500 orbits randomly chosen in the Astorb database.

fitting planetary ephemerides are in good agreement with those inferred from helioseismology (Pijpers 1998; Antia et al. 2008) with a value close to 2.2×10^{-7}. We emphasize that our value using the very fast core (e.g. following Fossat et al.) is smaller than the $J_2^{\odot} \approx 2.6 \times 10^{-7}$ found in Scherrer and Gough (2019). The reason of this difference is due to their large extent of the fast rotating core. Our differences in $J_2^{\odot}$ coming from the different core rotations are much smaller than our error bars, which prevents us to disentangle these core rotations with the current planetary ephemerides. We will see that with the inclusion of the Bepi-Colombo simulations (BC), this conclusion could be different (Sect. 4.2.2).

2.3. *Trans-Neptunian objects*

In INPOP19a, a modeling based on three circular rings representing the perturbations of Trans-Neptunian objects (TNO) located at 39.4, 44.0 and 47.5 AU has been introduced and outer planet orbits have been clearly improved, especially Saturn orbit (Fienga et al. 2019; Di Ruscio et al. 2020). However, with this circular ring modeling, the impact of the eccentricities of the TNO orbits was not included in the computation of the perturbing accelerations. The global mass of these rings appears also to be too important in comparison with theoretical estimations (see (Di Ruscio et al. 2020) for the full discussion). As TNO orbits tend to be more eccentric compared to main belt asteroid orbits, we implement an alternative representation by considering directly observed orbits extracted from the Astorb database (Moskovitz et al. 2018). In order to limit the number of objects to consider for not increasing too much the time of computation, on the total of 2225 objects with semi-major axis between 39.3 and 47.6 AU, we operated random selections of 500 of them that we integrated as individual objects with the same mass spread over the 500. Thanks to this approach the representation of the TNOs is more realistic, in particular, regarding the distributions in eccentricities and in semi-major axis (see Fig. 2) without increasing too much the integration time. For each random sampling of 500 objects a full fit was operated in adjusting the global mass of the 500, in addition to the regular planetary ephemeris parameters. Results being very similar from one random selection to another, one selection was chosen arbitrary for the rest of this study. After fit, the global mass for 500 TNOs is found to be $(1.91 \pm 0.05) \times 10^{23}$ kg which corresponds to (0.031 ± 0.001) $M_{\oplus}$. This mass is about two times smaller than the one proposed by (Di Ruscio et al. 2020). This difference can be explained by the differences in the dynamical modeling between this work and (Di Ruscio et al. 2020). While (Di Ruscio et al. 2020) used circular rings, we include here a real distribution of orbits with various eccentricities as one can see on Fig. 2.

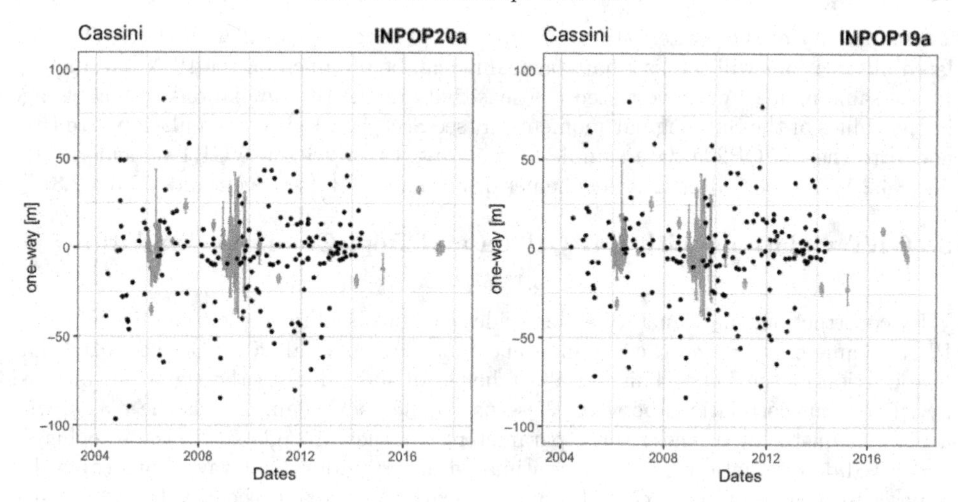

Figure 3. Cassini postfit residuals for INPOP19a and INPOP20a. The black dots are residuals obtained for JPL range sample when the orange dots indicate the residuals obtained for the Grand Finale (in 2017), the Navigation and the TGF range samples. For a more complete description of these sample, see (Di Ruscio et al. 2020).

2.4. *Postfit residuals and comparisons with INPOP19a*

One can find on Table 1 the weighted root mean squares (WRMS) for the INPOP19a and INPOP20a postfit residuals together with a brief description of the corresponding data sets (type of observations, number of observations, time coverage). The WRMS is defined as $WRMS = \sqrt{\sum_{i=1}^{N} \frac{(O_i - C_i)^2}{\sigma_i^2}}$, where $(O_i - C_i)$ is the postfit residual for the observation i, σ_i is the *a priori* instrumental uncertainty of the observation i given in Column 4 of Table 1. Where for the inner planets, the differences between INPOP19a and INPOP20a are not clearly visible, the improvement is more effective for outer planets, especially for Jupiter and Saturn. For Jupiter, the addition of 5 new perijove obtained up to P23 improves the residuals from about 18 m with INPOP19a to 14 m with INPOP20a.

For Saturn, one can note the significant improvement for Cassini samples, in particular for the Grand Final residuals obtained in 2017 for which the INPOP20a residuals is about 2.4 times smaller than the INPOP19a one. This result is a direct consequence of the introduction of the new TNO ring modele presented in Sect. 2.3 and is linked to the removal of a secular trend clearly visible in INPOP19a residuals but not in INPOP20a (see Fig. 3).

2.5. *Bepi-Colombo simulations*

In using INPOP20a as a reference planetary ephemerides, we have simulated possible range bias for the Bepi-Colombo mission between Mercury and the Earth. These simulations are used for estimating the impact for the INPOP construction of using very accurate Mercury-Earth distances as they should be obtained by the radio science MORE experiment. Based on the assumption that the radio tracking in KaKa-Band keeps the 1 cm accuracy that has been monitored during the commissioning phase of the Bepi-Colombo mission in 2019 and 2020 (Iess et al. 2021), we suppose a daily acquisition of range tracking data (Thor et al. 2020) during a period of 2.5, from 2026 to 2028.5. The simulated residuals obtained in using INPOP20a as reference ephemerides are plotted in Fig 4. These GRT residuals provide a reference against which can be tested the epheremides integrated with non-GRT parameters (i.e., PPN $\beta \neq \gamma \neq 1$ and $\dot{\mu}/\mu \neq 0$).

The capability of these alternative ephemerides to have a good fit with the GRT simulated observations will tell us what constraints can be obtained on the PPN parameters. In the same manner we have tested the sensitivity of the BC simulations to any change in the values of the Sun angular momentum (see Sect 4.2.2). These simulations are then added to the INPOP20a data sample for building a new reference GRT ephemeris (see Sect. 4.2.1) and new alternatives ephemerides in non-GRT (see Sect. and 4.2.3 4.2.3).

3. GRT violations with INPOP20a and Bepi-Colombo simulations

3.1. *Method*

By construction, the planetary ephemerides cannot disentangle the contribution of the PPN parameters β, γ and the Sun oblateness $J_2^\odot$ (Fienga et al. 2015; Bernus 2020). The introduction of the Lense-Thirring effect helps for individualize the signature induced by PPN γ but correlations between these parameters stay high. This is the reason why a direct adjustment of these three parameters together in a global fit leads to highly correlated determinations and under-estimated uncertainties. One way to overcome this issue is to fix one of these contributors, for example in fixing the γ value to the one estimated by (Bertotti et al. 2003) with the Cassini experiment in 2003. However, as with the Bepi-Colombo mission, far more accurate constraints are planned to be obtained with the same solar conjunction techniques (Imperi and Iess 2017), we decide not to fix γ but to add helioseismological limits for the Sun $J_2^\odot$ (see Sect 2.1). These thresholds are applicable when the Lense-Thirring effect is included in the dynamical modeling. Additionally, as explained in Sect 2.1, in the GRT case when no limits are applied to the $J_2^\odot$ determinations, INPOP20a gives a very consistent value, included in the uncertainties of the heliosismology. We note that the helioseismology limits are obtained based on the analysis of time variations of the Sun angular momentum and its kinetical energy in the Newtonian framework. Even if a strict approach would require a complete reanalysis of the helioseismology measurements in a non GRT-frame, one can expect a negligeable effect (Soffel and Frutos 2016). As in (Fienga et al. 2015), we introduce the parameter $\dot{G}/G$ through the secular variations of the gravitational mass of the Sun $\dot{\mu}/\mu$ with the equation

$$\dot{\mu}/\mu = \dot{G}/G + \dot{M}_\odot/M_\odot \tag{3.1}$$

where $\dot{M}_\odot/M_\odot$ is the Sun mass loss. This quantity is fixed by (Fienga et al. 2015) such as

$$\frac{\dot{M}_\odot}{M_\odot} = (-0.92 \pm 0.61) \times 10^{-13} \quad (3\sigma)\,\mathrm{yr}^{-1} \tag{3.2}$$

$\dot{G}/G$ is also accounting for the update of the masses of the planets and asteroids at each step of the integration of the body equations of motion as well as in the time-scale transformation TT-TDB and the correction of the Shapiro delay in the range computation. At each step t of the numerical integration of the INPOP equations of motion, we then estimate :

$$\begin{aligned} M_\odot(t) &= M_\odot(J2000) + (t - J2000) \times \dot{M}_\odot & (3.3)\\ G(t) &= G(J2000) + (t - J2000) \times \dot{G} & (3.4)\\ \mu(t) &= G(t) \times M_\odot(t) & (3.5) \end{aligned}$$

where $M_\odot(J2000)$, $G(J2000)$ are the mass of the sun and the constant of gravitation at the date J2000 and $(t - J2000)$ is the time difference between the date of the integration t and J2000 in years. The value of the Sun gravitational mass $\mu(t)$ corresponding to the date of the observation t is computed with Equation 3.5 and is then re-introduced in the

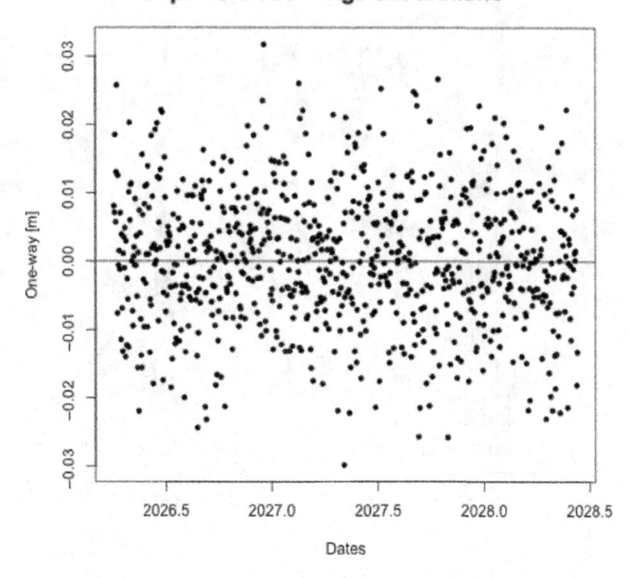

Figure 4. Bepi-Colombo 1-cm simulated residuals obtained in using INPOP20a as reference ephemeris in GRT and an 8-hour temporal resolution.

Table 3. Intervals for the uniform distributions for PPN parameters β and γ and $\dot{\mu}/\mu$.

Parameters	INPOP20a	with BC
β	$\pm 20 \times 10^{-5}$	$\pm 20 \times 10^{-5}$
γ	$\pm 20 \times 10^{-5}$	$\pm 1 \times 10^{-5}$
$\dot{\mu}/\mu$	$\pm 6 \times 10^{-13}\,\mathrm{yr}^{-1}$	$\pm 1 \times 10^{-13}\,\mathrm{yr}^{-1}$

Shapiro delay equation (8-38) given in (Moyer 2000). In this context, the strategy chosen for this study is the same as in (Fienga et al. 2015; Bernus et al. 2019, 2020): we built full planetary ephemerides by integrating and adjusting to observations Einstein-Imfeld-Hoffman equations of motion for planetary orbits, timescale transformation and Shapiro delay computation (Moyer 2000; Fienga et al. 2009) together with the Lense-Thirring effect (Eq. 2.1) and time varying G (Eq. 3.5). For the non-GRT parameters (PPN $\beta \neq 1$, $\gamma \neq 1$ and $\dot{\mu}/\mu \neq 0$), we take samples of random values following uniform distributions. Table 3 gives the intervals used for these uniform distributions. We have chosen random distributions instead of regular grids because we plan to use these results in a Bayesian context for a forthcoming study. Finally, these intervals were chosen in order to encompass the larger published limits (Fienga et al. 2015). For the BC simulation, the intervals were optimized according to the sensitivity of the simulated observations but still encompass the limits proposed by the literature in the frame of the BC mission (Imperi et al. 2018; De Marchi and Cascioli 2020). In this work, for filtering out the ephemerides built with the non GRT parameters (alternative ephemerides), we consider a study of the WRSS distribution.

3.2. *Construction of the Weighted Residual Sum of Squares distribution*

Before considering GRT violations, we study the instrumental noise variability of the INPOP adjustment. To do so, we use Monte Carlo simulations where we generate fake observations obtained by adding a Gaussian noise to the true observations. The standard deviation of the noise added to each observation is taken as the value of the INPOP

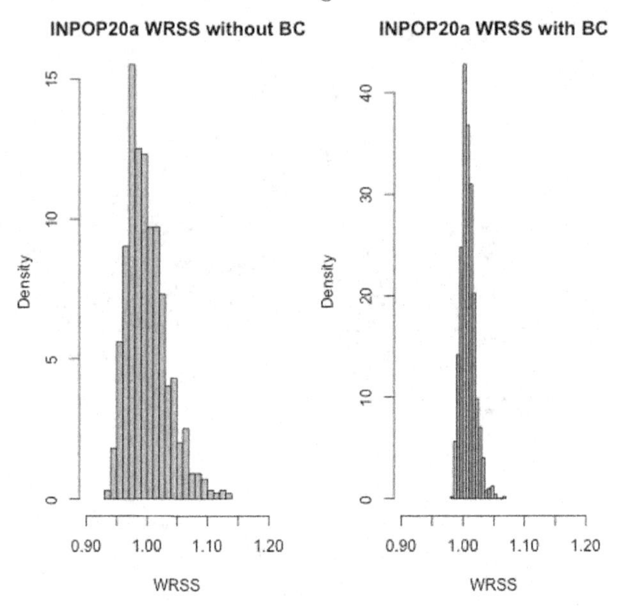

Figure 5. Distribution of INPOP20a WRSS including instrumental noise without BC simulations (left-hand side) and with BC simulations (right-hand side).

residual for the corresponding observation. We operate 1000 samplings and for each of them we refit the INPOP ephemerides and we compute the Weighted Residual Sum of Squares (WRSS) as followed:

$$WRSS = \frac{1}{N}\sum_{i=1}^{N}\frac{((O-C)_i)^2}{\sigma_i^2} \tag{3.6}$$

where $(O-C)_i$ is the difference between the observation O and the observable C computed with INPOP (postfit residual) for the observation i, σ_i is the *a priori* instrumental uncertainty of the observation i and N is the number of observations. We obtain a experimental WRSS distribution as presented in Fig. 5. From this empirical distribution, we can estimate the probability of a postfit WRSS to be explained by the instrumental uncertainties. We derive the quantiles corresponding to the 3-σ of the WRSS distribution after fitting a log-normal profile to this latest and we can estimate a confidence interval [WRSS $_{max}$:WRSS $_{min}$] that contains 99.7% of the distribution. We used these WRSS $_{max}$ and WRSS $_{min}$ as thresholds for the selection of alternative ephemerides (estimated with non-GRT parameters) compatible at 99.7% with the observations. With the INPOP20a data sampling, this leads to the definition of an interval of WRSS of about ±0.09 around the INPOP20a WRSS . We proceed in the same manner with the Bepi-Colombo (BC) simulations. We add BC simulations to the INPOP20a data sample in taking 1 centimeter as instrumental uncertainty. Fig. 5 gives the distribution of the WRSS including the BC simulations. The obtained profile for the WRSS distribution is clearly more narrow compared with the one obtained without. This may indicate the improvement of the fit quality and consequently, the increase of the constraint for the tested parameters. In considering the quantiles corresponding to the 3-σ WRSS distribution, we deduce the WRSS $_{min}$ and WRSS $_{max}$ to be used for selecting alternative ephemerides. The WRSS interval is then of ±0.03 around the WRSS of INPOP20a including BC.

Table 4. PPN β and γ confidence intervals given at 99.7 % of the reference WRSS distribution, with and without Bepi-Colombo simulations. The first column indicates the method being used for obtaining the results given in Columns 4 (with INPOP20a) and 6 (with INPOP20a and BC simulations).LS stand for Least squares, MC for Monte Carlo and GA for Genetic Algorithm.

	$(\beta - 1)$ $\times 10^5$	$(\gamma - 1)$ $\times 10^5$	$\dot{\mu}/\mu$ $\times 10^{13}$ yr^{-1}	$J_2^{\odot}$ $\times 10^7$
3-σ WRSS INPOP20a	-1.12 ± 7.16	-1.69 ± 7.49	-1.03 ± 2.28	2.206 ± 0.03
non-GRT LS INPOP20a	-1.9 ± 6.28	2.64 ± 3.44	-0.37 ± 0.32	2.165 ± 0.12
3-σ WRSS INPOP20a + BC	0.32 ± 5.00	0.09 ± 0.40	-0.19 ± 0.19	2.206 ± 0.009
non-GRT INPOP20a + BC	± 1.06	± 0.23	± 0.01	± 0.013
(Fienga et al. 2015) LS 3-σ	-6.7 ± 6.9	-0.8 ± 5.7	-0.50 ± 0.29	2.27 ± 0.25
(Fienga et al. 2015) MC	-0.8 ± 8.2	0.2 ± 8.2	-0.63 ± 1.66	1.81 ± 0.29
(Fienga et al. 2015) GA	0.0 ± 6.9	-1.55 ± 5.01	-0.43 ± 0.74	2.22 ± 0.13

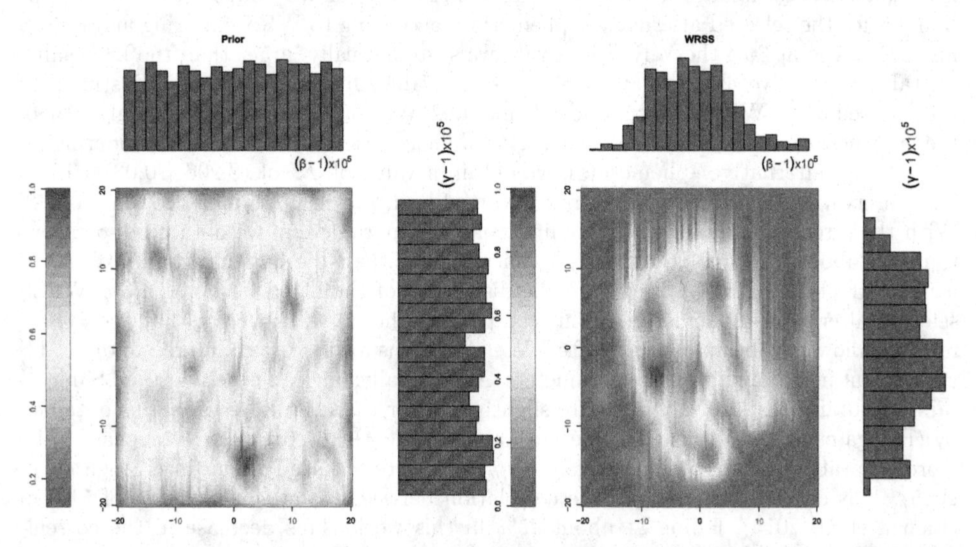

Figure 6. 2D-histograms of PPN β and γ. The first left-hand side plot shows the uniform distribution of the prior when the middle histogramms give the distribution of the selected values of β and γ according to the 3-σ INPOP20a WRSS distribution without considering the Bepi-Colombo simulations. The color-scale indicates the normalised probability.

4. Results

4.1. *with INPOP20a*

4.1.1. PPN parameters β and γ

3800 runs are estimated with values of PPN β and γ different from unity, as described in Sect .3.1. For each of these runs, alternate planetary ephemerides are integrated and fitted over the same data sample as INPOP20a in an iterative process, equivalent to the one used for INPOP20a construction. When the iterations converged, the WRSS is compared to the INPOP20a WRSS distribution as discussed in Sect. 3.2. The selection described above is then operated according to the estimated WRSS and only 23% of the runs have been kept. In Fig. 6, are plotted the 2-D histogram for the initial (uniform) distributions of PPN parameters β and γ together with the 2-D histogram of the selected ephemerides, compatible at 3-σ with the INPOP WRSS distribution, without BC simulations. Table 4 gives the deduced intervals for the two quantities based on the WRSS filtering. These intervals are all compatible with GRT and can be compared with the one obtained by direct least square fit (non-GRT LS), based on the INPOP20a data sample. These results are obtained by adding the non-GRT parameters to the full INPOP20a adjustment together with the 402 other parameters (including the mass of the Sun and its oblateness, constrained by helioseismology values). As for the WRSS filtering, these estimations are also consistent with GRT. Nevertheless, there are major correlations obtained from LS covariance analysis, between β, γ and the other parameters of fit. In particular we note 75% of correlation between β and $J_2^\odot$ and between $J_2^\odot$ and γ as well as between γ and the semi-major axis of the inner planets. As a first consequence of these correlations, in the global fit including β and γ, the value of the $J_2^\odot$ decreases significantly and could escape from the (Pijpers 1998) interval considering the LS uncertainties. The WRSS of this non-GRT fit is in the 3-σ quantiles of the INPOP20a WRSS distribution, showing that the WRSS filtering is more conservative than a direct least squares determination. This is also clearly visible when we compare the intervals for β and γ for the selected alternative ephemerides according to WRSS filtering and the LS intervals given at 3-σ: the WRSS intervals are systematically larger than the LS results. In Table 4, we give also the average of the $J_2^\odot$ values fitted during the adjustment of the selected $3-\sigma$ WRSS alternative ephemerides. Without BC simulations, only 12% of the ephemerides have $J_2^\odot$ values reaching the heliocentric boundaries. In considering all the selected alternative ephemerides, we obtain a value of $J_2^\odot$ of $(2.206 \pm 0.03) \times 10^{-7}$. This interval is consistent with the one obtained by LS fit in GRT presented in Table 2. With the direct LS adjustment including non-GRT parameters, we obtain a 3-σ uncertainty of about 0.12×10^{-7}, which is 4 times greater than the 3-σ uncertainty obtained for the direct LS fit in GRT or than the dispersion of the fitted $J_2^\odot$ for the 3-σ WRSS selected alternative ephemerides. This comparison shows, as expected, that the WRSS filtering allows to obtain less correlated determinations for the $J_2^\odot$ compared to direct LS adjustment including non-GRT parameters. Additionally, one can note that the LS uncertainties obtained with INPOP20a are slightly smaller than the LS estimations obtained by (Fienga et al. 2015), also provided in Table 4 (line 4). In particular, γ appears to be more accurately constrained relatively to β in the present study than in (Fienga et al. 2015). This can be explained by the correlation between β and γ which was of 51% in (Fienga et al. 2015), falling at about 25% in this work. This decrease in the correlation is induced by the Lense-Thirring acceleration, introducing an additional constraint on γ, independently from β. In comparisons with (Fienga et al. 2015) and the results obtained with a genetic algorithm (labelled "GA" in Table 4) and the Monte Carlo runs (labelled "MC" in Table 4), one can be surprised that the limits for the β and γ intervals have not been more reduced as more accurate planetary tracking observations were used for the construction of INPOP20a. Several reasons can be proposed. Firstly, as already mentioned, the reference ephemerides used in (Fienga et al. 2015), INPOP15a, has built using a different dynamical modeling (without Lense-Thirring effect nor TNO ring). The INPOP15a data sample did also not account for Juno data nor Cassini recent re-analysed observations. Secondly the selection criteria used in (Fienga et al. 2015) were also different from the one used in this work. Finally, regarding the MC runs, the intervals of randomly selected β and γ values are larger in this present work $(\pm 20 \times 10^{-5})$ compared to (Fienga et al. 2015) $(\pm 15 \times 10^{-5})$. Additionally, with the genetic algorithm method, one can not demonstrate that the (Fienga et al. 2015) convergency had reached the unique extremum (and not a local extremum), or even if such unique extremum does exist (Katoch and Chauhan 2021). Using a GA approach is then problematic and has been not used for this work. Comparisons with the present results are then difficult.

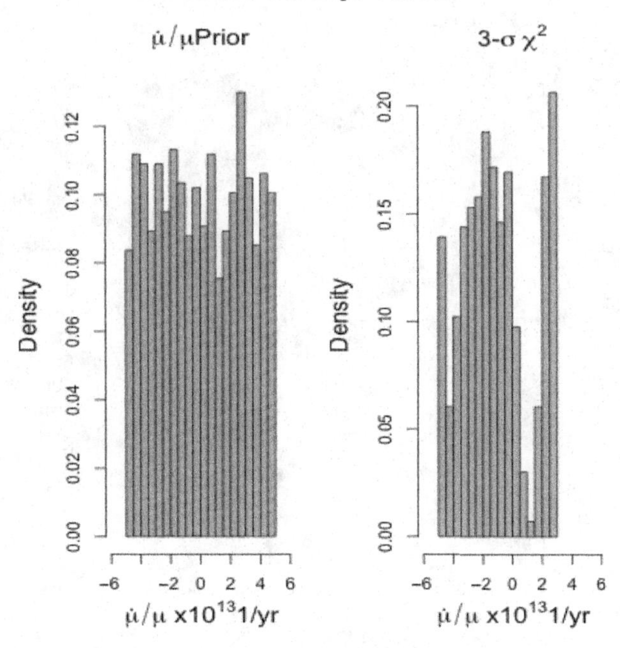

Figure 7. 1D-histograms of $\dot{\mu}/\mu$ in yr^{-1}. The first left-hand side plot shows the uniform distribution of the prior when the middle histograms give the distribution of the selected values of η according to the 3-σ INPOP20a WRSS distribution.

4.1.2. $\dot{\mu}/\mu$

The same procedure has been used for $\dot{\mu}/\mu$ using the INPOP20a datasets. Fig. 7 gives the 1-D histograms of the initial distribution of $\dot{\mu}/\mu$ and the distribution of $\dot{\mu}/\mu$ corresponding to alternative ephemerides selected according to the 3-σ WRSS method. A limit of about 2.3 $\times 10^{-13}$ yr^{-1} is obtained with 60% of the runs selected. If one supposes a constant Sun mass loss of about $(-0.92 \pm 0.61) \times 10^{-13}$ yr^{-1}, this result leads to $\dot{G}/G = (-0.08 \pm 2.84) \times 10^{-13}$ yr^{-1}, still consistent with GRT. In terms of LS, $\dot{\mu}/\mu$ shows small correlations (less than 0.5) with the rest of the parameters involved in the INPOP20a adjustment.The LS covariance, $\pm 0.32 \times 10^{-13}$ yr^{-1}, is still smaller than the interval obtained with 3-σ WRSS filtering demonstrating again that this latest approach is more conservative then the direct LS. Besides, these results are consistent with the 3-σ LS uncertainties given by (Fienga et al. 2015) but higher than their Monte Carlo estimates. As for β and γ, this can be explained by the larger interval of random values explored in this work, $\pm 6 \times 10^{-13}$ yr^{-1}, in comparison with the interval used in (Fienga et al. 2015), $\pm 4 \times 10^{-13}$ yr^{-1}.

4.2. *Adding Bepi-Colombo simulations*

4.2.1. Planetary orbits and other fitted parameters

The first aspect to consider when one introduces BC simulations in planetary adjustments is the impact on the determination of planetary orbits by the means of the evolution of the covariance matrix of the planetary orbit initial conditions and other parameters of the fit. On Fig. 8, we plotted the ratio between the standard deviations (defined as the square root of the diagonal terms of the covariance matrix deduced from the least squares adjustment) for the 402 parameters of INPOP20a in GRT obtained in including the BC simulation (σ w BC) and without the BC simulation (σ wo BC). As one can see on this

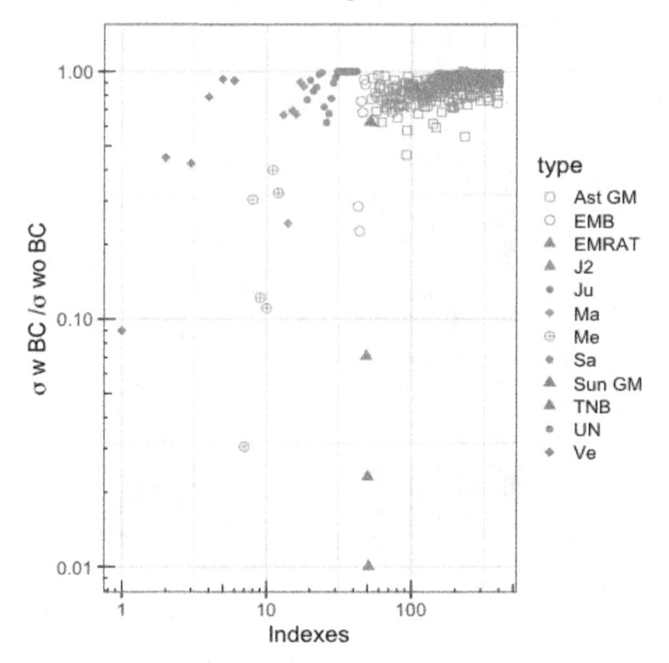

Figure 8. Distribution of the ratio between the parameter uncertainties obtained with and without BC simulations in log scale.The colors and shapes indicate the different types of parameters considered in the INPOP adjustment : Me,V,Ma,Ju,Sa,UN, EMB represent the ratio of the uncertainties for the 6 orbital initial conditions for Mercury, Venus, Mars, Jupiter, Saturn, Uranus, Neptune and the Earth-Moon barycenter respectively. $J_2^\odot$, Sun GM, EMRAT and TNB give the ratios for the Sun oblateness and mass, the ratio between the Earth and the Moon masses and the mass of the TNO ring respectively. Finally Ast GM indicate the ratio for the 343 Main Belt asteroid masses.

figure, the introduction of the BC simulation does not introduce any degradation of the parameter uncertainties as no ratio is greater than 1. The highly improved parameters, besides the Mercury and the Earth-Moon barycenter orbits, are the Earth-Moon mass ratio, the mass of the Sun and its oblateness. The ratio between the variances obtained with and without BC are of at least of one order of magnitude for these parameters thanks to a better constraint on the Mercury geocentric orbit perturbed by the sun, provided by the BC observations. As secondary perturbers of the Mercury-Earth distance, Venus sees also its orbit improved as well as Mars. At a lower level, Jupiter and Saturn orbits are also better estimated when the other outer planet orbits are almost insensitive to the BC introduction. The determination of Main Belt asteroid masses does not seem to be drastically improved even if a noticeable increase of the ratio from 0.75 to 1 is visible. This indicates a slight reduction of the mass uncertainties for some of the perturbers.

4.2.2. Sun core rotation

In GRT, the results of the $J_2^\odot$ LS adjustment including BC simulations for different models of Sun core rotations are given in Table 5. One can notice a significant reduction of the 3-σ LS uncertainty from 3×10^{-9} with INPOP20a to 2×10^{-10} when including the BC simulations (as noticeable in Fig. 8 as well). At this level of accuracy, the differences between solar core rotation hypothesis appear to be detectable thanks to BC. More precisely, at the first glance, considering the four Sun core rotation modeles, no significant

Table 5. Sun Angular Momentum and oblateness obtained in GRT (with PPN parameters fixed to unity) considering INPOP20a data samples and BC simulations. In Column 4, we give the differences $\Delta J_2^{\odot}$ between the reference $J_2^{\odot}$ fitted using (Pijpers 1998) angular momentum value and the $J_2^{\odot}$ obtained after fit using different values for the amplitude of the angular momentum given in Column 2. Different models of Sun core rotation (identified in Column 1) are used for estimating S. The fourth column gives the differences between the reference WRSS and the WRSS obtained for different Sun core rotations. The uncertainties are given at 3-σ. See Sect 2.2 for the significance of the different rotation modeling.

Type of rotation	**S $\times 10^{48}$ g.cm^{-2}.s^{-1}**	**$\Delta\chi^2$ $\times 10^4$**	**$\Delta J_2^{\odot}$ $\times 10^{10}$**
INPOP20a + BC	1.90 ± 1.5	0	0.0 ± 2.3
Slow rotation	1.896	-4	-2 ± 2.3
Uniform at 435 nHz	1.926	-2	0.0 ± 2.3
Fast rotation	1.976	1	2 ± 2.3
Very fast rotation	1.998	3	3 ± 2.3

differences are noticeable in terms of WRSS (Column 3 of Table 5) as they remain smaller than the interval of 3-σ χ^2, ±0.03, defined in Sect 3.2. This means that these ephemerides are acceptable for the WRSS filtering whatever the model for Sun core rotation. However, despite the fact that the estimated values for the Sun oblateness are still consistent at 2σ with the (Antia et al. 2008) value, they differ from one Sun core rotation to another by a maximum of 5×10^{-10} (between the slow and the very fast rotations) , which is more than 2 times bigger than the 3-σ LS uncertainty. This could indicate a possible detection of the Sun core rotation thanks to the addition of the BC data. In non-GRT (see Table 4), when we consider the 3-σ WRSS filtering , the addition of BC simulations induces that only 1 % of the computed runs reach the helioseismological limits. The interval of the fitted $J_2^{\odot}$ deduced from the selected WRSS alternative ephemerides is about $\pm 1 \times 10^{-9}$. This corresponds to an improvement of a factor 3 relative to the WRSS fitted $J_2^{\odot}$ obtained without BC, but it is 5 times larger than the LS 3-σ uncertainty, $\pm 2.3 \times 10^{-10}$, obtained in Table 5 by direct adjustment in GRT. If we consider the direct fit of $J_2^{\odot}$ together with the non-GRT parameters, the obtained 3-σ uncertainty is improved relative to the fit without BC of about a factor almost 10, but it remains 5 times larger than the uncertainty obtained in GRT. In this context, the detection of the different models for the Sun core rotation appears then to be out of reach when we consider a simultaneous estimation of non-GRT parameters and $J_2^{\odot}$. We can conclude that, there are some indications of a possible detection of the Sun angular momentum from future BC observations in GRT with direct LS adjustment. Such detections seem to be difficult if tests of non-GRT are done simultaneously, even when considering the WRSS filtering.

4.2.3. PPN parameters β, γ and $\dot{\mu}/\mu$

On Table 4 and Fig 9 are given the results obtained by adding the BC simulations to the INPOP20a data sample. A first striking result is that BC will improve drastically the constraint on the possible violation of GRT through the PPN parameters β and γ. For the 3-σ WRSS filtering, the most spectacular is the estimation of the γ parameter which gains a factor 19 in comparison with the INPOP20a results (see also Fig. 6). The constraint on β is less improved, of about a factor 1.5. We also note an improvement of the LS results with and without BC of about a factor 6 for β and 15 for γ. These differences between β and γ can again be explained by the introduction of the Lense-Thirring acceleration into the dynamical modeling of the planetary motion, that allows for a more efficient disentangling of the two parameters (see Sect. 4.1.1 and the correlation discussion). The introduction of the BC simulations also reduces the number of selected

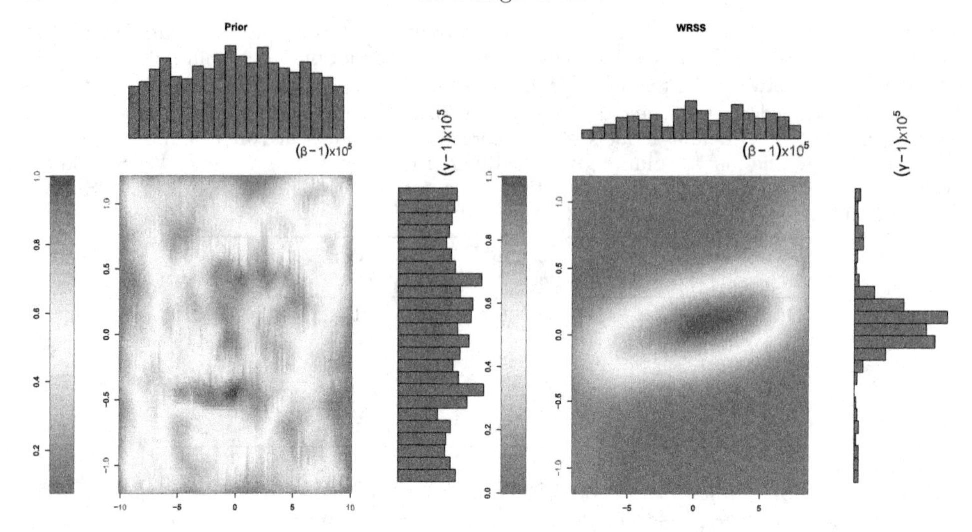

Figure 9. 2D-histograms of PPN β and γ selected according to the 3-σ INPOP20a WRSS distribution in considering the Bepi-Colombo simulations. The color-scale indicates the normalised probability.

alternative ephemerides as only 8% of the alternative ephemerides have been selected with the 3-σ WRSS filtering. On Fig. 10, are plotted the distributions of $\dot{\mu}/\mu$ before and after the 3-σ WRSS filtering including BC simulations. 27% of the runs are selected leading to a reduction of the interval of possible $\dot{\mu}/\mu$ values of a factor 12. For the direct LS estimate, the improvement is even more important, of about a factor 30. With such a constraint, in the perspective of measuring $\dot{G}/G$, it will be important to have independent and accurate constraints for the Sun mass loss which has currently a higher uncertainty (0.61×10^{-13} yr^{-1}).

5. Discussion and conclusion

5.1. *Helioseismology and variations of the Sun mass*

Limits for the possible variations of non-GRT parameters (β, γ and $\dot{\mu}/\mu$) have been obtained with Sun oblateness thresholds based on helioseismology measurements (see Sect 2.1). These measures were obtained in considering the variations of the Sun angular momentum and its kinematic energy for a fixed value of the gravitational mass of the Sun, μ. So it is interesting to address the question of how could change the value of the helioseismological Sun oblateness for different values of μ. From (Gough 1981; Pijpers 1998) we see that the $J_2^\odot$ measurement deduced from helioseismology relies on the following equation

$$J_2^\odot = -\frac{R_\odot}{\mu}\phi_{12}(R_\odot),$$

where $R_\odot$ is the radius of the Sun, μ its gravitational mass and ϕ_{12} is the quadrupole component of the gravitational potential, deduced from the Poisson equation. ϕ_{12} depends on $R_\odot$ and consequently, we can estimate the impact on the estimation of $J_2^\odot$, $\delta(J_2^\odot)$, of introducing a change in the gravitational mass $\delta\mu$ for a fixed value of $R_\odot$ by

$$\frac{\delta(J_2^\odot)}{J_2^\odot} = -\frac{1}{2}\frac{\delta\mu}{\mu}. \tag{5.1}$$

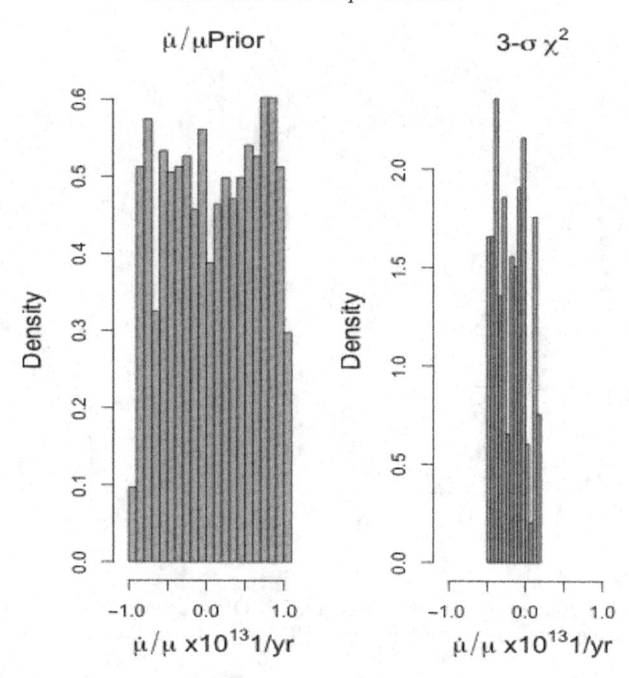

Figure 10. 1D-histograms of $\dot{\mu}/\mu$ in yr^{-1} including Bepi-Colombo simulations. The first left-hand side plot shows the uniform distribution of the prior when the middle histogramms give the distribution of the selected values of η according to the 3-σ INPOP20a WRSS distribution considering the Bepi-Colombo simulations.

In Sec 2.1, we consider possible variations of $J_2^{\odot}$ into the range of $\pm 1.5 \times 10^{-8}$. For such an interval of $\delta(J_2^{\odot})$ the equation 5.1 gives a corresponding variation of μ, $\delta\mu$, of about 4×10^{-5} UA^3.d^{-2}. This means that for inducing a change in the helioseismic estimation of $J_2^{\odot}$ greater than the interval considered in Sec 2.1, we need to introduce variations of μ of the order of 10^{-5} UA^3.d^{-2}. However, as one can see on the histogram of Fig. 11, the distribution of the differences between the INPOP20a μ and the values estimated for the alternative ephemerides are clearly under this threshold, the maximum difference being of about 4×10^{-15} UA^3.d^{-2}. With such a difference, the impact on the helioseismic $J_2^{\odot}$ is of about 10^{-18}. We can also note that the differences between 1998-published DE405 gravitational mass (Standish 2001) and INPOP20a is about $3{\times}10^{-11}$ UA^3.d^{-2}, again below 10^{-5} UA^3.d^{-2}. We can then conclude that even if we consider different values of μ compared to the value used by (Pijpers 1998), the impact on the $J_2^{\odot}$ determinations is clearly encompassed in the interval of uncertainty used in this work.

5.2. *Comparisons with previous results*

In 2015, a similar approach based on Monte Carlo sampling of PPN parameters, was proposed by (Fienga et al. 2015) using INPOP15a. Since then, improvements have been brought to the adjustment method, especially regarding asteroid mass determination (Fienga et al. 2020), and the addition of accurate data for Mercury, Mars, Jupiter and Saturn. Values obtained by the (Fienga et al. 2015) Monte Carlo sampling are presented for comparison in Table 4. The intervals obtained with INPOP20a show a clear improvement for parameters estimated by (Fienga et al. 2015). In 2020, (De Marchi and Cascioli 2020) presented a covariance analysis of least square determinations of the same non-GRT parameters. They consider different sets of tracking data used for planetary ephemerides

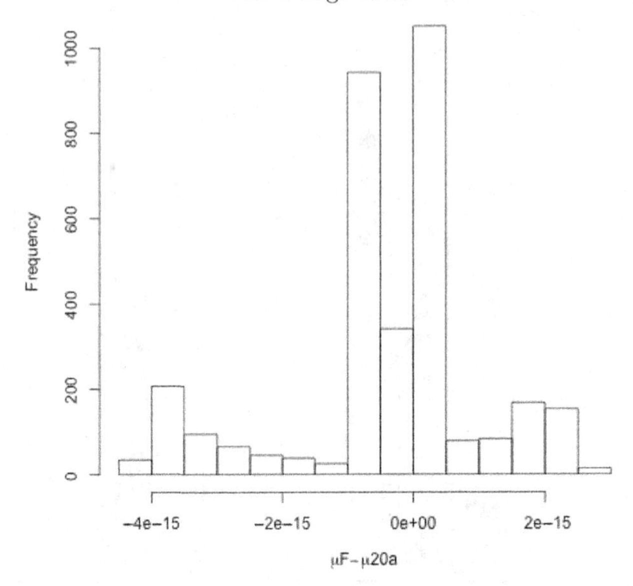

Figure 11. Differences in $UA^3.d^{-2}$ between INPOP20a Sun μ and μ fitted over 4500 alternative ephemerides built with non-unity PPN β and γ and with non-zero η and $\dot{\mu}/\mu$.

construction (Messenger, Mars orbiters, Juno and Cassini missions) as well as future missions including BC. Before comparing our results with theirs, it is important to note that (De Marchi and Cascioli 2020) introduce a linear relation (Nordvedt relation) between the PPN parameters β, γ and η as $\eta = 4(\beta - 1) - (\gamma - 1) - \alpha_1 - \frac{2}{3}\alpha_2$ where η is the ratio between inertial mass and gravitational mass, accounting for the strong equivalence test and α_1 and α_2 testing preferred-frame hypothesis, can be consider as 0. By introducing such a relation, it has been demonstrated that the uncertainty on β estimations is severely diminished (see for example (Imperi et al. 2018)) but at the cost of generality in terms of possible theories to be tested. This is for this reason that we do not introduce this relation in our work. With the present INPOP20a data sets, our LS results are in good agreement with (De Marchi and Cascioli 2020) covariances. In comparison, the intervals produced by the 3-σ WRSS filtering are larger for all parameters, showing again that this method is more conservative relative to the LS estimation or the covariance analysis. The improvement brought by BC simulations on the 3-σ WRSS determinations of the non-GRT parameters are very close to the improvements proposed by (De Marchi and Cascioli 2020). We find that the ratio between the 3-σ WRSS acceptable intervals with and without BC for $\gamma - 1$, Sun $J_2^{\odot}$, $\beta - 1$ are quite close or of the same order of magnitude than (De Marchi and Cascioli 2020) despite the fact that we did not include the Nordvedt relation. For $\dot{\mu}/\mu$, however, we note an improvement induced by BC simulations of about a factor 10 when (De Marchi and Cascioli 2020) indicate a factor 5.

5.3. *Conclusion*

In this study, we give a detailed description of the latest INPOP planetary ephemerides INPOP20a as well as new evaluations of possible GRT violations with the PPN parameters β,γ and $\dot{\mu}/\mu$. With a new method for selecting acceptable alternative ephemerides we provide conservative limits of about 7.16×10^{-5} and 7.49×10^{-5} for $\beta - 1$ and $\gamma - 1$ respectively using the present day planetary data samples. We show, as already stated in (Imperi et al. 2018; De Marchi and Cascioli 2020), that the use of future BC range observations should improve these estimates, in particular γ. Limits of possible secular variations of the Sun gravitational mass are given with a limit of about $2.28 \times 10^{-13}\ yr^{-1}$ without BC simulations and $0.19 \times 10^{-13}\ yr^{-1}$ with. Finally, interesting perspectives for the detection of the Sun core rotation seem to be reachable thanks to the BC mission and its accurate range measurements in the GRT frame.

References

H. M. Antia, S. M. Chitre, and D. O. Gough. Temporal variations in the Sun's rotational kinetic energy. *A&A*, 477(2):657–663, January 2008.

T. Appourchaux and T. Corbard. Searching for g modes. II. Unconfirmed g-mode detection in the power spectrum of the time series of round-trip travel time. *A&A*, 624:A106, April 2019.

B. A. Archinal, C. H. Acton, M. F. A'Hearn, A. Conrad, G. J. Consolmagno, T. Duxbury, D. Hestroffer, J. L. Hilton, R. L. Kirk, S. A. Klioner, D. McCarthy, K. Meech, J. Oberst, J. Ping, P. K. Seidelmann, D. J. Tholen, P. C. Thomas, and I. P. Williams. Report of the IAU Working Group on Cartographic Coordinates and Rotational Elements: 2015. *Celestial Mechanics and Dynamical Astronomy*, 130(3):22, February 2018.

L. Bernus. *Tests de graviation à l'échelle du systeme solaire.* PhD thesis, Observatoire de Paris, 2020.

L. Bernus, O. Minazzoli, A. Fienga, M. Gastineau, J. Laskar, and P. Deram. Constraining the Mass of the Graviton with the Planetary Ephemeris INPOP. *Phys. Rev. Lett.*, 123(16):161103, October 2019.

L. Bernus, O. Minazzoli, A. Fienga, M. Gastineau, J. Laskar, P. Deram, and A. Di Ruscio. Constraint on the Yukawa suppression of the Newtonian potential from the planetary ephemeris INPOP19a. *Phys. Rev. D*, 102(2):021501, July 2020.

B. Bertotti, L. Iess, and P. Tortora. A test of general relativity using radio links with the Cassini spacecraft. *Nature*, 425(6956):374–376, September 2003.

W. J. Chaplin, J. Christensen-Dalsgaard, Y. Elsworth, R. Howe, G. R. Isaak, R. M. Larsen, R. New, J. Schou, M. J. Thompson, and S. Tomczyk. Rotation of the solar core from BiSON and LOWL frequency observations. *MNRAS*, 308(2):405–414, September 1999.

J. Christensen-Dalsgaard. Private communication, 2021.

F. De Marchi and G. Cascioli. Testing general relativity in the solar system: present and future perspectives. *Classical and Quantum Gravity*, 37(9): 095007, May 2020.

M. P. Di Mauro. *Helioseismology: A Fantastic Tool to Probe the Interior of the Sun*, volume 599, pages 31–67. 2003.

A. Di Ruscio, A. Fienga, D. Durante, L. Iess, J. Laskar, and M. Gastineau. Analysis of Cassini radio tracking data for the construction of INPOP19a: A new estimate of the Kuiper belt mass. *A&A*, 640:A7, August 2020.

A. Fienga, J. Laskar, T. Morley, H. Manche, P. Kuchynka, C. Le Poncin-Lafitte, F. Budnik, M. Gastineau, and L. Somenzi. INPOP08, a 4-D planetary ephemeris: from asteroid and time-scale computations to ESA Mars Express and Venus Express contributions. *A&A*, 507:1675–1686, December 2009.

A. Fienga, J. Laskar, P. Exertier, H. Manche, and M. Gastineau. Numerical estimation of the sensitivity of INPOP planetary ephemerides to general relativity parameters. *Celestial Mechanics and Dynamical Astronomy*, 123: 325–349, November 2015.

A. Fienga, P. Deram, V. Viswanathan, A. Di Ruscio, L. Bernus, D. Durante, M. Gastineau, and J. Laskar. INPOP19a planetary ephemerides. *Notes Scientifiques et Techniques de l'Institut de Mecanique Celeste*, 109, December 2019.

A. Fienga, C. Avdellidou, and J. Hanuš. Asteroid masses obtained with INPOP planetary ephemerides. *MNRAS*, 492(1):589–602, February 2020.

E. Fossat and F. X. Schmider. More about solar g modes. *A&A*, 612:L1, April 2018.

E. Fossat, P. Boumier, T. Corbard, J. Provost, D. Salabert, F. X. Schmider, A. H. Gabriel, G. Grec, C. Renaud, J. M. Robillot, T. Roca-Cortés, S. Turck-Chièze, R. K. Ulrich, and M. Lazrek. Asymptotic g modes: Evidence for a rapid rotation of the solar core. *A&A*, 604:A40, August 2017.

R. A. García, T. Corbard, W. J. Chaplin, S. Couvidat, A. Eff-Darwich, S. J. Jiménez-Reyes, S. G. Korzennik, J. Ballot, P. Boumier, E. Fossat, C. J. Henney, R. Howe, M. Lazrek, J. Lochard, P. L. Pallé, and S. Turck-Chièze. About the rotation of the solar radiative interior. *Sol. Phys.*, 220(2):269–285, April 2004.

R. A. García, S. Turck-Chièze, S. J. Jiménez-Reyes, J. Ballot, P. L. Pallé, A. Eff-Darwich, S. Mathur, and J. Provost. Tracking Solar Gravity Modes: The Dynamics of the Solar Core. *Science*, 316(5831):1591, June 2007.

E. Gavryuseva, V. Gavryusev, and M. P. Di Mauro. Rotational Split of Solar Acoustic Modes from GONG Experiment. In S. Korzennik, editor, *Structure and Dynamics of the Interior of the Sun and Sun-like Stars*, volume 418 of *ESA Special Publication*, page 193, January 1998.

A. Genova, E. Mazarico, S. Goossens, F.G. Lemoine, G. A. Neumann, D. E. Smith, and M. T. Zuber. Solar system expansion and strong equivalence principle as seen by the NASA MESSENGER mission. *Nature Communications*, 9:289, January 2018.

D. O. Gough. A new measure of the solar rotation. *MNRAS*, 196:731–745, September 1981.

D. O. Gough. Some Glimpses from Helioseismology at the Dynamics of the Deep Solar Interior. *Space Sci. Rev.*, 196(1-4):15–47, December 2015.

A. Hees. Private communication, 2015.

L. Iess, S. W. Asmar, P. Cappuccio, G. Cascioli, F. De Marchi, I. di Stefano, A. Genova, N. Ashby, J. P. Barriot, P. Bender, C. Benedetto, J. S. Border, F. Budnik, S. Ciarcia, T. Damour, V. Dehant, G. Di Achille, A. Di Ruscio, A. Fienga, R. Formaro, S. Klioner, A. Konopliv, A. Lemaître, F. Longo, M. Mercolino, G. Mitri, V. Notaro, A. Olivieri, M. Paik, A. Palli, G. Schettino, D. Serra, L. Simone, G. Tommei, P. Tortora, T. Van Hoolst, D. Vokrouhlický, M. Watkins, X. Wu, and M. Zannoni. Gravity, Geodesy and Fundamental Physics with BepiColombo's MORE Investigation. *Space Sci. Rev.*, 217(1):21, February 2021.

L. Imperi and L. Iess. The determination of the post-Newtonian parameter γ during the cruise phase of BepiColombo. *Classical and Quantum Gravity*, 34(7): 075002, April 2017.

L. Imperi, L. Iess, and M. J. Mariani. An analysis of the geodesy and relativity experiments of BepiColombo. *Icarus*, 301:9025, February 2018.

S. Katoch and S. .S. Chauhan. A review on genetic algorithm: past, present, and future. *Multimedia Tools and Applications*, 80(5): 8091–8126, February 2021.

R. Komm, R. Howe, B. R. Durney, and F. Hill. Temporal Variation of Angular Momentum in the Solar Convection Zone. *ApJ*, 586(1):650–662, March 2003.

M. Lazrek, A. Pantel, E. Fossat, B. Gelly, F. X. Schmider, D. Fierry-Fraillon, G. Grec, S. Loudagh, S. Ehgamberdiev, I. Khamitov, J. T. Hoeksema, P. L. Pallé, and C. Régulo. Is the Solar Core Rotating Faster of Slower Than the Envelope? *Sol. Phys.*, 166(1):1–16, June 1996.

J. Lense and H. Thirring. Über den Einfluß der Eigenrotation der Zentralkörper auf die Bewegung der Planeten und Monde nach der Einsteinschen Gravitationstheorie. *Physikalische Zeitschrift*, 19:156, January 1918.

N. Moskovitz, R. Schottland, B. Burt, M. Bailen, and L. Wasserman. astorb at Lowell Observatory: A comprehensive system to enable asteroid science. In *AAS/Division for Planetary Sciences Meeting Abstracts #50*, volume 50 of *AAS/Division for Planetary Sciences Meeting Abstracts*, page 408.08, October 2018.

T.D. Moyer. Formulation for observed and computed values of deep space network data types for navigation. Monography of DEEP SPACE COMMUNICATIONS AND NAVIGATION Series 2, JPL, 2000.

R. S. Park, W. M. Folkner, A. S. Konopliv, J. G. Williams, D. E. Smith, and M. T. Zuber. Precession of Mercury's Perihelion from Ranging to the MESSENGER Spacecraft. *AJ*, 153:121, March 2017.

F. P. Pijpers. Helioseismic determination of the solar gravitational quadrupole moment. *MNRAS*, 297(3):L76–L80, July 1998.

T. Roca Cortés, M. Lazrek, L. Bertello, S. Thiery, F. Baudin, R. A. Garcia, and GOLF Team. The Solar Acoustic Spectrum as Seen by GOLF. III. Asymmetries, Resonant Frequencies and Splittings. In S. Korzennik, editor, *Structure and Dynamics of the Interior of the Sun and Sun-like Stars*, volume 418 of *ESA Special Publication*, page 329, January 1998.

I. W. Roxburgh. Gravitational multipole moments of the Sun determined from helioseismic estimates of the internal structure and rotation. *A&A*, 377:688–690, October 2001.

P. H. Scherrer and D. O. Gough. A Critical Evaluation of Recent Claims Concerning Solar Rotation. *ApJ*, 877(1):42, May 2019.

Hannah Schunker, Jesper Schou, Patrick Gaulme, and Laurent Gizon. Fragile Detection of Solar g-Modes by Fossat et al. *Sol. Phys.*, 293(6):95, June 2018.

M. Soffel and F. Frutos. On the usefulness of relativistic space-times for the description of the Earth's gravitational field. *Journal of Geodesy*, 90(12):1345–1357, December 2016.

E.M. Standish. The jpl de405 planetary and lunar ephemerides. 2001.

M. J. Thompson, J. Christensen-Dalsgaard, M. S. Miesch, and J. Toomre. The Internal Rotation of the Sun. *ARA&A*, 41:599–643, January 2003.

R. N. Thor, R. Kallenbach, U. R. Christensen, A. Stark, G. Steinbrügge, A. Di Ruscio, P. Cappuccio, L. Iess, H. Hussmann, and J. Oberst. Prospects for measuring mercury´s tidal love number h2 with the bepicolombo laser altimeter. *A&A*, 633:A85, 2020. URL `https://doi.org/10.1051/0004-6361/201936517`.

Multi-scale (time and mass) dynamics of space objects
Proceedings IAU Symposium No. 364, 2022
A. Celletti, C. Galeş, C. Beaugé, A. Lemaitre, eds.
doi:10.1017/S1743921321001423

The Lidov-Kozai resonance at different scales

Anne-Sophie Libert

naXys, Department of Mathematics, University of Namur,
61 Rue de Bruxelles, 5000 Namur, Belgium
email: anne-sophie.libert@unamur.be

Abstract. The Lidov-Kozai (LK) resonance is one of the most widely discussed topics since the discovery of exoplanets in eccentric orbits. It constitutes a secular protection mechanism for systems with high mutual inclinations, although large variations in eccentricity and inclination are observed. This review aims to illustrate how the LK resonance influences the dynamics of the three-body problem at different scales, namely i) for two-planet extrasolar systems where the orbital variations occur in a coherent way such that the system remains stable, ii) for inclined planets in protoplanetary discs where the LK cycles are produced by the gravitational force exerted by the disc on the planet, iii) for migrating planets in binary star systems, whose dynamical evolution is strongly affected by the LK resonance even without experiencing a resonance capture, and iv) for triple-star systems for which the migration through LK cycles combined with tidal friction is a possible explanation for the short-period pile-up observed in the distribution of multiple stars.

Keywords. Celestial mechanics, planetary systems, planetary systems: formation, planetary systems: protoplanetary discs, planets and satellites: dynamical evolution and stability, binaries: close, stars: kinematics and dynamics, stars: formation

---

1. Introduction

When two bodies gravitationally interact, the problem is integrable and its solution, which consists of two fixed elliptic orbits, is known since the works of Kepler in the 17th century. However, the perturbations caused by additional bodies produce changes in the shape and orientation of the Keplerian orbits on secular timescales. In particular, the planetary perturbations induce the precession of the pericenter argument of the orbits. In the 18th century, Laplace and Lagrange built a linear approximation for the secular motion of the planets of the Solar system, showing that the semi-major axes present no secular variation while the eccentricities and inclinations of the orbits suffer from limited variations inducing no possible orbit crossing or planet collision. This is no longer true when considering planetary systems with large eccentricities and / or mutual inclinations.

Lidov (1962) and Kozai (1962) investigated the secular motion of small bodies under the effect of a pertuber on a circular orbit. Lidov (1962) focused on the orbital motion of artificial Earth satellites perturbed by the Moon, while Kozai (1962) studied the influence of Jupiter on the motion of asteroids. They showed that, for inclined orbits of the small body (considered here as a massless particle), the dynamics can be characterized by the libration of the pericenter argument ω of the small body around 90° or 270°, as well as large periodic coupled variations of its eccentricity e and inclination i. These variations are bounded by the conservation of the *Kozai constant*, $H = \sqrt{1-e^2}\cos i$ (i.e., the adimensional z-component of the small body angular momentum). As a result, the orbit of the small body will be more eccentric for smaller inclinations and less eccentric for larger inclinations. This particuliar dynamics is known as the *LK resonance* or more

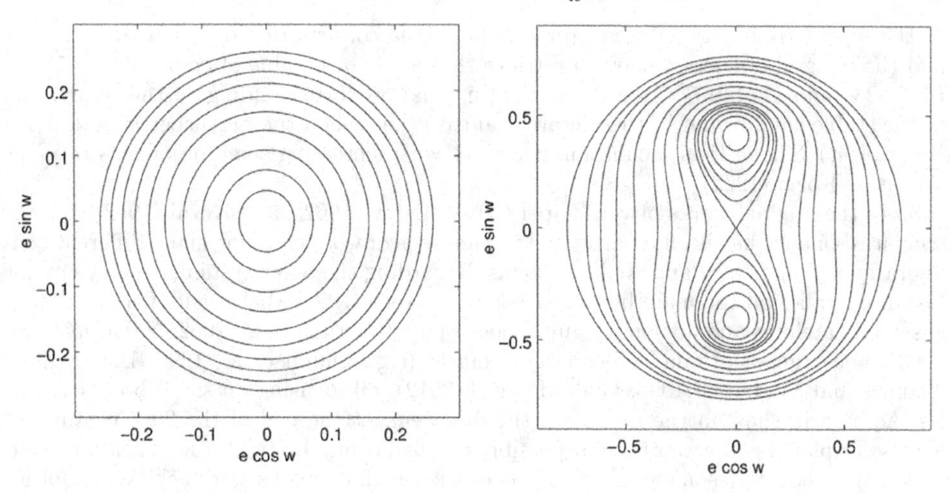

Figure 1. Level curves of the Hamiltonian $\mathcal{H}_{\rm QUAD}$ in the $(e\cos\omega, e\sin\omega)$ phase space, for two values of the Kozai constant: $H = 0.97$ (left panel) and $H = 0.64$ (right panel).

recently *ZLK resonance*, as it turns out that the mechanism was originally discussed by von Zeipel (1910) for the motion of (long-period) comets (see e.g., Ito and Ohtsuka 2019, for a comparison between the works of von Zeipel (1910), Lidov (1962), and Kozai (1962)).

More precisely, Kozai (1962) showed that the Hamiltonian formulation of the three-body problem in the test particle approach can be reduced to two degrees of freedom after short-period averaging (i.e., averaging with respect to the mean anomalies to obtain the secular evolution of the bodies whose semi-major axis is then considered as constant) and node reduction (i.e., adoption of the invariant plane as reference plane, see e.g., Poincaré 1892). Assuming that the orbit of the perturber is circular, the problem becomes integrable. The one-degree of freedom Hamiltonian formulation at quadrupole level of the expansion in Legendre polynomials writes (e.g., Kinoshita and Nakai 1999; Naoz et al. 2013)

$$\mathcal{H}_{\rm QUAD} = -\tfrac{Gm_0m_2}{16a_2(1-e_2^2)^{3/2}}\left(\tfrac{a}{a_2}\right)^2\left((2+3e^2)(3\cos^2 i - 1) + 15e^2\sin^2 i\cos(2\omega)\right), \quad (1.1)$$

where the index 2 refers to the perturber and with a the semi-major axis of the particle, m_0 the mass of the central star, and m_2 the mass of the perturber.

The dynamics can be represented, for a given value of the Kozai constant H (i.e., bounded values of the eccentricity and inclination), in the phase space $(e\cos\omega, e\sin\omega)$, where the level curves of the Hamiltonian correspond to the trajectories of the test particle (e.g., Thomas and Morbidelli 1996). Such a representation is shown in Fig. 1. For large values of H, or equivalently for small inclination values of the massless body (for instance, for $i = 15°$ at $e = 0$, left panel), the phase space shows a stable equilibrium at zero eccentricity of the massless body. The eccentricity (and thus the inclination) of the test particle at low inclination suffers from limited variations during the precession of the orbit. However, for higher inclinations (for instance, for $i = 50°$ at $e = 0$, right panel), the equilibrium becomes unstable and a separatrix divides the phase space in three regions: two libration islands around $\omega = 90°$ and $\omega = 270°$ and a region characterized by the circulation of ω. These two equilibria are referred to as *LK equilibria*. The bifurcation of the central equilibrium induces large eccentricity variations for the massless body initially in a circular orbit since its real motion (short periods included) will remain close to the separatrix of the reduced problem. From the conservation of the Kozai constant and the integral of energy, one can retrieve the well-known value of $\cos i = \pm\sqrt{3/5}$, i.e., $i = 39.23°$ and $i = 140.77°$, for the critical inclination corresponding to the bifurcation of the stable equilibrium in the hierarchical case (i.e., when the perturber is on a much wider orbit). The critical inclination decreases with increasing semi-major axis ratio, as shown by Kozai (1962).

Since the original works of von Zeipel (1910), Lidov (1962), and Kozai (1962), the LK secular resonance has been investigated in hundreds of works and for many different configurations of planetary and stellar systems. Regarding the Solar system, we may cite for instance studies on the main belt asteroids (e.g., Kozai 1985; Michel and Thomas 1996), gas giant satellites and Jovian irregular moons (e.g., Carruba et al. 2002; Nesvorný et al. 2003), and trans-neptunian objects and comets (e.g., Quinn et al. 1990; Bailey 1992; Thomas and Morbidelli 1996; Gallardo et al. 2012), all focusing on small body dynamics. As we will show in the following, the discovery, at the end of the 20th century, of extrasolar planets on eccentric and possibly inclined orbits brought a new field of applications for the LK resonance. Extensions of the previous works were achieved in order to take into account eccentric orbits for the perturber as well as different mass ratios among the bodies, in the context of planetary systems (e.g., Michtchenko et al. 2006; Libert and Henrard 2007; Migaszewski and Goździewski 2009) and multiple-star systems (e.g., Innanen et al. 1997; Naoz et al. 2013). In particular, the formation through the LK resonance of hot Jupiters with orbital periods of a few days only has been deeply investigated (e.g., Fabrycky and Tremaine 2007; Wu et al. 2007; Naoz et al. 2011).

In this work I aim to show how the LK resonance influences the dynamics at four different scales of the three-body problem: two-planet extrasolar systems in Section 2, planets perturbed by their protoplanetary disc in Section 3, circumprimary planets in binary stars in Section 4, and finally triple-star systems in Section 5. This review has not the ambition to draw an exhaustive picture of the latest results in the four different fields discussed here, but rather it aims to take a closer look at how the LK dynamics can be easily transposed among different fields and provide valuable contribution in all of them.

2. Two-planet systems

We first consider the planetary three-body problem. Many two-planet extrasolar systems were detected via the radial velocity (RV) method in the last 25 years. This detection method measures only the line of sight component of the star velocity, and thus gives no information on the orbital inclinations of the planets. The planetary masses are also unclear since they have to be scaled by the sinus of the unknown orbital inclination. It means that three-dimensional (3D) configurations for the detected planetary systems cannot be uncovered by the RV technique alone. Recent years have seen the emergence of a number of observational evidence on the existence of 3D planetary systems. A well-known example of a mutually inclined system is υ And for which the mutual inclination between the orbital planes of planets *c* and *d* was estimated to 30°, by combining different detection methods (Deitrick et al. 2015).

Because of the large eccentricities (and possibly large inclinations) of many detected exoplanets, the classical analytical theories, such as the Laplace-Lagrange linear perturbation theory, are unable to describe correctly the motion of these planets. As a result, several authors carried out analytical works on the secular evolution of the 3D planetary three-body problem (e.g., Michtchenko et al. 2006; Libert and Henrard 2007; Migaszewski and Goździewski 2009; Naoz et al. 2013). A development commonly used is the Hamiltonian expansion in Legendre polynomials to high order in the semi-major axis ratio (e.g., Kozai 1962; Ford et al. 2000; Lee and Peale 2003; Migaszewski and

Goździewski 2009; Naoz et al. 2013), which is valid for hierarchical systems. Another well-known analytical approach consists in the generalization of the Laplace-Lagrange Hamiltonian expansion to high order in the eccentricities and inclinations, as given here (e.g., Libert and Henrard 2007)

$$\mathcal{H} = -\frac{Gm_0m_1}{2a_1} - \frac{Gm_0m_2}{2a_2} - \frac{Gm_1m_2}{a_2} \sum_{k,i_l,j_l,l\in\underline{4}} A_{i_l}^{k,j_l} \sqrt{\frac{2P_1}{L_1}}^{|j_1|+2i_1} \sqrt{\frac{2P_2}{L_2}}^{|j_2|+2i_2} \sqrt{\frac{2Q_1}{L_1}}^{|j_3|+2i_3} \sqrt{\frac{2Q_2}{L_2}}^{|j_4|+2i_4} \cos\Phi, \tag{2.2}$$

with

$$\Phi = [(k+j_1+j_3)\lambda_1 - (k+j_2+j_4)\lambda_2 + j_1p_1 - j_2p_2 + j_3q_1 - j_4q_2] \tag{2.3}$$

and where subscript 1 refers to the planet closer to the star (inner planet) and 2 to the outer one. The expansion is expressed in the classical modified Delaunay's elements:

$$\begin{aligned} \lambda_i &= \text{mean longitude of } m_i, & L_i &= m_i\sqrt{Gm_0a_i} \\ p_i &= \text{- the longitude of the pericenter of } m_i, & P_i &= L_i\left[1-\sqrt{1-{e_i}^2}\right] \\ q_i &= \text{- the longitude of the node of } m_i, & Q_i &= L_i\sqrt{1-e_i^2}\,[1-\cos i_i]\,. \end{aligned} \tag{2.4}$$

The coefficients $A_{i_l}^{k,j_l}$ depend only on the ratio a_1/a_2 of the semi-major axes.

As we are interested in the long-term evolution of extrasolar systems, we can average the expansion over the short period terms by simply removing from the Hamiltonian the terms depending on the mean anomalies (first-order averaging), if the system is not close to a mean-motion resonance (a similar analytical development for planetary systems in mean-motion resonance can be found in Sansottera and Libert 2019). Moreover, by adopting the invariant Laplace plane which is orthogonal to the total angular momentum vector (i.e., elimination of the nodes, see Jacobi 1842), the Hamiltonian can be reduced to two degrees of freedom, namely $\mathcal{H}(e_1, e_2, \omega_1, \omega_2)$. For a fixed value of the total angular momentum

$$C = (L_1 - P_1)\cos i_1 + (L_2 - P_2)\cos i_2 \tag{2.5}$$

or equivalently for a fixed value of the angular momentum deficit (Laskar 1997)

$$AMD = L_1 + L_2 - C, \tag{2.6}$$

one can easily determine the values of the inclinations as functions of the eccentricities. It was shown that the averaged expansion limited to order 12 in the eccentricities and inclinations is accurate enough to describe precisely the secular evolution of extrasolar systems with moderate to high eccentricities and inclinations (Libert and Henrard 2007).

To visualize the dynamics, we can define, for a given value of the AMD (i.e., bounded values of the eccentricities and inclinations), a *representative plane* $(e_1 \sin\omega_1, e_2 \sin\omega_2)$ where both arguments of the pericenters are fixed to $\pm 90°$. This plane is neither a phase portrait nor a surface of section, since the problem is four dimensional. However, nearly all the orbits will cross the representative plane at several points of intersection on the same energy curve. The maximal value of the mutual inclination between the two orbital planes, $i_{\rm mut} = i_1 + i_2$, is reached at the origin of the representative plane and is given by

$$\max i_{\rm mut} = \arccos\left(\frac{C^2 - L_1^2 - L_2^2}{2L_1L_2}\right). \tag{2.7}$$

Representative planes are shown in Fig. 2 for different AMD values. Depending on the value of the AMD, one or three equilibria are visible: the central equilibrium at $e_1 = e_2 = 0$ which is stable for small mutual inclination values (for instance, for $\max i_{\rm mut} = 23°$, left panel) and, for higher mutual inclination values (for instance, for $\max i_{\rm mut} = 53°$, middle

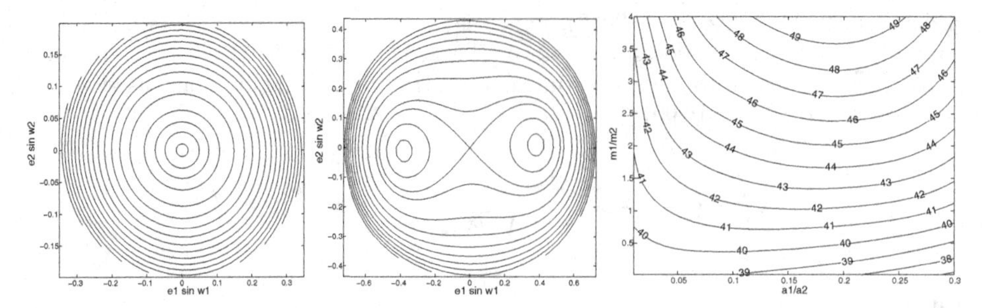

Figure 2. Level curves of the Hamiltonian $\mathcal{H}$ in the $(e_1 \sin\omega_1, e_2 \sin\omega_2)$ representative plane with both arguments of pericenters fixed to $\pm 90°$, for two AMD values corresponding to a maximal mutual inclination of 23° (left panel) and 53° (middle panel). Planetary parameters are fixed to $a_1/a_2 = 0.1$ and $m_1/m_2 = 1$. In the right panel, critical mutual inclinations as a function of the semi-major axis ratio and mass ratio. Adapted from Libert and Henrard (2007).

panel), the stable LK equilibria generated by the bifurcation of the central equilibrium. The critical value of the maximal mutual inclination corresponding to the change in the stability of the equilibrium at the origin depends on the planetary mass ratio and semi-major axis ratio, as shown in the right panel of Fig. 2. As expected, when the two ratios are small, namely $a_1/a_2 = 0.01$ and $m_1/m_2 = 0.01$, we retrieve the critical mutual inclination of 39.23° found by Kozai (1962).

From Fig. 2, we deduce that non-coplanar two-planet systems can be long-term stable, either at low mutual inclination between the orbital planes, whatever the planetary eccentricities (in this case the eccentricities and inclinations will stay close to their initial values), or at high mutual inclination for elliptic orbits only, since planets on quasi-circular orbits will feel some instability in the neighborhood of the unstable equilibrium. In the latter case, the LK equilibria provide stability islands for highly mutually inclined systems. In the invariant Laplace plane reference frame, LK resonant systems are characterized by large coupled variations of the eccentricity and the inclination of the inner planet as well as the libration of the argument of the pericenter of the same planet around $\pm 90°$ (e.g., Libert and Tsiganis 2009).

Volpi et al. (2019) showed that many RV-detected extrasolar systems have orbital parameters compatible with a LK resonant state when varying the (unknown) orbital inclinations with respect to the plane of the sky, denoted here I_1 and I_2, and the mutual inclination i_{mut}. For simplicity, although the inclinations I_1 and I_2 could differ, the authors fixed them to the same value I and therefore used the same scaling factor $\sin I$ for both planetary masses. This scaling factor has an impact on the long-term evolution of the planetary system, since the dynamics of the three-body problem depends on the individual planetary masses, and not only on their mass ratio (see Eq. (2.2)). The extent of the LK resonant region is shown in the left panel of Fig. 3 for the HD11506 system considered with different I and i_{mut} values (dark blue region characterized by the libration of ω_1). The LK resonance is essential to ensure long-term regular evolutions for the highly mutually inclined systems, as shown in the right panel of Fig. 3, where chaotic regions are unveiled by the MEGNO chaos indicator (Cincotta et al. 2003). For moderate to high values of the eccentricities, such as the ones reported by the RV detections, significant chaos (yellow) is generally observed around the islands of the LK resonance (purple), which are the only regular regions in the case of a mutual inclination higher than 40°.

For exoplanets which are very close to the star with orbital periods of a few days only, additional effects such as tides and general relativity can greatly affect the long-term evolution of the system. Several authors found that the precession induced by the tides

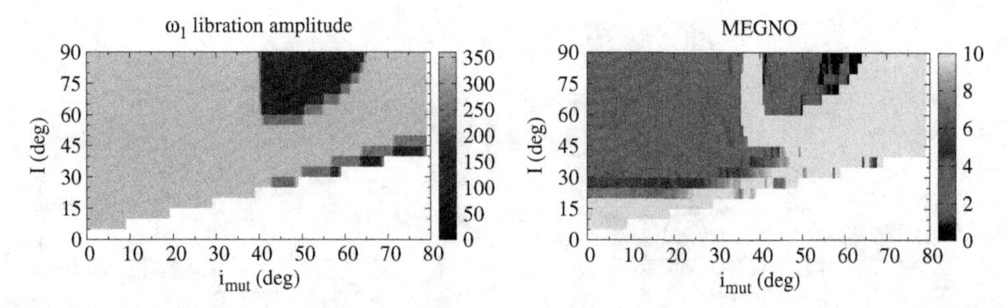

Figure 3. Long-term evolution of HD11506 (parameters from Tuomi and Kotiranta (2009)). Left panel: Libration amplitude of the argument of the pericenter ω_1. Right panel: Values of the (mean) MEGNO chaos indicator. The white region corresponds to mutual inclination values incompatible with a given value of I in the relation $\cos i_{\rm mut} = \cos^2 I + \sin^2 I \cos(\Delta\Omega)$ (i.e., $i_{\rm mut} \geq 2I$). Adapted from Volpi et al. (2019).

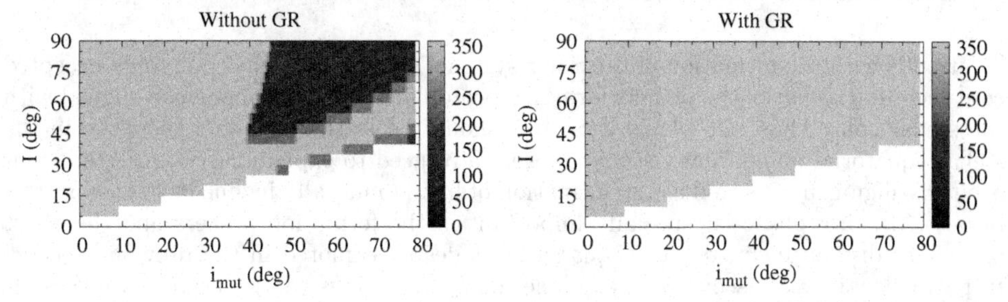

Figure 4. Libration amplitude of the pericenter argument ω_1 for GJ649 (parameters from Wittenmyer et al. (2013)), without (left) and when (right) considering relativistic corrections.

is generally negligible with respect to the one caused by the relativistic corrections (e.g., Migaszewski and Goździewski 2009; Veras and Ford 2010; Sansottera et al. 2014; Naoz 2016). The general relativity causes an advance of the pericenter which can be damaging for mutually inclined systems given that the LK resonance acts on the argument of the pericenter of the inner planet. More precisely, the secular perturbation caused by the general relativity on the pericenter argument of the inner planet is

$$\mathcal{H}_{\rm GR} = \frac{-3(Gm_0)^4 m_1^5}{c^2 L_1^3 G_1} \tag{2.8}$$

with c the speed of light (e.g., Migaszewski and Goździewski 2009). This corresponds to the following relativistic advance of the inner pericenter

$$\dot{\omega}_{1,\rm GR} = \frac{3(Gm_0)^{3/2}}{c^2 a^{5/2}(1-e_1^2)}. \tag{2.9}$$

Volpi and Libert (in preparation) investigated the possibility for RV-detected exoplanetary systems with close-in planets to be in a 3D configuration and found that, for the majority of the systems they examined, the LK resonance region disappears when the relativistic corrections are considered. An example is shown in Fig. 4 for GJ649 system.

3. Planets in protoplanetary discs

Besides the existence of non-coplanar extrasolar systems, the discovery of giant planets very close to their parent star (hot Jupiters) and the strong spin-orbit misalignment (i.e., high angle between the sky projection of the stellar spin axis and the orbit normal) of a significant fraction of them (Albrecht et al. 2012) are several characteristics that appear to

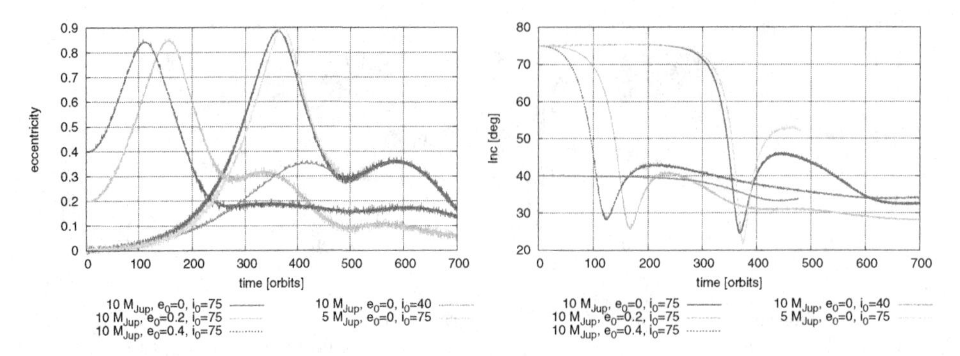

Figure 5. LK cycles of a planet evolving in the protoplanetary disc during the gas phase. The mass of the disc is $0.01 M_\odot$. The curves correspond to different planetary mass, initial eccentricity and initial inclination values. Reproduced from Bitsch et al. (2013).

be at odds with the formation of the Solar system. According to the commonly accepted core accretion scenario, the planets form in a protoplanetary disc supposedly aligned with the stellar spin. Thus, the planetary orbits should lie in the disc, and therefore in the stellar equatorial plane. Many scenarios were advanced to explain the unexpected spin-orbit misalignments, as well as the formation of highly mutually inclined systems. In this respect, the disc phase is especially important in the formation process and the effect of the gas disc on inclined (giant) planets was deeply explored in the previous decade. In particular, it was discovered that planet-disc interactions can generate LK cycles for the planet evolving in the gas disc (e.g., Terquem and Ajmia 2010; Xiang-Gruess and Papaloizou 2013; Teyssandier et al. 2013; Bitsch et al. 2013).

An example of LK evolution during the disc phase is shown in Fig. 5. It is issued from 3D hydrodynamical simulations of protoplanetary discs with embedded high-mass single planets (1 to 10 $M_{\rm Jup}$) achieved by Bitsch et al. (2013), in which they let the planet evolve freely under the influence of the disc forces. For a low initial inclination of the planet, the planetary eccentricity and inclination are both generally damped by the disk (i.e., the dynamical friction that the disc exerts on the planet when it crosses the disc), except for very massive planets (above $\sim 5 M_{\rm Jup}$) where eccentricity can increase at low inclination. For initial planetary inclinations larger than a critical value, the gravitational force exerted by the disc on the planet leads to LK cycles during which the orbital eccentricity reaches high values (~ 0.9 for an initial planetary inclination of $75°$, as shown in Fig. 5), in antiphase with the inclination. Note that the critical inclination value depends on the mass ratio between the planet and the disc, and it can be as small as $\sim 20°$ in specific cases according to Teyssandier et al. (2013).

In Fig. 5, we observe that the LK cycles are damped through time by the dynamical friction exerted by the disc. If the dissipation of the disc is not rapid, the LK cycles only delay the alignment of the planetary orbit with the disc and its circularization caused by the damping forces of the disc on the planet. Otherwise, the LK cycles make it hard to predict the exact evolution of the planet and its orbital parameters at the dispersal of the disc.

4. Planets in binary star systems

Gravitational perturbations from a planetary companion and a protoplanetary disc were considered in Sections 2 and 3, respectively. We now focus on the gravitational influence of a binary companion. It is believed that half of the Sun-like stars are part of multiple-star systems (Duquennoy and Mayor 1991; Raghavan et al. 2010). More than 100

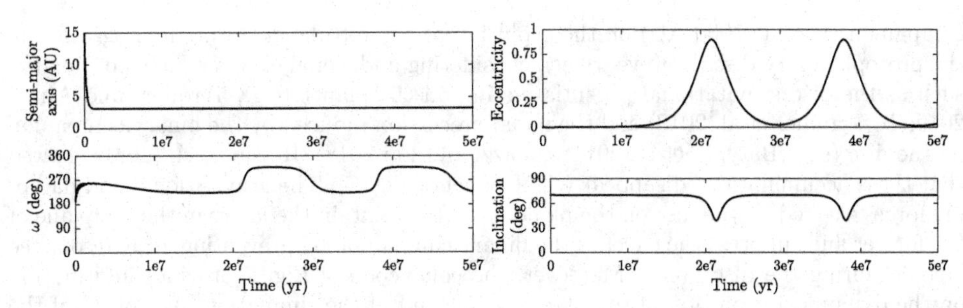

Figure 6. Evolution of a planet in LK resonance with an inclined binary companion, as shown by the libration of the argument of the pericenter in the invariant plane reference frame. The initial parameters for the binary are $m_B = 1M_\odot$, $a_B = 500$ AU, $e_B = 0.3$, and $i_B = 70°$. The ones for the planet are $m = 2.02\ M_{\rm Jup}$, $a = 15$ AU, $e \sim 0$, $i \sim 0$, and $\omega = 16°$.

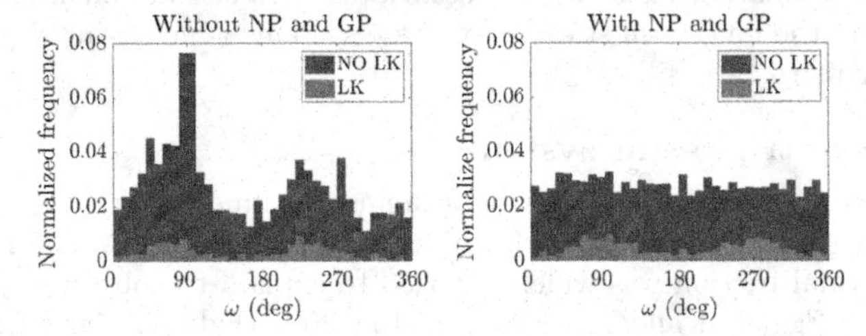

Figure 7. Normalized distribution of the pericenter argument of the planet (in the invariant Laplace plane reference frame) at the dispersal of the disc, as found in the simulations of Roisin et al. (2021), when the disc gravitational potential (GP) acting on the planet and the nodal precession (NP) induced by the binary companion on the disc are included (right panel) or not (left panel). The simulations follow the evolution of 3200 single giant planets migrating in a S-type configuration with a wide binary companion at 1000 AU and with a mass of $1M_\odot$. The initial planetary parameters are $m \in [1,5]\ M_{\rm Jup}$, $a = 20$ AU, $e = 10^{-3}, 0.1, 0.3,\ 0.5$, and $i \in [0°, 70°]$.

circumprimary planets (also called S-type planets) have been discovered in multiple-star systems (Schwarz et al. 2016), mostly in wide binaries with separation of at least 500 AU. The stellar companion can strongly impact the formation and long-term evolution of these planets (see e.g., Thébault and Haghighipour 2015, for a review).

The LK resonance was widely discussed for planets in binary star systems. In particular, it was identified as a possible origin for the hot Jupiters via the LK migration scenario (e.g., Wu and Murray 2003; Naoz et al. 2011; Petrovich 2015; Anderson et al. 2016; Naoz 2016, for a review; see also Section 5). In this section, we show how the LK resonance induced by an inclined binary companion can interfer with the planetary migration driven by the protoplanetary disc during the late-stage formation process.

An example of a planetary evolution in the LK resonance with a binary companion is shown in Fig. 6. The binary companion has an inclination of 70° with respect to the disc plane (where the planet is embedded). After the migration phase (~ 1 Myr, top left panel), the planet is rapidly captured in the LK resonance, as indicated by the libration of the pericenter argument (bottom left panel) and the high eccentricity and inclination variations (right panels).

Using a symplectic N-body integrator designed for binary star systems, in which the dissipation for the planet due to the disc is implemented following the formulas of Bitsch et al. (2013), Roisin and Libert (2021) noted that accumulations of the pericenter arguments around 90° and 270° are clearly visible at the dispersal of the disc (left panel of Fig. 7). This reflects the significant dynamical influence of a wide binary companion (here at 1000 AU) on the evolution of a giant planet during its migration in the protoplanetary disc. However, when considering additional effects related to the disc, such as the disc gravitational potential acting on the planet (e.g., Terquem and Ajmia 2010; Teyssandier et al. 2013) and the nodal precession induced by the binary companion on the disc (e.g., Batygin et al. 2011; Zanazzi and Lai 2018), Roisin et al. (2021) showed that these accumulations disappear (right panel of Fig. 7). The gravitational and damping forces exerted by the disc on the planet tend to maintain the latter in the midplane of the former and suppress the effect of the binary companion by preventing a LK resonance locking during the disc phase. Moreover, the color code of Fig. 7 provides information on the dynamical evolution of the planet at the end of the simulation. A capture of the planet in the LK resonance is far from being automatic, since only 30% of the systems with an inclined binary companion (inclination above 40°) are found in a LK-resonant state in the long term (red color). The non-resonant evolutions which also present high eccentricity and inclination variations are associated with circulation around the LK stability regions and the long-term stability of these systems is not assured (see Roisin and Libert 2021, for more details).

5. Hierarchical triple-star systems

For a significant proportion of triple-star systems, the inner binary has a period smaller than six days (e.g., Duquennoy and Mayor 1991; Tokovinin 2014). The LK mechanism combined with tidal friction was widely invoked to explain the observed pile-up around three days (e.g., Eggleton and Kiseleva-Eggleton 2001; Fabrycky and Tremaine 2007; Naoz and Fabrycky 2014; Liu et al. 2015; Toonen et al. 2016; Anderson et al. 2017). Due to the gravitational perturbation of an (outer) highly-inclined stellar companion, LK cycles are observed for the inner binary star. As shown in Fig. 8 (top panel), the eccentricity of the inner orbit can reach values close to unity during the LK cycles. As a result, the periastron distance can reach very small values. Tidal dissipation comes then into play and causes the inner orbit to shrink. This mechanism is often called *LK migration*. The evolution of Fig. 8 was obtained when using vectorial secular octupole-order equations for both the spin and orbital evolutions with general relativity corrections, tidal effects, stellar oblateness, and magnetic spin-down braking (Correia et al. 2016).

The evolution of the inner binary can also be displayed on the phase portrait of a simplified Hamiltonian formulation, namely the quadrupole approximation (given by Eq. (1.1)) with general relativity corrections, as done in Bataille et al. (2018). For hierarchical stellar systems as the ones considered here (i.e., an inner binary with masses m_0 and m_1 perturbed by a distant companion star with mass m_2 and mutual inclination i), the adimensional angular momentum of the inner orbit $H = \sqrt{1-e_1^2}\cos i$ can be considered as constant (i.e., the Kozai constant) and the trajectory of the system can be followed, for a given value of H, on the level curves of the one-degree Hamiltonian in the plane (ω_1, i). In the bottom panel of Fig. 8, the time evolution of the system is indicated by the color code, from its initial configuration (in blue) to the final one (in yellow). The phase portrait clearly shows the LK equilibria at $\omega_1 = 90°$ and $\omega_1 = 270°$. The system initially evolves around one of these equilibria, then, due to tidal dissipative effects, moves away from it by crossing different Hamiltonian level curves, before the final locking of the inner orbit in a quasi-circular orbit with orbital period of a few days only.

Fabrycky and Tremaine (2007) studied the secular evolution of hierarchical triple-star systems using the quadrupole approximation with general relativity and tidal effects. They observed an excess of short-period binaries in their simulations, which suggests the possible existence of distant tertiary companions for the observed population of short-period binaries, preferably with a mutual inclination of $\sim 40°$ or $\sim 140°$. In their extension of the previous study to the octupole order, Naoz and Fabrycky (2014) showed that the

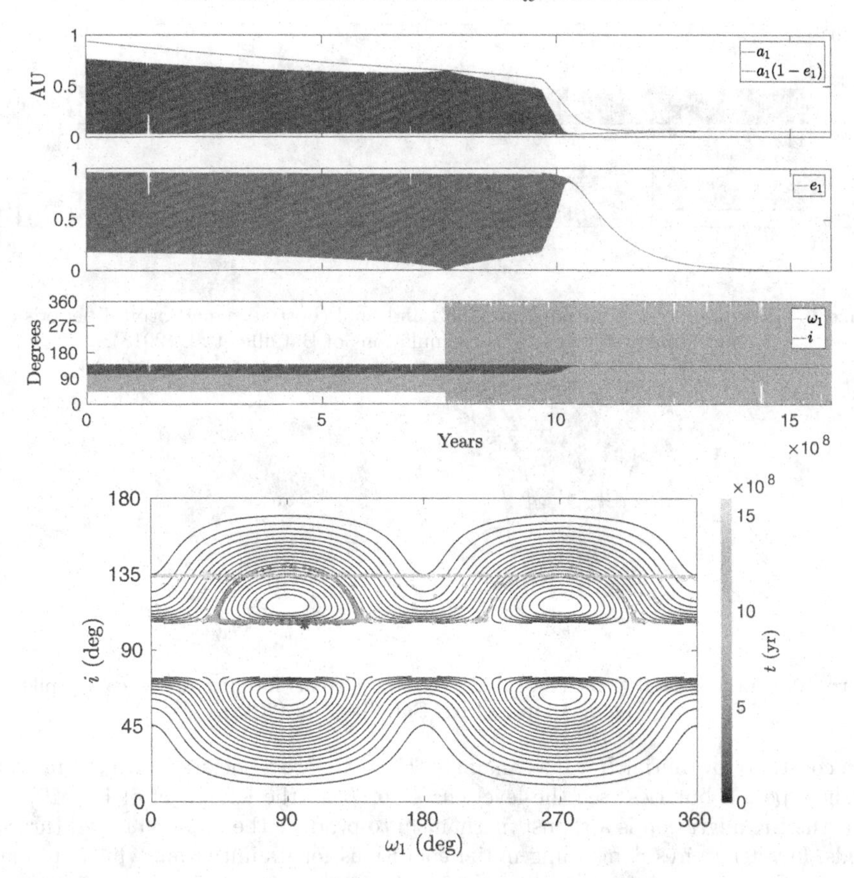

Figure 8. Triple-star system undergoing LK migration. Both the orbital and the spin evolutions are considered (octupole order approximation with general relativity corrections, tidal effects, stellar oblateness, and magnetic spin-down braking). Top: Time evolution of the orbital elements of the inner binary. Bottom: Same evolution displayed on the phase portrait of the quadrupole level Hamiltonian with general relativity corrections. Adapted from Bataille et al. (2018).

LK migration can be induced by a larger range of initial mutual inclinations. Further refinements were proposed by Anderson et al. (2017), Moe and Kratter (2018), and Bataille et al. (2018). In this last work, the authors aimed to identify the initial values of the orbital elements leading to the LK migration, by using uniform distributions of the initial elements in their simulations. They highlighted that the probability to initiate the migration, and thus the formation of the short-period pile-up around three days visible in Fig. 9 (bottom panel), is higher for the following initial orbital elements: $e_1 \sim 0.9$, $\omega_1 \sim 0°$ or $180°$, and mutual inclination $i \sim 90°$, as it can be deduced from the accumulations shown in Fig. 10.

Since very high eccentricities are needed for the migration caused by the tidal effects, either the inner binary is initially formed on very eccentric orbit, or its eccentricity has to be pumped to high values via the LK cycles (like in the evolution of Fig. 8). In the latter case, the initial mutual inclination preferably lies in the range $[40°; 140°]$. However, this condition on the mutual inclination is not sufficient to initiate the migration because when the system is very close to a LK equilibrium, the eccentricity variations are too limited for the tidal effects to operate. These findings can also be interpreted in terms of the

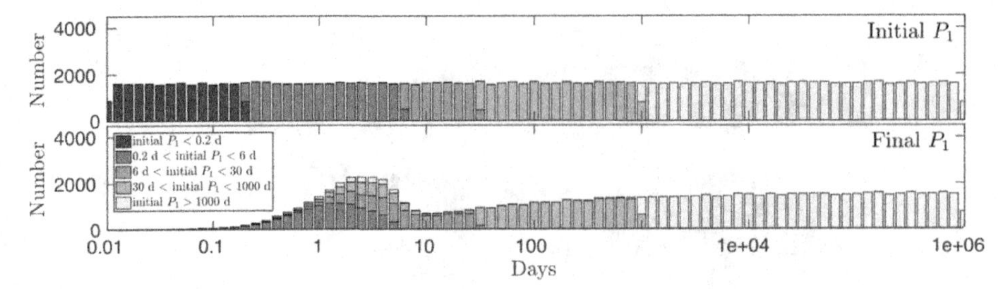

Figure 9. Histogram of the initial (top panel) and final (bottom panel) orbital periods of the inner binary, as given by the simulations of Bataille et al. (2018).

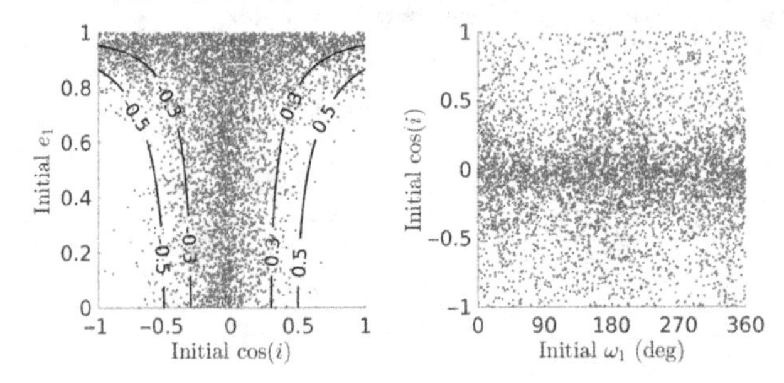

Figure 10. Initial orbital elements of the triple-star systems forming the three-day pile-up in Fig. 9. Adapted from Bataille et al. (2018).

Kozai constant: low initial H-values, namely $|H| \lesssim 0.5$, are associated with the migration into short-period binaries (see the level curves of H in the left panel of Fig. 10). As a result, the LK migration is a robust mechanism to produce the pile-up around three-day periods, but it requires demanding initial conditions for its initiation, which are maybe not encountered during the formation of the triple-star systems.

6. Conclusions

After the original works of von Zeipel (1910), Lidov (1962), and Kozai (1962), the LK resonance recently attracted renewed interest among very diverse scientific fields, from planetary and stellar systems to galaxies and black holes. This is due to the universal nature of the mechanism which clearly seems to operate at different scales. The list of the applications developed here is far from being exhaustive. The present contribution aimed to give a fairly accurate description of the LK resonance and to show how this mechanism can provide answers to current research questions in planetary and stellar systems, such as the formation of inclined planetary orbits and the existence of the hot Jupiters and short-period binaries.

The coming years will give us the opportunity to carry out deeper observations and confirm as well as quantify the occurence of the LK mechanism, a mechanism whose initial conditions in eccentricities and/or inclinations are quite demanding. This will provide valuable clues for a better understanding of the formation and long-term stability of planetary and stellar systems.

Acknowledgements

The author would like to warmly thank A. Lemaître and J. Teyssandier for their careful reading and suggestions. This work is supported by the Fonds de la Recherche Scientifique - FNRS under Grant No. F.4523.20 (DYNAMITE MIS-project).

Computational resources have been provided by the PTCI (Consortium des Equipements de Calcul Intensif CECI), funded by the FNRS-FRFC, the Walloon Region, and the University of Namur (Conventions No. 2.5020.11, GEQ U.G006.15, 1610468 and RW/GEQ2016).

References

Albrecht, S., Winn, J. N., Johnson, J. A., Howard, A. W., Marcy, G. W., Butler, R. P., Arriagada, P., Crane, J. D., Shectman, S. A., Thompson, I. B., Hirano, T., Bakos, G., & Hartman, J. D. 2012, *ApJ*, 757(1), 18.

Anderson, K. R., Lai, D., & Storch, N. I. 2017, *Monthly Notices of the Royal Astronomical Society*, 467, 3066–3082.

Anderson, K. R., Storch, N. I., & Lai, D. 2016, *MNRAS*, 456(4), 3671–3701.

Bailey, M. E. 1992, *Celestial Mechanics and Dynamical Astronomy*, 54(1-3), 49–61.

Bataille, M., Libert, A.-S., & Correia, A. C. M. 2018, *Monthly Notices of the Royal Astronomical Society*, 479(4), 4749–4759.

Batygin, K., Morbidelli, A., & Tsiganis, K. 2011, *A&A*, 533, A7.

Bitsch, B., Crida, A., Libert, A. S., & Lega, E. 2013, *A&A*, 555, A124.

Carruba, V., Burns, J. A., Nicholson, P. D., & Gladman, B. J. 2002, *Icarus*, 158(2), 434–449.

Cincotta, P. M., Giordano, C. M., & Simó, C. 2003, *Physica D Nonlinear Phenomena*, 182(3–4), 151–178.

Correia, A. C. M., Boué, G., & Laskar, J. 2016, *Celestial Mechanics and Dynamical Astronomy*, 126, 189–225.

Deitrick, R., Barnes, R., McArthur, B., Quinn, T. R., Luger, R., Antonsen, A., & Benedict, G. F. 2015, *ApJ*, 798(1), 46.

Duquennoy, A. & Mayor, M. 1991, *A&A*, 248, 485–524.

Eggleton, P. P. & Kiseleva-Eggleton, L. 2001, *The Astrophysical Journal*, 562, 1012–1030.

Fabrycky, D. & Tremaine, S. 2007, *ApJ*, 669(2), 1298–1315.

Ford, E. B., Kozinsky, B., & Rasio, F. A. 2000, *ApJ*, 535(1), 385–401.

Gallardo, T., Hugo, G., & Pais, P. 2012, *Icarus*, 220(2), 392–403.

Innanen, K. A., Zheng, J. Q., Mikkola, S., & Valtonen, M. J. 1997, *AJ*, 113, 1915.

Ito, T. & Ohtsuka, K. 2019, *Monographs on Environment, Earth and Planets*, 7(1), 1–113.

Jacobi, M. 1842, *Astronomische Nachrichten*, 20(6), 81.

Kinoshita, H. & Nakai, H. 1999, *Celestial Mechanics and Dynamical Astronomy*, 75(2), 125–147.

Kozai, Y. 1962, *AJ*, 67, 591.

Kozai, Y. 1985, *Celestial Mechanics*, 36(1), 47–69.

Laskar, J. 1997, *A&A*, 317, L75–L78.

Lee, M. H. & Peale, S. J. 2003, *ApJ*, 592(2), 1201–1216.

Libert, A.-S. & Henrard, J. 2007, *Icarus*, 191(2), 469–485.

Libert, A.-S. & Tsiganis, K. 2009, *A&A*, 493(2), 677–686.

Lidov, M. L. 1962, *Planetary Space Science*, 9, 719–759.

Liu, B., Mu noz, D. J., & Lai, D. 2015, *Monthly Notices of the Royal Astronomical Society*, 447, 747–764.

Michel, P. & Thomas, F. 1996, *A&A*, 307, 310.

Michtchenko, T. A., Ferraz-Mello, S., & Beaugé, C. 2006, *Icarus*, 181(2), 555–571.

Migaszewski, C. & Goździewski, K. 2009, *MNRAS*, 395(4), 1777–1794.

Moe, M. & Kratter, K. M. 2018, *ApJ*, 854(1), 44.

Naoz, S. 2016, *ARAA*, 54, 441–489.

Naoz, S. & Fabrycky, D. C. 2014, *The Astrophysical Journal*, 793, 137.

Naoz, S., Farr, W. M., Lithwick, Y., Rasio, F. A., & Teyssandier, J. 2011, *Nature*, 473(7346), 187–189.

Naoz, S., Farr, W. M., Lithwick, Y., Rasio, F. A., & Teyssandier, J. 2013, *MNRAS*, 431(3), 2155–2171.

Nesvorný, D., Alvarellos, J. L. A., Dones, L., & Levison, H. F. 2003, *AJ*, 126(1), 398–429.

Petrovich, C. 2015, *ApJ*, 799(1), 27.

Poincaré, H. 1892, *Les méthodes nouvelles de la mécanique céleste*. Gauthier-Villars et fils.
Quinn, T., Tremaine, S., & Duncan, M. 1990, *ApJ*, 355, 667.
Raghavan, D., McAlister, H. A., Henry, T. J., Latham, D. W., Marcy, G. W., Mason, B. D., Gies, D. R., White, R. J., & ten Brummelaar, T. A. 2010, *The Astrophysical Journal Supplement*, 190, 1–42.
Roisin, A. & Libert, A. S. 2021, *A&A*, 645, A138.
Roisin, A., Teyssandier, J., & Libert, A.-S. 2021, *MNRAS*, 506(4), 5005–5014.
Sansottera, M., Grassi, L., & Giorgilli, A. On the relativistic Lagrange-Laplace secular dynamics for extrasolar systems. In *Complex Planetary Systems, Proceedings of the International Astronomical Union* 2014,, volume 310, pp. 74–77.
Sansottera, M. & Libert, A. S. 2019, *Celestial Mechanics and Dynamical Astronomy*, 131(8), 38.
Schwarz, R., Funk, B., Zechner, R., & Bazsó, Á. 2016, *MNRAS*, 460(4), 3598–3609.
Terquem, C. & Ajmia, A. 2010, *MNRAS*, 404(1), 409–414.
Teyssandier, J., Terquem, C., & Papaloizou, J. C. B. 2013, *MNRAS*, 428(1), 658–669.
Thébault, P. & Haghighipour, N. 2015, *Planet Formation in Binaries*, pp. 309–340. Springer.
Thomas, F. & Morbidelli, A. 1996, *Celestial Mechanics and Dynamical Astronomy*, 64(3), 209–229.
Tokovinin, A. 2014, *AJ*, 147, 87.
Toonen, S., Hamers, A., & Portegies Zwart, S. 2016, *Computational Astrophysics and Cosmology*, 3, 6.
Tuomi, M. & Kotiranta, S. 2009, *A&A*, 496(2), L13–L16.
Veras, D. & Ford, E. B. 2010, *ApJ*, 715(2), 803–822.
Volpi, M., Roisin, A., & Libert, A.-S. 2019, *A&A*, 626, A74.
von Zeipel, H. 1910, *Astronomische Nachrichten*, 183(22), 345.
Wittenmyer, R. A., Wang, S., Horner, J., Tinney, C. G., Butler, R. P., Jones, H. R. A., O'Toole, S. J., Bailey, J., Carter, B. D., Salter, G. S., Wright, D., & Zhou, J.-L. 2013, *The Astrophysical Journal Supplement*, 208(1), 2.
Wu, Y. & Murray, N. 2003, *ApJ*, 589(1), 605–614.
Wu, Y., Murray, N. W., & Ramsahai, J. M. 2007, *ApJ*, 670(1), 820–825.
Xiang-Gruess, M. & Papaloizou, J. C. B. 2013, *MNRAS*, 431(2), 1320–1336.
Zanazzi, J. J. & Lai, D. 2018, *MNRAS*, 478(1), 835–851.

Multi-scale (time and mass) dynamics of space objects
Proceedings IAU Symposium No. 364, 2022
A. Celletti, C. Galeş, C. Beaugé, A. Lemaitre, eds.
doi:10.1017/S1743921322000461

A numerical criterion evaluating the robustness of planetary architectures; applications to the υ Andromedæ system

Ugo Locatelli[1], Chiara Caracciolo[2], Marco Sansottera[2] and Mara Volpi[1]

[1]Dipartimento di Matematica dell'Università degli Studi di Roma "Tor Vergata",
via della ricerca scientifica 1, 00133 Roma, Italy
emails: `locatell@mat.uniroma2.it`, `volpi@mat.uniroma2.it`

[2]Dipartimento di Matematica dell'Università degli Studi di Milano,
via Saldini 50, 20133 Milano, Italy
emails: `chiara.caracciolo@unimi.it`, `marco.sansottera@unimi.it`

Abstract. We revisit the problem of the existence of KAM tori in extrasolar planetary systems. Specifically, we consider the υ Andromedæ system, by modelling it with a three-body problem. This preliminary study allows us to introduce a natural way to evaluate the robustness of the planetary orbits, which can be very easily implemented in numerical explorations. We apply our criterion to the problem of the choice of a suitable orbital configuration which exhibits strong stability properties and is compatible with the observational data that are available for the υ Andromedæ system itself.

Keywords. Planetary Systems, Celestial Mechanics.

---

1. Introduction

From the very beginning of their history, physical sciences have been an inexhaustible source of problems and inspiration for mathematics. In particular, the orbital characteristics of more and more extrasolar systems are raising very challenging questions which concern the modern theory of stability for planetary Hamiltonian systems.

Since the announcement of the discovery of the first one (Mayor & Queloz 1995), thousands of exoplanets have been detected. Systems hosting more than one planet show a rather surprising variety of configurations which can be remarkably different with respect to that of the Solar System, which presents planetary orbits that are well separated, quasi-circular and nearly coplanar. The situation is made even more complex by the fact that none of the detection methods nowadays available to discover extrasolar planets is able to measure all their orbital elements. In this regard, the Radial Velocity (hereafter, RV) method is the most effective observation technique, because it provides values for the semi-major axis a, the eccentricity e, and the argument of the pericentre ω of an exoplanet (see, e.g., Perryman 2018). Moreover, the RV method is able to evaluate the so-called minimal mass $m\sin(\iota)$, where m and ι are the mass and the inclination† of the observed exoplanet, respectively. Indeed, this is as a very serious limitation of the currently available detection techniques, since they are often unable to completely determine

† More precisely, ι refers to the inclination of the Keplerian ellipse with respect to the plane orthogonal to the line of sight (i.e., the direction pointing to the object one is observing), which is usually said to be "tangent to the celestial sphere".

such an important parameter like the mass of an exoplanet (which is crucial to draw conclusions about, e.g., its habitability). In particular, the three-dimensional architecture of a multi-planetary system eludes the observational measures, when they are made using the RV method. However, this can be determined by crossing the results provided by multiple detection techniques, when different methods can be applied to the same system. Since the transit photometry is the most prolific technique in the discovering of exoplanets, its joined use with the RV method is expected to be very promising for what concerns their orbital characterisation. For instance, the combination of the transit and the RV method allowed to measure the inclination of three exoplanets orbiting around the L 98-59 star and so to determine rather narrow ranges for the values of their masses (Cloutier et al. 2019). Although the data obtained through astrometry are less precise with respect to the aforementioned detection techniques, they can be joined with the measures provided by the RV method to evaluate both the inclination ι and the longitude of the node Ω for some massive-enough exoplanets (e.g., in the case of HD 128311 c, see McArthur et al. 2014). Moreover, combining astrometry and RV methods it was possible to determine ranges of values for all the orbital elements except the mean anomalies M for the two exoplanets that are expected to be the most massive ones among those orbiting the υ Andromedæ A star† (McArthur et al. 2010). On one hand, this allowed to describe rather carefully the 3D structure of the main part of this extrasolar system, with an instantaneous value of the mutual inclination of $29.9^\circ \pm 1^\circ$; on the other hand, the uncertainty on the knowledge of a few orbital elements is so large that the estimated error on the mass of one of the exoplanets is quite relevant (i.e., $\simeq 30\%$), which is also due to the fact that its orbital plane is very inclined with respect to the line of sight.

According to the approach designed by Morbidelli & Giorgilli (1995), the stability of quasi-integrable systems can be efficiently analysed by combining the KAM theorem with the Nekhoroshev's one. In fact, their joint application can ensure the effective stability (that is valid for interval of times larger than the estimated age of the universe) for Hamiltonian systems of physical interest. This strategy has been successfully applied to a pair of non-trivial planetary models describing the dynamics of the two or three innermost Jovian planets of our Solar System; in both those cases, upper bounds on the diffusion speed have been provided by suitable estimates on the remainder of the Birkhoff normal form which is preliminarily constructed in the neighbourhood of an invariant torus (Giorgilli et al. 2009 & 2017). The so-called Arnold diffusion is a phenomenon which cannot take place in Hamiltonian systems having two degrees of freedom (hereafter, d.o.f.), because 2D invariant tori act as topological barriers separating the orbits. Nevertheless, reverse KAM theory can be applied in a way that is far from being trivial for what concerns the secular dynamics of extrasolar systems including three bodies (which can be described by a Hamiltonian model with 2 d.o.f.). In fact, in Volpi et al. (2018) the explicit construction of invariant KAM tori is used to infer information on the possible ranges of values of the mutual inclinations between the orbital planes of the two exoplanets hosted in the three following systems: HD 141399, HD 143761 and HD 40307. However, such an approach suffers serious limitations, mainly due to the fact that is based on an algorithm which was designed to construct suitable normal forms for the secular dynamics of our Solar System (Locatelli & Giorgilli 2000). Firstly, this computational procedure is apparently unable to deal with the case of eccentricities larger than 0.1, which is quite frequent

† Indeed, υ Andromedæ is a binary star. Since the companion is a red dwarf that is quite far (about 750 AU) from the primary star, the former is expected to not appreciably affect the planetary system orbiting the latter one. For the sake of simplicity, with the name of υ Andromedæ hereafter we will refer to both its primary star (which is, more precisely, υ Andromedæ A) and the extrasolar system hosting the exoplanets that have been discovered around it.

for exoplanets discovered by the RV detection method. Moreover, although the algorithm constructing the normal forms can work with bunches of initial conditions at the same time (if the implementation is made by using interval arithmetics, see Volpi et al. 2018), this kind of procedures can be rather demanding from a computational point of view, if they are not tailored carefully to the model under consideration. Therefore, the possibility to apply extensively such an approach to the study of many extrasolar systems looks rather doubtful. The so-called criterion of the Angular Momentum Deficit (hereafter AMD, see Laskar & Petit 2017, and Petit et al. 2017 for its reformulation adapted to planetary systems in mean motion resonance) gives an elegant answer to the need of a "coarse-graining" method for quickly studying the stability of many extrasolar planetary systems. However, also the AMD criterion does not cover all the extrasolar planetary systems that are known up to now, in the sense that is unable to ensure the stability for some of them. In particular, the AMD criterion can become inapplicable to systems where the orbital plane of (at least) one exoplanet is highly inclined with respect to the line of sight; for instance, this is exactly what occurs in the case of υ Andromedæ, which is very challenging. On the one hand, the 2D three-body model which includes the star and its two exoplanets with the largest minimal masses looks stable according to the AMD criterion, when the line of sight lies in their common orbital plane (Fig. 7 of Laskar & Petit 2017). On the other hand, when also the inclinations and the longitudes of the nodes are taken into account, then there is a remarkable fraction of the possible initial conditions that generates motions which are evidently unstable (McArthur et al. 2010; Deitrick et al. 2015). This is mainly due to the fact that the actual value of the mass of υ And c should be larger than 5 times the minimal one, while the increasing factor affecting the value of υ And d's mass is about 2.5 . Therefore, the perturbation of the Keplerian orbits (that is mainly due to the mutual gravitation) due to the updated values of the exoplanetary masses is one order of magnitude larger than the perturbation in the two-dimensional models of the υ Andromedæ system considering the data derived by the first observational measures provided by the RV detection method.

In Caracciolo et al. 2022, we have studied the secular dynamics of the υ Andromedæ system by adopting the so called averaged model at order two in the masses. In that framework we have shown how to construct an invariant (KAM) manifold which is a very accurate approximation of the orbit originating from initial conditions that are within the range of the observed values. Moreover, we have also shown rigorously the existence of such a KAM torus, by adopting a suitable technique based on a computer-assisted proof. Let us recall that this ensures that there is a small region around those initial conditions (and so, consistent with the observational data) for which the secular dynamics is effectively stable (see again the aforementioned paper by Morbidelli & Giorgilli 1995). Indeed, we have carefully selected those initial conditions by using a numerical criterion to evaluate the *robustness* of the corresponding orbit. The present work is devoted to the description of such a criterion. As it will be discussed in the next sections, the concept of robustness actually refers to the eventually existing torus which covers the orbit. The key remark which allows us to introduce such a criterion can be shortly summarised as follows: for what concerns the secular dynamics of the υ Andromedæ system, a KAM torus is as more persistent to the perturbing terms as it is closer to a periodic orbit which corresponds to the anti-alignment of the pericentre arguments of υ And c and υ And d. Thus, it is natural to apply our robustness criterion in situations where the exoplanets are in a librational regime with respect to the difference of their pericentre arguments. Let us recall that υ And c and υ And d were conjectured to be in such an apsidal locking state just a few years after their discovery (Chiang et al. 2001, see also Michtchenko & Malhotra 2004 for an explanation of such a dynamical mechanism within the framework of a secular model). Although our robustness criterion

Table 1. Orbital elements and minimal masses of the exoplanets υ And c and υ And d. All the data appearing in the following first three columns are reported from Table 13 of McArthur et al. (2010). In the rightmost column we have included also the relative errors for each quantity. In all our numerical integrations the stellar mass of υ Andromedæ is assumed to be $m_0 = 1.31\, M_\odot$. As usual, $M_\odot$ and M_J denote the solar mass and the Jupiter one, respectively.

	υ **And** c	υ **And** d	**rel. err.**
$a(0)$ [AU]	0.829 ± 0.043	2.53 ± 0.014	$\simeq 5\,\%$
$e(0)$	0.245 ± 0.006	0.316 ± 0.006	$\simeq 5\,\%$
$\iota(0)$ [°]	7.868 ± 1.003	23.758 ± 1.316	1.4/180
$\omega(0)$ [°]	247.66 ± 1.76	252.99 ± 1.31	1.8/360
$\Omega(0)$ [°]	236.85 ± 7.53	4.07 ± 3.31	7.6/360
$m \sin(\iota(0))$ $[M_J]$	1.96 ± 0.05	4.33 ± 0.11	$\lesssim 3\,\%$

simply applies in combination with numerical integrations (any averaging procedure is not strictly necessary), it is somehow related with the dynamical phenomenon we have described by adopting the language of the normal forms and the refined computational procedure which is fully detailed in Caracciolo et al. 2022. Therefore, our robustness numerical indicator does not aim to be as general as the AMD stability criterion, at least in its first formulation we are going to introduce; eventual extensions to contexts different with respect to the librations in an apsidal locking regime (or in the anti-apsidal one) could need some nontrivial adaptations.

2. The orbital dynamics of the exoplanets in the υ Andromedæ system: a short overview

The discovery of three exoplanets orbiting around υ Andromedæ was made at the end of the last century, by applying the RV detection method (Butler et al. 1999). Moreover, McArthur et al. (2010) remarked that a long-period trend in the analysis of the signals is an indication of the presence of a fourth planet (named υ And e). The long-term stability of a planetary system which includes υ And b, υ And c and υ And d has been studied in Deitrick et al. 2015, by performing many numerical integrations; let us also recall that several of them have shown unstable motions. In the present work, we are going to further restrict the model by limiting us to consider the two exoplanets that are expected to be the largest ones. There are good reasons to assume that the influence exerted by υ And b and υ And e is negligible: the latter is known quite poorly (and the RV method is rather sensitive to more massive bodies), while the former is very tightly close to the star and its minimal mass is one order of magnitude smaller than the ones of υ And b and υ And c†.

The initial values of the orbital elements (except the mean anomalies, that are unknown) for the pair of exoplanets υ And c, υ And d and their minimal masses are reported in Table 1. Hereafter, in our three-body model of the υ Andromedæ extrasolar system the indexes 1, 2 will be used to refer to the inner planet and the outer one, respectively, while m_0 will denote the stellar mass. Looking at Table 1, one can appreciate that all the reported data are given with a relative uncertainty that is not larger than a few percentage units. Due to the occurrence of the increasing factor $1/\sin(\iota_j(0))$ (with $j = 1,\, 2$), this is no more true for the exoplanetary masses. A straightforward evaluation starting from the data reported in Table 1 gives $m_1 = 14.6 \pm 2.2$ and $m_2 = 10.8 \pm 0.9$. Therefore, the relative uncertainty of at least one parameter (which plays a crucial role

† The ratio between the semi-major axes of two consecutive planets is $\simeq 14$ in the case of the pair υ And c – υ And b, while it is a bit more than 3 in the case of υ And d – υ And c. In the case of υ And b the value of the quantity $m \sin(\iota(0))$ is known to be $0.0594 \pm 0.0003\, M_J$ (see Table 13 of McArthur et al. (2010), whose data concerning υ And c and υ And d are reported in our Table 1 above). Let us also recall that the initial inclination $\iota(0)$ of υ And b is unknown and, thus, the minimal value is the only information available about its mass.

Table 2. List of the values of the parameters that are kept fixed in all our numerical explorations. In the first two rows, the initial conditions concerning with semi-major axes and eccentricities of the exoplanets υ And c and υ And d are reported. In the last row, their values of the minimal masses are given.

	υ **And** c	υ **And** d
$a(0)$ [AU]	0.829	2.53
$e(0)$	0.239	0.310
$m\sin(\iota(0))$ $[M_J]$	1.91	4.22

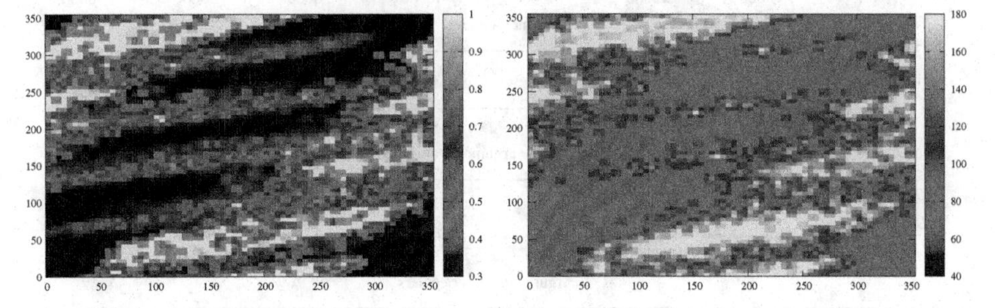

Figure 1. On the left, a colour-grid plot of the maximal value reached by the eccentricity e_2 of υ And d. On the right, the same for the maximal value of the difference between the pericentres, i.e., $\max_t |\omega_1(t) - \omega_2(t)|$. In both panels, the plots are made as a function of the initial values of the mean anomalies $M_1(0)$ and $M_2(0)$. See the text for more details about the choice of the initial conditions.

in the discussion about the stability of this extrasolar planetary system) can reach† 15 % of the corresponding mid value; this is the case of the mass of υ And c.

We emphasise that an extensive study of the possible motions with an homogeneous and accurate covering of all the possible initial conditions and parameters gets immediately far too complex from a computational point of view, because it would require to deal with a fourteen-dimensional grid. For the sake of simplicity, we started to reduce the complexity by fixing some of the parameters that are determined rather precisely by the observational measures made by using the RV method; in detail, they are two pairs of orbital elements, i.e., the initial values of semi-major axes and eccentricities, and the minimal masses. The latter two pairs have been fixed so to be equal to the lowest possible values of the corresponding ranges given in Table 1. This choice has been made in order to increase the fraction of the orbital motions that are apparently stable. All the values of the parameters that have been so fixed by us are reported in Table 2.

We start our numerical explorations by investigating the dependence on the pair of the orbital elements that are unknown, namely the mean anomalies M_1 and M_2 ‡. For this purpose, we decide to consider sets of initial conditions such that $\iota_j(0)$, $\omega_j(0)$ and $\Omega_j(0)$ are set to be equal to the corresponding mid values reported in Table 1, $\forall\, j = 1, 2$. Moreover, the initial conditions are complemented with the data reported in Table 2, while the initial values of the mean anomalies are taken from a regular 2D grid covering all the set $[0°, 360°] \times [0°, 360°]$ with a grid-step of 5°. Hereafter, the mass of each exoplanet is always determined by multiplying its minimal value (appearing in Table 2) by the

† Taking into account all the uncertainties due to the observational measures, the errors ranges are even wider. Indeed, in Table 13 of McArthur et al. (2010) the following values for the exoplanetary masses are given: $m_1 = 13.98^{+2.3}_{-5.3}$ and $m_2 = 10.25^{+0.7}_{-3.3}$.

‡ The observational data reported in both online catalogues and published papers determine values for the orbital period and the epoch of periastron. From these two values it is possible to infer the values of the mean anomalies, but they are never explicitly determined. In this respect, we then consider them as *unknown*.

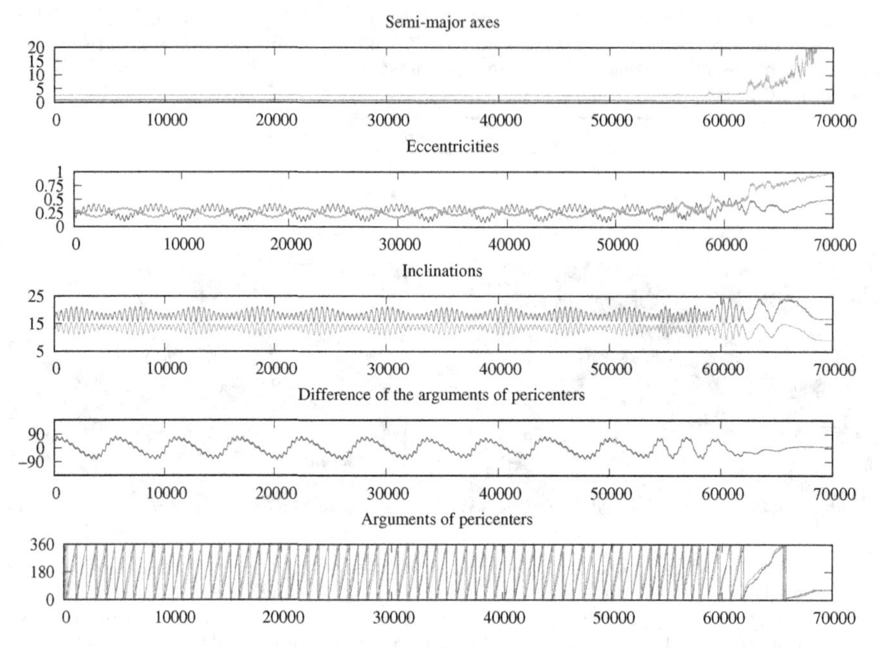

Figure 2. Orbital evolution of the exoplanets υ And c and υ And d in the case of one single set of initial conditions among those that have been considered also in Fig. 1 with the particular choice $M_1(0) = 0°$ and $M_2(0) = 120°$ for what concerns the initial values of the mean anomalies. From top to bottom, the five graphs include the plots of the evolution for the following quantities: semi-major axes a_1 and a_2, eccentricities e_1 and e_2, inclinations ι_1 and ι_2, difference of the arguments of the pericentres $\omega_2 - \omega_1$, arguments of the pericentres ω_1 and ω_2. The plots in green refer to the orbital motion of υ And *d*. The inclinations are evaluated with respect to the direction of the total angular momentum. In all the abscissas, the year is assumed as unit of measure of time.

corresponding increasing factor $1/\sin(\iota(0))$. Starting from each of the initial conditions defined just above, we have numerically integrated the Hamilton equations describing our three-body planetary model, by using the symplectic method $\mathcal{SBAB}_{C3}$ as it is defined in Laskar & Robutel (2001) for a timespan of 10^5 yr, with an integration step of 0.02 yr. The main results so obtained are summarised in Fig. 1, which highlights that the choice of the initial values of the mean anomalies affects the orbital dynamics in a very remarkable way. In fact, the regions that appear with lighter colours in the left panel correspond to motions that can experience close encounters. Let us recall that the threshold value of the eccentricity of the outer planet on top of which collisions with the inner planet are possible can be roughly evaluated as $1 - (a_1(0)/a_2(0)) \simeq 0.67$. On the other hand, about 50 % of the colour-grid plot in the left panel is in dark; this means that the maximum value of the eccentricity of the outer planet looks to be safely below that threshold allowing close encounters with the inner one. The strong similarity between the two panels of Fig. 1 clearly suggests that stable configurations are possible when the difference of the pericentre arguments is in a librational regime, i.e., the orbital motions are such that the maximum of the half-width of the oscillations concerning with $\omega_1(t) - \omega_2(t)$ is less than $180°$. Let us emphasise that this kind of phenomena has already been observed in the last few years. In fact, the relevance of the impact due to the mean anomalies on the orbital dynamics of extrasolar systems that are close or in mean motion resonance has been shown, e.g., in Libert & Sansottera (2013) and Sansottera & Libert (2019).

In order to make clear the ideas, it can be convenient to have a close look at the dynamical evolution of most of the orbital elements, for a motion starting form one single set of initial conditions, that is selected among those considered in Fig. 1. In particular, the orbital evolution described by the plots included in Fig. 2 refers to $M_1(0) = 0°$ and $M_2(0) = 120°$; let us recall that the other values of the initial conditions are taken from the mid values of Table 1 (for what concerns $\iota(0)$, $\omega(0)$ and $\Omega(0)$ only) and from Table 2 (for the remaining data). Looking at the plots of the orbital elements one can appreciate that the orbit is unstable; for any of those plots, the lack of quasi-periodicity is particularly evident after 50000 yr. From the behaviour of the semi-major axes and the eccentricities, it is obvious that the outer planet is ejected from the system at the end of the numerical simulation. Let us also stress that our standard implementation of the symplectic method $\mathcal{SBAB}_{C3}$ usually crashes for all the motions starting from initial conditions which correspond to regions of lighter colour in the plot of the left panel in Fig. 1.

Hereafter, we will refer to the three-body planetary problem that has been described in the present section as the *complete* model, in order to distinguish it with respect to the *secular* one. The latter is an Hamiltonian system which is defined by a suitable procedure of averaging that will be briefly discussed in the next section.

3. The construction of invariant tori in the secular dynamics of the υ Andromedæ system as a source of inspiration

The present section is devoted to recall some of the ideas we recently used in order to successfully construct KAM tori, that are invariant for the secular dynamics of the υ Andromedæ planetary system and are also in librational regime with respect to the difference of the pericentre arguments (Caracciolo et al. 2022). Our aim is to explain in a rather natural way the reasons to introduce our numerical criterion evaluating the robustness of planetary configurations, that will be properly defined in the next section.

In the case of the secular dynamics of the υ Andromedæ planetary system, the preliminary construction of the normal form for a particular elliptic torus is essential to be performed before the one constructing the final KAM torus. These two constructive procedures can be described in an unified way, as we explained in Locatelli et al. (2022). We defer the reader to those pedagogical notes for all the details about this kind of (so-called) semi-analytic algorithms, that can be summarised as follows for our goals.

The proof scheme of the KAM theorem can be formulated in terms of a constructive algorithm whose convergence is ensured if some suitable hypotheses are satisfied. This procedure starts by considering an analytic Hamiltonian function $H^{(0)} : \mathcal{A} \times \mathbb{T}^n \mapsto \mathbb{R}$ (being $\mathcal{A} \subseteq \mathbb{R}^n$ an open set) of the form $H^{(0)}(\boldsymbol{p}, \boldsymbol{q}) = \boldsymbol{\nu} \cdot \boldsymbol{p} + h^{(0)}(\boldsymbol{p}, \boldsymbol{q}) + \varepsilon f^{(0)}(\boldsymbol{p}, \boldsymbol{q})$, where n denotes the number of degrees of freedom, $\boldsymbol{\nu} \in \mathbb{R}^n$ is an angular velocity vector and $h^{(0)}$ is at least quadratic with respect to the actions $\boldsymbol{p}$, i.e., $h^{(0)}(\boldsymbol{p}) = \mathcal{O}(\|\boldsymbol{p}\|^2)$ for $\boldsymbol{p} \to \boldsymbol{0}$. The term $\varepsilon f^{(0)}(\boldsymbol{p}, \boldsymbol{q})$ appearing in $H^{(0)}$ is usually called the perturbing term and it is made smaller and smaller by the normalisation procedure, which is defined by an infinite sequence of canonical transformations. This entails that we have to introduce a sequence of Hamiltonians $H^{(r)}$ that are iteratively defined so that

$$H^{(r)} = \exp\left(\mathcal{L}_{\chi_2^{(r)}}\right) \exp\left(\mathcal{L}_{\chi_1^{(r)}}\right) H^{(r-1)} \qquad \forall\, r \geq 1\,, \tag{3.1}$$

where the generating functions $\chi_1^{(r)}$ and $\chi_2^{(r)}$ are determined in such a way to remove the part of the perturbation term that is both $\mathcal{O}(\varepsilon^r)$ and not dependent on $\boldsymbol{p}$ or linear in $\boldsymbol{p}$, respectively. We then say that formula (3.1) defines the r-th normalization step. We stress that the Lie series operators $\exp\left(\mathcal{L}_{\chi_2^{(r)}}\right)$ and $\exp\left(\mathcal{L}_{\chi_1^{(r)}}\right)$ define canonical transformations when they are applied to the whole set of variables $(\boldsymbol{p}, \boldsymbol{q})$. This is due to the fact that they are given in terms of the Lie derivatives $\mathcal{L}_{\chi_2^{(r)}}\,\mathcal{L}_{\chi_1^{(r)}}$ (which in turn are expressed as Poisson brackets, i.e., $\mathcal{L}_g f = \{f, g\}$ for any pair of dynamical functions f and g that are defined on the phase space). The statement of the KAM theorem (see Kolmogorov 1954, Arnold 1963 and Moser 1962) can be shortly formulated as follows:

if $\boldsymbol{\nu}$ is non-resonant enough, $h^{(0)}$ is non-degenerate with respect to the actions $\boldsymbol{p}$ and the parameter ε is small enough, then there is a canonical transformation $(\boldsymbol{p},\boldsymbol{q}) = \Psi(\boldsymbol{P},\boldsymbol{Q})$, leading $H^{(0)}$ in the so called Kolmogorov normal form

$$\mathcal{K}(\boldsymbol{P},\boldsymbol{Q}) = \nu\cdot\boldsymbol{P} + \mathcal{O}(\|\boldsymbol{P}\|^2)\ , \tag{3.2}$$

being $\mathcal{K} = H\circ\Psi$.

Indeed, the final canonical transformation Ψ is obtained by composing all the canonical transformations induced by $\exp\left(\mathcal{L}_{\chi_1^{(1)}}\right)$, $\exp\left(\mathcal{L}_{\chi_2^{(1)}}\right)$, ... $\exp\left(\mathcal{L}_{\chi_1^{(r)}}\right)$, $\exp\left(\mathcal{L}_{\chi_2^{(r)}}\right)$ ... Moreover, one can easily verify that the quasi-periodic motion law $t\mapsto(\boldsymbol{P}(t)=\boldsymbol{0}\ ,\ \boldsymbol{Q}(t) = \boldsymbol{Q}_0+\boldsymbol{\nu}t)$ is the unique solution for the Hamilton equations related to the Kolmogorov normal form (3.2) with initial conditions $(\boldsymbol{P}(0)\ ,\ \boldsymbol{Q}(0)) = (\boldsymbol{0}\ ,\ \boldsymbol{Q}_0)$. Since the canonical transformations have the property of preserving solutions, then the n-dimensional KAM torus $\left\{(\boldsymbol{p},\boldsymbol{q}) = \Psi(\boldsymbol{0},\boldsymbol{Q})\ ,\ \forall\,\boldsymbol{Q}\in\mathbb{T}^n\right\}$ is invariant with respect the flow induced by the initial Hamiltonian $H^{(0)}$.

3.1. *Preliminaries*

As it has been first explained in Locatelli & Giorgilli (2000), the so-called secular model at order two in the masses can be properly introduced by performing a first step of normalization, which aims at removing the perturbation terms depending on the fast revolution angles. In order to set the ideas let us recall that a three-body Hamiltonian problem has nine degrees of freedom. Three of them can be easily separated because they describe the uniform motion of the centre of mass in an inertial frame. The nontrivial part of the dynamics is represented in astrocentric canonical coordinates and its degrees of freedom can be further reduced by two using the conservation of the total angular momentum $\boldsymbol{C}$. As it is shown in section 6 of Laskar (1989), this allows us to write the Hamiltonian as a function of four pairs of Poincaré canonical variables, that are

$$\begin{aligned}
\Lambda_j &= \frac{m_0 m_j\sqrt{G(m_0+m_j)a_j}}{m_0+m_j}\,, & \xi_j &= \sqrt{2\Lambda_j}\sqrt{1-\sqrt{1-e_j^2}}\,\cos(\omega_j)\,, & \\
\lambda_j &= M_j+\omega_j\,, & \eta_j &= -\sqrt{2\Lambda_j}\sqrt{1-\sqrt{1-e_j^2}}\,\sin(\omega_j)\,, & \forall\, j=1,2\,.
\end{aligned} \tag{3.3}$$

We also recall that the reduction of the total angular momentum makes implicit the dependence on the orbital elements that are missing in formula (3.3). They are the inclinations and the longitudes of the nodes, which are conveniently expressed with respect to the so-called Laplace invariant plane, that is orthogonal to the total angular momentum $\boldsymbol{C}$. However, also the instantaneous values of these two pairs of orbital elements can be recovered by the knowledge of all the others and the euclidean norm of $\boldsymbol{C}$. The actions Λ_1 and Λ_2 (that are conjugate with respect to the mean anomalies λ_1 and λ_2 , respectively) are usually expanded around a pair of reference values, namely Λ_1^* and Λ_2^*. These values are obtained by replacing the semi-major axes appearing in the corresponding definition included in formula (3.3) with their initial values $a_1(0)$ and $a_2(0)$ reported in Table 2. Thus, after the reduction of the constants of motion, the Hamiltonian describing the three-body planetary problem can be expressed as a function of four pairs of canonical variables: $\boldsymbol{L} = \boldsymbol{\Lambda} - \boldsymbol{\Lambda}^*$, λ, ξ and η. We can introduce the secular model at order two in the masses thanks to the following three operations: we perform a first step of normalization aiming to reduce the perturbing part that does not depend on $\boldsymbol{L}$ and does depend on the angles λ; we put $\boldsymbol{L}=\boldsymbol{0}$ (this is made because we expect that the oscillations of the semi-major axes close to their initial values have negligible effects); we finally average over the mean anomalies $\boldsymbol{\lambda}$ (as it is usual, when the analysis is focused on the long-term evolution of a planetary system). Therefore, we can write our secular Hamiltonian model as follows:

$$H^{(\mathrm{sec})}(\boldsymbol{\xi},\boldsymbol{\eta}) = \sum_{s=1}^{N_S/2} h_{2s}^{(\mathrm{sec})}(\boldsymbol{\xi},\boldsymbol{\eta}) \,, \tag{3.4}$$

where h_{2s} is an homogeneous polynomial of degree $2s$. This means that the expansion contains just terms of even degree, as a further consequence of the well known D'Alembert rules. Let us stress that the canonical variables $(\boldsymbol{\xi},\boldsymbol{\eta})$ appearing in formula (3.4) are not the ones defined in (3.3), by abuse of notation. Indeed, the former variables are obtained from the latter ones, by performing the canonical transformation defined by the normalization step introducing the secular model at order two in the masses. Since this change of variables differs from the identity, because of a small correction that is of order one in the masses, then the values of the canonical variables $(\boldsymbol{\xi},\boldsymbol{\eta})$ appearing in formula (3.4) are quite close to the corresponding ones that are defined in (3.3). These last comments joined with the remark that both ξ_j and η_j are $\mathcal{O}(e_j)$ for $e_j \to 0\ \forall\, j=1,2$ (as it can be easily checked by looking at the definition (3.3)) allow us to give a meaning to the parameter N_S , in the sense that $H^{(\mathrm{sec})}$ provides an approximation of the secular dynamics up to order N_S in the eccentricities. On the one hand, in practical applications one is interested in expansions up to high order in eccentricities†; on the other hand, the computational effort critically increases with respect to N_S . To fix the ideas, in the case of the υ Andromedæ planetary system we have found that setting $N_S=8$ is a good balance between these two different needs that are in opposition to each other.

We have explicitly performed all the computations of Poisson brackets (required by Lie series formalism to express canonical transformations) and all the expansions briefly described in the present section, by using *Χρόνος*. It is a software package especially designed for doing computer algebra manipulations into the framework of Hamiltonian perturbation theory (see Giorgilli & Sansottera 2012 for an introduction to its main concepts). Such computations also allow an easy visualisation of the secular dynamics by adopting a classical tool in the context of the numerical investigations: the Poincaré sections. In fact, we have performed many numerical integrations of the secular model $H^{(\mathrm{sec})}$ that is defined in (3.4) by simply applying the RK4 method‡.

A few dynamical features of the Hamiltonian model defined by $H^{(\mathrm{sec})}$ are summarised in the plots reported in Fig. 3. The orbit plotted in red in both panels refers to a set of initial conditions of the same type with respect to those considered in the previous Section 2. In detail, the initial values of the mean anomalies have been set so that $M_1(0)=M_2(0)=0°$, while the other initial conditions are taken from the mid values of Table 1 (for what concerns $\iota(0)$, $\omega(0)$ and $\Omega(0)$ only) and from Table 2 (for the remaining data). Since the Poincaré sections are plotted in correspondence to the hyperplane $\eta_2=0$ (with the additional condition $\xi_2>0$) and the canonical variables $(\boldsymbol{\xi},\boldsymbol{\eta})$ appearing in formula (3.4) are close to those defined in (3.3), then we can assume that on the surface of section $\omega_2 \simeq 0$. In the left panel of Fig. 3, therefore, the difference of the pericentre

† However, it must be taken into account that too large expansions of the secular model introduced here can be meaningless, because the high quality of the approximation in the eccentricities can be shadowed by the lack of precision with respect to the masses.

‡ It is very well known that long-term numerical integrations of secular models are much less computationally expensive than those dealing with the corresponding complete planetary system (see, e.g., Laskar 1988 and the references therein).

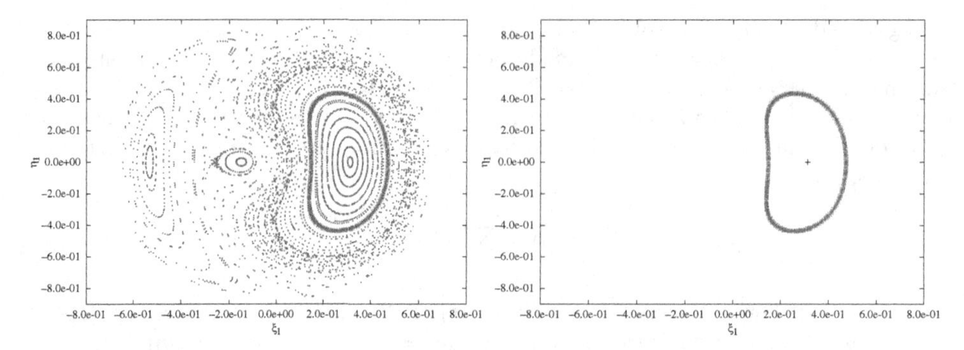

Figure 3. On the left, Poincaré sections that are corresponding to the hyperplane $\eta_2 = 0$ (with the additional condition $\xi_2 > 0$) and are generated by the flow of the Hamiltonian secular model $H^{(\rm sec)}$, given in (3.4) at order two in the masses for the exoplanetary system υ Andromedæ; the orbit in red refers to the motion starting from initial conditions of the same type of those considered in Figs. 1–2 with the additional choice of the initial mean anomalies, that have been fixed so that $M_1(0) = M_2(0) = 0°$. On the right panel, the same orbit in red is shown: approximately at its centre the symbol + represents the orbit of a one-dimensional elliptic torus (that reduces to a fixed point in these Poincaré sections). See the text for more details.

arguments $\omega_2 - \omega_1$ is evaluated by the polar angle, whose width is measured, as usual, with respect to the set of the positive abscissas, i.e., $\{(\xi_1 > 0, \eta_1 = 0)\}$. Thus, we can easily appreciate that this angle is librating around 0° also in the case of the secular model, in agreement with the corresponding plots reported in Figs. 1–2, that refer to the dynamics of the complete planetary system. By taking into account the fact that the nodes are opposite in the Laplace frame, this means that the pericentres of υ And c and υ And d are in the so-called apsidal locking regime in the vicinity of the anti-alignment of the pericentres. It is easy to remark that the Poincaré sections plotted in red (that are corresponding to the motion starting from the initial conditions we have chosen to consider) are orbiting around a fixed point, whose presence is also highlighted in the right panel of Fig. 3. Let us recall that all the Poincaré sections reported in Fig. 3 refer to the same level of energy, say E, corresponding to the set of the initial conditions we have previously described. Since $H^{(\rm sec)}$ is a two degrees of freedom Hamiltonian, the manifold labelled by such a value of the energy will be three-dimensional; in other words, by plotting the Poincaré sections, we automatically reduce by one the dimensions of the orbits. This is the reason why a fixed point actually corresponds to a periodic orbit. Since such a fixed point with positive value of the abscissa is surrounded by closed curves, then we can argue that such a periodic orbit is linearly stable for what concerns the transverse dynamics. This means that it can be seen as a one-dimensional elliptic torus. Therefore, we can conclude that the orbit which intersects the hyperplane $\eta_2 = 0$ in correspondence with the red dots is actually winding around a linearly stable periodic orbit, by remaining in its vicinity. This explains why it can be convenient to adopt a strategy based on two different algorithms: the first one refers to the elliptic torus (that corresponds to a fixed point in the Poincaré sections) and provides a good enough approximation to start the second computational procedure that constructs the final KAM torus (which shall include also the points marked in red in Fig. 3).

3.2. *Construction of the normal form for an one-dimensional elliptic torus*

It is now convenient to introduce a new set of canonical coordinates by including among them also an angle which describes the libration of the difference of the pericentre arguments, i.e., $\omega_2 - \omega_1$. For such a purpose, we first introduce a set of action-angle variables $(\boldsymbol{\mathcal{J}}, \boldsymbol{\psi})$ via the canonical transformation

$$\xi_j = \sqrt{2\mathcal{J}_j}\cos\psi_j\ , \qquad \eta_j = \sqrt{2\mathcal{J}_j}\sin\psi_j\ , \qquad \forall\ j=1,2, \tag{3.5}$$

being $(\boldsymbol{\xi}, \boldsymbol{\eta})$ the variables appearing as arguments of the secular Hamiltonian $H^{(\mathrm{sec})}$ defined in (3.4). Then, we define a new set of variables $(\boldsymbol{I}, \boldsymbol{\vartheta})$ such that

$$\vartheta_1 = \psi_1 - \psi_2\ , \quad \vartheta_2 = \psi_2\ , \quad I_1 = \mathcal{J}_1\ , \quad I_2 = \mathcal{J}_2 + \mathcal{J}_1\ . \tag{3.6}$$

In view of the discussion included in the previous subsection, we have that the angle $\vartheta_1 \simeq \omega_2 - \omega_1$ is expected to librate in the model under consideration. We now move to (new) canonical polynomial variables $(\boldsymbol{x}, \boldsymbol{y})$ defined as

$$x_j = \sqrt{2I_j}\cos\vartheta_j\ , \qquad y_j = \sqrt{2I_j}\sin\vartheta_j\ , \qquad \forall\ j=1,2\ . \tag{3.7}$$

Let us also remark that making Poincaré sections with respect to the hyperplane $\eta_2 = 0$, when $\xi_2 > 0$ is equivalent to impose $\psi_2 = 0$, because of the definitions in (3.5). Therefore, looking at formulæ (3.6)–(3.7), one can easily realise that the drawing in the left panel of Fig. 3 can be seen as a plot of the Poincaré sections in coordinates $(x_1\ , y_1)$ with respect to $y_2 = 0$ and with the additional condition $x_2 > 0$. By a simple numerical method, we can easily determine the initial condition $(\boldsymbol{x}^\star, \boldsymbol{y}^\star)$ that is in correspondence with a Poincaré section and generates a periodic solution. We can now subdivide the variables in two different pairs. The first one is given by $(p, q) \in \mathbb{R}\times\mathbb{T}$, i.e., the action-angle pair describing the periodic motion. Thus, we rename the angle ϕ_2 as q, while the action is obtained by translating the origin of I_2 so that $p = I_2 - \big((x_2^\star)^2 + (y_2^\star)^2\big)/2$. For what concerns the second pair of canonical coordinates, we start from the polynomial variables (x_1, y_1) in order to describe the motion transverse to the periodic orbit. It is now convenient to rescale the transverse variables $(\bar{x}_1, y_1)$, being $\bar{x}_1 = x_1 - x_1^\star$, in such a way that the Hamiltonian part which is quadratic in the new variables (x, y) and does not depend on (p, q) is in the form $\Omega^{(0)}(x^2 + y^2)/2$. This rescaling can be done by a canonical transformation as the quadratic part does not have any mixed term $\bar{x}_1 y_1$ and the coefficients of $\bar{x}_1^2$ and y_1^2 have the same sign, because of the proximity to an elliptic equilibrium point. Thus, since such a quadratic part is in the preliminary form $a\bar{x}_1^2 + by_1^2$, it suffices to define the new variables (x, y) as $x = \sqrt[4]{\frac{a}{b}}\,\bar{x}_1$, $y = \sqrt[4]{\frac{b}{a}}\,y_1$. Finally, we introduce the second pair of canonical coordinates $(J, \phi) \in \mathbb{R}_+ \cup \{0\} \times \mathbb{T}$ so that $x = \sqrt{2J}\cos\phi$ and $y = \sqrt{2J}\sin\phi$.

After having performed all the canonical transformation described above, the Hamiltonian can be written in the following way:

$$\mathcal{H}^{(0)}(p, q, J, \phi) = \mathcal{E}^{(0)} + \nu^{(0)}p + \Omega^{(0)}J + h(p, J, \phi) + \varepsilon f^{(0)}(p, q, J, \phi)\ , \tag{3.8}$$

where $\mathcal{E}^{(0)}$ is constant (that is close to the energy value of the wanted periodic orbit), $\nu^{(0)}$ and $\Omega^{(0)}$ are angular velocities, the function $h(p, J, \phi) = \mathcal{O}(\|(p, J)\|^{3/2})$ when the action vector† $(p, J) \to \mathbf{0}$ and $\varepsilon f^{(0)}(p, q, J, \phi)$ is a generic perturbing term, with ε playing the role of the small parameter. If such a perturbation is small enough, then it is possible to successfully perform a normalization algorithm, which allows to construct another canonical transformation Φ that conjugates the initial Hamiltonian $\mathcal{H}^{(0)}$ to $\mathcal{H}^{(\infty)} = \mathcal{H}^{(0)} \circ \Phi$ having the following (normal) form:

$$\mathcal{H}^{(\infty)}(P, Q, X, Y) = \mathcal{E}^{(\infty)} + \nu^{(\infty)}P + \frac{\Omega^{(\infty)}}{2}(X^2 + Y^2) + \mathcal{R}(P, Q, X, Y)\ , \tag{3.9}$$

where $\mathcal{E}^{(\infty)}$ is constant, $\nu^{(\infty)}$ and $\Omega^{(\infty)}$ are angular velocities and the remainder $\mathcal{R}$ is such that $\mathcal{R}(P, Q, X, Y) = o\big(|P| + \|(X, Y)\|^2\big)$, when $(P, X, Y) \to (0, 0, 0)$. Therefore, one

† Because of the change of coordinates which introduces the canonical pair of variables (J, ϕ), i.e., $x = \sqrt{2J}\cos\phi$ and $y = \sqrt{2J}\sin\phi$, also semi-integer powers of J can appear in the expansion (3.8) of the Hamiltonian $\mathcal{H}^{(0)}$.

can easily check that

$$(P(t), Q(t), X(t), Y(t)) = \big(0, Q(0) + \nu^{(\infty)} t, 0, 0\big) \tag{3.10}$$

is a solution of the Hamilton equations, since the function $\mathcal{H}^{(\infty)}$, contains terms of type $\mathcal{O}(P^2)$, $\mathcal{O}(|P| \|(X,Y)\|)$ and $\mathcal{O}(\|(X,Y)\|^3)$ only, except for its main part (that is made by a constant, a linear term in P and another quadratic in both X and Y). Because of this remark, it is evident that the 1D manifold $\big\{(P,Q,X,Y) : P=0,\ Q \in \mathbb{T},\ X = Y = 0\big\}$ is invariant. The energy level of such a solution is equal to $\mathcal{E}^{(\infty)}$. The elliptical character is given by the fact that, in the remaining degree of freedom, the transverse dynamics is given by an oscillatory motion whose period tend to the value $2\pi/\Omega^{(\infty)}$, in the limit of $(P, X, Y) \to (0,0,0)$. Of course, this is due to the occurrence of the term $\Omega^{(\infty)}(X^2+Y^2)/2$ which overwhelms the effect of the remainder $\mathcal{R}$ in the so-called limit of small oscillations.

In the case under study, dealing with the exoplanetary system υ Andromedæ, the normalization algorithm can be adapted so as to construct the 1D elliptic torus with a value of the parameter $\mathcal{E}^{(\infty)}$ equal to the energy level of the Poincaré sections (Caracciolo et al. 2022). In the right panel of Fig. 3 all the intersections of the corresponding orbit with the Poincaré surface $\eta_2 = 0$ are marked with a black cross. Of course, they perfectly superpose each other in a single fixed point corresponding to the wanted periodic orbit. In Caracciolo (2021), the normalization algorithm we have adopted to construct elliptic tori is fully described and its convergence is thoroughly analysed from a theoretical point of view. In short, such a procedure can be made in strict analogy with the construction of the Kolmogorov normal form. In fact, it can be formulated in such a way to introduce a sequence of Hamiltonians $H^{(r)}$ that are iteratively defined by a normalization step that is mainly composed by three Lie series: the first aims to reduce the perturbation that is not depending on the actions (p, J); the second achieves the same with the terms proportional to $\sqrt{J}$; also the third has the same goal for what concerns the terms that are linear in p or in J. We remark that in the normal form Hamiltonian $\mathcal{H}^{(\infty)}$ written in (3.9) we have expressed the dynamics that is transverse to the 1D elliptic torus in terms of the normalised canonical coordinates (X, Y) of polynomial type (instead of using action–angle variables), in order to highlight the existence of the periodic solution (3.10).

3.3. *Final construction of the invariant KAM torus*

It is now convenient to express also the second pair of canonical coordinates appearing in the normalised Hamiltonian (3.9) in the form of action–angle variables, i.e., we introduce (I, Θ) so that $X = \sqrt{2I}\cos\Theta$ and $Y = \sqrt{2I}\sin\Theta$. A very simple canonical change of variables, i.e.,

$$p_1 = P\,, \quad q_1 = Q\,, \quad p_2 = I - I^\star\,, \quad q_2 = Q\,, \tag{3.11}$$

is now enough in order to transform the Hamiltonian $\mathcal{H}^{(\infty)}$ (introduced at the end of the previous subsection) to $H^{(0)}(\boldsymbol{p},\boldsymbol{q}) = \boldsymbol{\nu}\cdot\boldsymbol{p} + h^{(0)}(\boldsymbol{p},\boldsymbol{q}) + \varepsilon f^{(0)}(\boldsymbol{p},\boldsymbol{q})$, that is in a suitable form to start the classical normalization algorithm that is the base of the proof scheme of KAM theorem. In a first approximation, the translation constant can be determined as $I^\star = (X_0^2 + Y_0^2)/2$, where $(X_0\,, Y_0)$ are values of the canonical coordinates (X, Y) corresponding to the initial conditions. Moreover, as a preliminary step we determine the angular velocity vector $\boldsymbol{\nu}$ by using the frequency analysis method. The choice of $I^\star$ can be optimised by applying a Newton method; so as to approach as much as possible the vector $\boldsymbol{\nu}$ (Caracciolo et al. 2022). Fig. 4 highlights that the algorithm constructing the Kolmogorov normal form (3.2) is successful also for the initial conditions considered in the present section, i.e., with mean anomalies fixed so that $M_1(0) = M_2(0) = 0°$, while

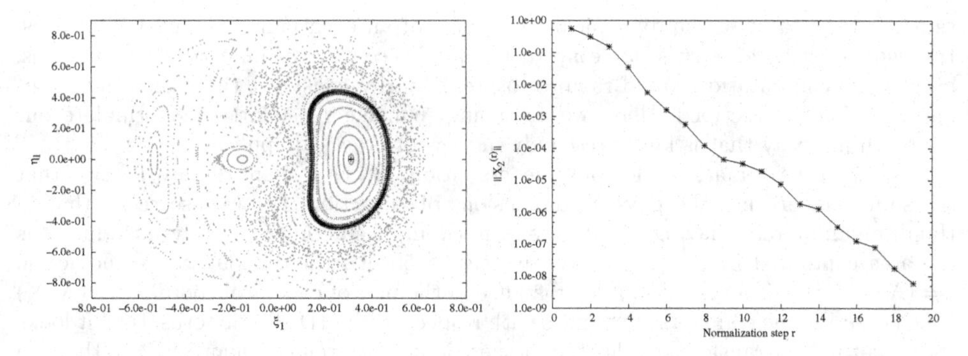

Figure 4. On the left, comparisons between the Poincaré sections generated by two different initial conditions. The first ones are marked in red and are exactly the same as those appearing in the left panel of Fig. 3 (where they are plotted in red as well). The second ones are marked in black and correspond to the orbit on the invariant KAM torus. The other "background" Poincaré sections are defined in the same way as those reported in Fig. 3; in particular, the dots plotted in blue there are located exactly in the same positions as those marked in orange here. The black symbol + refers to the orbit of the 1D elliptic torus also here. On the right, the behaviour of $\|\chi_2^{(r)}\|$ is plotted as a function of the normalization step r.

the other initial conditions are taken from the mid values of Table 1 (for what concerns $\iota(0)$, $\omega(0)$ and $\Omega(0)$ only) and from Table 2 (for the remaining data). In the left panel the Poincaré sections that are plotted (in red) during the numerical integration of the equations of motion related to the Hamiltonian (3.4) perfectly superpose to the orbit produced by composing all the canonical transformations briefly described in the present section, which is marked in black. The right panel of Fig. 4 clearly shows the regularity of the decrease of the norms of the generating functions (which are computed by simply adding up the absolute values of all the Taylor–Fourier coefficients). This gives a clear numerical indication of the convergence of the computational procedure in the case under study dealing with the exoplanetary system υ Andromedæ.

We stress that the importance of the translation constant $I^\star$ is crucial. Indeed, the abundance of the KAM manifolds surrounding an invariant torus is an increasing function of the inverse of the distance from said torus (as it has been shown, e.g., in Morbidelli & Giorgilli 1995). This is in agreement with the rather well known fact that the small parameter ε, which enters in the definition of the Hamiltonian $H^{(0)}$, is proportional to the shift value $I^\star$ (see, e.g., Giorgilli et al. 2017). Therefore, also the rate of the exponential decrease of the generating functions depends on $I^\star$: the smaller the latter the faster the former. In other words, we can also say that the invariant tori surrounding a reference one are more and more *robust* when the shift value $I^\star$ tends to zero. This means that larger and larger additional perturbing terms are needed in order to destroy this invariant structure for $I^\star \to 0$. Of course, all these remarks still hold true also when the reference torus (corresponding to $I^\star = 0$) is of elliptic type, as it is in the case of the periodic orbit $(P(t), Q(t), X(t), Y(t)) = \big(0, Q(0) + \nu^{(\infty)}t, 0, 0\big)$ that is obviously invariant with respect to the Hamiltonian flow of the normal form $\mathcal{H}^{(\infty)}$ written in (3.9).

4. The criterion of the minimal area as a robustness indicator

4.1. *Motivation and definition*

In the final discussion at the end of the previous section, we have explained why the shift value $I^\star$ appearing in the canonical transformation (3.11) can be considered as a good indicator of the dynamical robustness of an eventually existing KAM torus. However, such a concept is not easy to use in the context of numerical explorations, because its computation would require to preliminarily construct the normal forms we have previously described. Here, we are going to make the effort to reformulate our approach in a way that is far more handy in view of practical applications.

Firstly, let us remark that from the definition (3.11) it immediately follows that the shift value $I^\star$ has the physical dimensions of an action. Let us also recall that in Hamiltonian systems having one degree of freedom, the action is usually introduced as the area contoured by a closed orbit (see, e.g., § 50 of Arnold 1989). Since the action $I=(X^2+Y^2)/2$ is a sort of squared distance in the pair of canonical coordinates (X,Y) which describe the transverse dynamics with respect to the 1D elliptic torus, then it looks rather natural to transfer the role of robustness indicator from the quantity $I^\star$ to the area enclosed by an orbit in the Poincaré sections. Let us directly refer to Fig. 3 in order to fix the ideas. We recall that we have adopted the non-normalised canonical coordinates $(\boldsymbol{\xi},\boldsymbol{\eta})$ to plot those Poincaré sections, instead of (X,Y) that are much more expensive to compute. Nevertheless, in the hyperplane $\eta_2=0$ (after having fixed the energy level) the pair $(\xi_1\,,\,\eta_1)$ evidently describes a manifold that is transverse to the 1D elliptic torus, which is located by a fixed point marked with a black cross in the right panel. Since all the invariant tori winding around that periodic orbit describe Poincaré sections which are enclosing each other, then we can assume that the area embraced by the Poincaré sections is proportional to the distance (in action) from the elliptic torus. Therefore, by combining all the arguments explained at the end of the previous section with those discussed at the beginning of the present one, it is natural to assume that *an invariant torus is as more robust as smaller is the area contoured by the corresponding Poincaré sections.*

We now come to the approximated evaluation of such an area. By focusing our attention on the Poincaré sections marked in red in both panels of Fig. 3, we can say that the corresponding area is nearly equal to

$$\left(\max_t\left\{\xi_1(t)\right\}-\min_t\left\{\xi_1(t)\right\}\right)\left(\max_t\left\{\eta_1(t)\right\}-\min_t\left\{\eta_1(t)\right\}\right)\,. \tag{4.1}$$

Let us recall that ξ_1 and η_1 are proportional to $e_1\cos\omega_1$ and $e_1\sin\omega_1$, respectively, as determined by the definitions (3.3). Therefore, we can assume that also the area written in the formula above is proportional to

$$\mathcal{A}=\left[\left(e_{1;\max}\right)^2-\left(e_{1;\min}\right)^2\right]\max_t\left|\omega_1(t)-\omega_2(t)\right|\,, \tag{4.2}$$

where the meaning of the new symbols we have just introduced is $e_{1;\max}=\max_t\left\{e_1(t)\right\}$ and $e_{1;\min}=\min_t\left\{e_1(t)\right\}$. Moreover, in order to write the definition of the quantity $\mathcal{A}$ above as an approximation of the action surface written in formula (4.1), we have also assumed that (by symmetry reasons) both the extremals $\max_t\{\xi_1(t)\}$ and $\min_t\{\xi_1(t)\}$ are in correspondence with $\omega_1=0$, while we have evaluated the width $\max_t\{\eta_1(t)\}-\min_t\{\eta_1(t)\}$ with a circular arc centred in the origin $(\xi_1\,,\,\eta_1)=\mathbf{0}$ of the frame of the Poincaré surface. We remark that the half-width of that arc is evaluated by referring to $\left|\omega_1(t)-\omega_2(t)\right|$, because ω_2 is equal to zero in the region of the Poincaré surface with $\xi_1>0$ and we want to evaluate the quantity $\mathcal{A}$ for any motion in librational regime with respect to the difference of the pericentre arguments.

We can summarise all the discussion above by formulating the following *robustness criterion: we assume that a* **quasi-periodic Hamiltonian motion describing an invariant torus** *is as more robust as smaller is the corresponding quantity $\mathcal{A}$ defined in* (4.2).

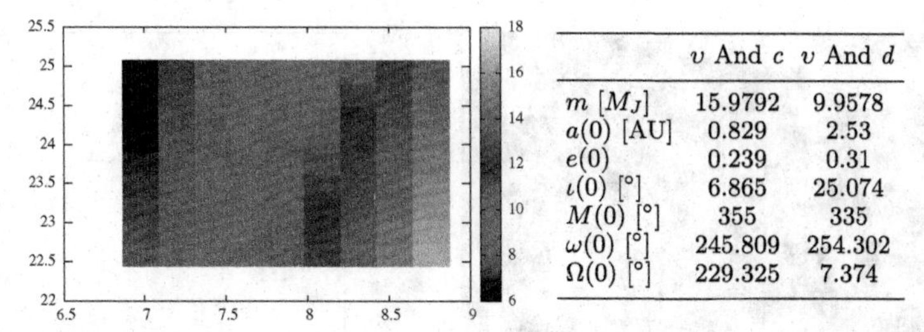

	υ And c	υ And d
m $[M_J]$	15.9792	9.9578
$a(0)$ [AU]	0.829	2.53
$e(0)$	0.239	0.31
$\iota(0)$ [°]	6.865	25.074
$M(0)$ [°]	355	335
$\omega(0)$ [°]	245.809	254.302
$\Omega(0)$ [°]	229.325	7.374

Figure 5. On the left, colour-code plot of the area $\mathcal{A}$ for different initial values of the inclinations $\iota_1(0)$ and $\iota_2(0)$. See formula (4.2) for the definition of the quantity $\mathcal{A}$. On the right, Table including the values of the masses and the initial conditions expected to correspond to the most robust planetary orbit compatible with the observed data available for υ And c and υ And d, according to the criterion of the minimal area.

It is quite evident that the statement above requires to minimise the area enclosed by the Poincaré sections, when we look for the most robust orbit originating from a set of possible initial conditions. For short, hereafter, we will refer to that as the criterion of the "minimal area".

4.2. *An application to the υ Andromedæ extrasolar planetary system*

In spite of the fact that we have constantly referred to a *secular* model in order to introduce and motivate our robustness criterion, we emphasise that its formulation is so flexible that it can be applied also to the study of the *complete* planetary dynamics of extrasolar systems. As we have claimed since the Introduction of the present work, we are going to select a set of initial conditions that is corresponding to an orbital configuration of the υ Andromedæ three-body model which is extremely stable. In our opinion, looking for *robust* invariant tori with a numerical criterion inspired by the secular dynamics has a twofold meaning. Firstly, they have more chances to persist when the perturbing effects due to the fast dynamics are taken into account; as we have shown in Section 2, chaotic motions compatible with the initial conditions are not rare in a probabilistic sense. Moreover, *robust* invariant tori describing the orbits υ And c and υ And d are expected to stay within a dynamically stable region of the phase space also when the effects due to υ And b and/or υ And e are included in the model.

In order to avoid the extensive study of a grid of initial conditions having a too high dimensionality, we will split our analysis in three different layers. As a first step, we consider initial conditions such that the mean anomalies are fixed so that $M_1(0) = M_2(0) = 0°$, while $\boldsymbol{\omega}(0)$ and $\boldsymbol{\Omega}(0)$ are taken from the corresponding mid values of Table 1; moreover, the assumed values of $\boldsymbol{a}(0)$, $\boldsymbol{e}(0)$ and minimal masses come from Table 2; the initial data are completed by covering the range of values of $\iota_1(0)$ and $\iota_2(0)$ which is reported in Table 1 with a regular grid of 10x10 points. For each of these 100 initial conditions, we numerically integrate the equations of motion, by using the symplectic method $\mathcal{SBAB}_{C3}$ (also here we adopt the same integrator as in Section 2, which is described in Laskar & Robutel 2001, with the same total timespan and integration step, that are 10^5 yr and 0.02 yr, respectively), and we compute the corresponding value of the numerical indicator $\mathcal{A}$. The results are reported in the left panel of Fig. 5. A straightforward application of the minimal area criterion allows us to conclude that the initial conditions that are expected to correspond to the most robust planetary orbit are such that

$$\iota_1(0) = 6.865° \ , \quad \iota_2(0) = 25.074° \ , \quad \Longrightarrow \quad m_1 = 15.9792\, M_J \ , \quad m_2 = 9.9578\, M_J \ . \quad (4.3)$$

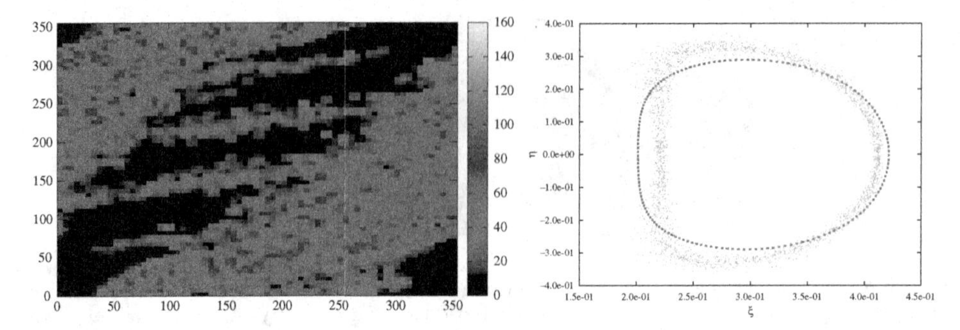

Figure 6. On the left, colour-code plot of the area $\mathcal{A}$ for different initial values of the mean anomalies $M_1(0)$ and $M_2(0)$. See formula (4.2) for the definition of the quantity $\mathcal{A}$. On the right, in orange, plot of the Poincaré sections that are corresponding to the hyperplane $\eta_2 = 0$ (with the additional condition $\xi_2 > 0$) and are generated by the flow of the complete Hamiltonian model of the exoplanetary system υ Andromedæ; such a motion is started from the initial conditions listed in the Table included on the right of Fig. 5. In red, plot of the Poincaré sections generated by the flow of the secular model $H^{(\mathrm{sec})}$, given in (3.4) and starting from initial conditions generated by those same values of the orbital elements.

The left panel of Fig. 5 clearly shows a rather surprising result: the most robust configurations correspond to the minimal value of the initial inclination $\iota_1(0)$ and, thus, to the maximal value of the υ And c mass (i.e., $\simeq 16\,M_J$). This conclusion is in agreement with a similar analysis that has been performed in Caracciolo et al. (2022), by studying the ratio between the norm of the last- and first-computed generating function, among those reported in a graph analogous to that appearing in the right panel of Fig. 4. The decrease rate of the sequence of the generating functions $\{\chi_2^{(r)}\}_{r\geq 1}$ (which are defined by the normalization algorithm eventually leading to the final Kolmogorov normal form) has been been firstly adopted as a robustness indicator starting from Volpi et al. (2018). We stress that this our new result looks to be rather unexpected when compared with the existing ones in the scientific literature: none of the four stable (and prograde) orbital configurations reported in Table 3 of Deitrick et al. (2015) is such that the υ And c mass is greater than $11\,M_J$, that is below the lowest possible value of $m_1 = 1.91/\sin(\iota_1(0))\,M_J$, where $1.91\,M_J$ is the minimal mass of υ And c taken from Table 2 and its initial inclination $\iota_1(0)$ is ranging in the corresponding interval reported in Table 1.

We continue our analysis by studying a second layer. We now consider initial conditions such that the mean anomalies are still fixed so that $M_1(0) = M_2(0) = 0°$, while $\boldsymbol{a}(0)$, $\boldsymbol{e}(0)$ are taken from Table 2 and the values of the initial inclinations and masses are as written in formula (4.3). In this second layer of analysis, the initial data are completed by covering the range of values of the angles $\boldsymbol{\omega}(0)$ and $\boldsymbol{\Omega}(0)$ with a regular 4D grid. Since the uncertainties on the knowledge of both the pericentre arguments and the longitudes of the node are not so large, we limit ourselves to define a grid which considers for each of the angles $\omega_1(0)$, $\omega_2(0)$, $\Omega_1(0)$, and $\Omega_2(0)$ just three possible values that are the minimum, the mid-point and the maximum of the corresponding values range reported in Table 1, respectively. For each of the so defined 81 initial conditions, we perform the same type of numerical integration we have described above. In this case, the application of the minimal area criterion leads to the conclusion that the initial values of $\boldsymbol{\omega}$ and $\boldsymbol{\Omega}$ that are expected to correspond to the most robust orbit are those reported in the Table included on the right of Fig. 5.

We come now to the description of the third layer of our analysis. In this last case, we consider the values of the planetary masses m_1 , m_2 and the initial conditions for the orbital elements $\boldsymbol{a}$, $\boldsymbol{e}$, $\boldsymbol{\iota}$, $\boldsymbol{\omega}$ and $\boldsymbol{\Omega}$ as they are given in the Table included on the right of

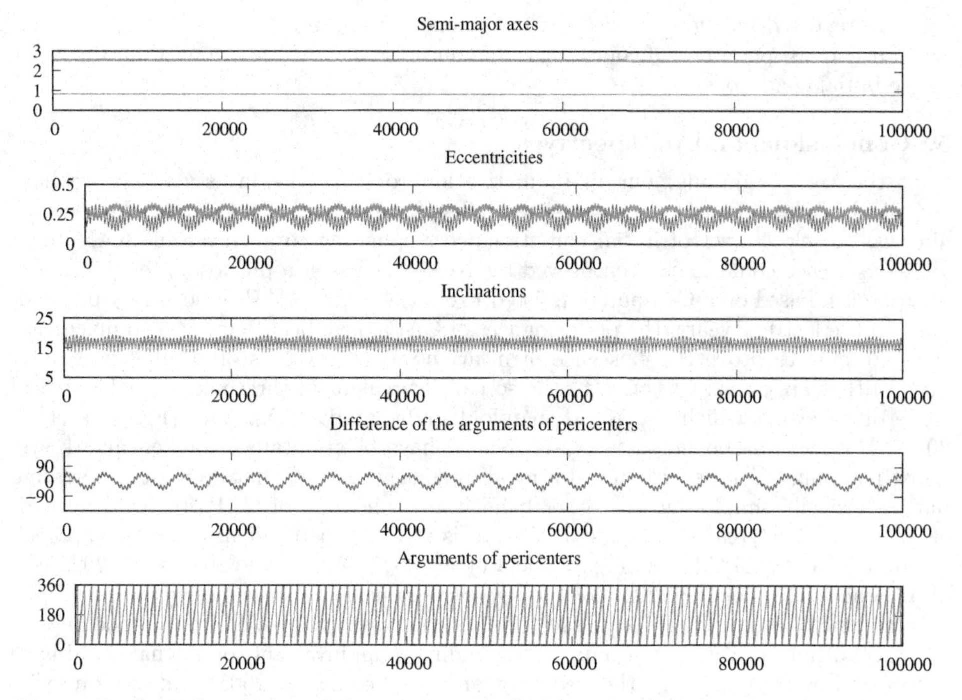

Figure 7. Orbital evolution of the exoplanets υ And c and υ And d in the case of the single set of initial conditions which is described in the Table included on the right of Fig. 5.

Fig. 5, while we make the coverage of all the possible initial values of the mean anomalies $\{(M_1(0), M_2(0))\} \in [0°, 360°] \times [0°, 360°]$ by means of a regular 2D grid with a grid-step of 5°. Once again, for each of these $72^2 = 5184$ different initial conditions, we perform the same type of numerical integration we have described above. For all of them, we compute the quantity $\mathcal{A}$, that is defined in (4.2). The results are reported in the left panel of Fig. 6. By comparing that colour-code plot with those included in Fig. 1, we can appreciate that there is good agreement between them: the most robust regions (according to the criterion of the minimal area) look also well apart from possible collisions (because the eccentricity of the outer planet does not reach large values) and fairly inside the librational regime with respect to the difference of the pericentre arguments. The initial values of the mean anomalies that are expected to correspond to the most robust orbit are the following ones:

$$M_1(0) = 355° \ , \qquad M_2(0) = 335° \ . \tag{4.4}$$

In the right panel of Fig. 6, we have plotted the intersections of the corresponding "most robust" orbit with respect to the Poincaré hypersurface $\eta_2 = 0$. Moreover, we have done the same also for the flow of the Hamiltonian $H^{(\mathrm{sec})}$, which is defined in (3.4), starting from the corresponding initial conditions $\big(\boldsymbol{\xi}(0), \boldsymbol{\eta}(0)\big)$ that are computed in terms of the secular canonical coordinates. The comparison of these two different kinds of Poincaré sections allows us to conclude that, for what concerns the most robust orbit, the behaviour of the eccentricities in the case of the complete planetary Hamiltonian should be rather close to that we can observe in the secular model at order two in the masses.

Fig. 7 describes the dynamical evolution of the exoplanets υ And c and υ And d in the case of the orbit that we consider as the most robust, according to the analysis we have widely discussed in the present section. The comparison with the corresponding graphs that are reported in Fig. 2 allows us to appreciate that the behaviour has now become pleasantly quasi-periodic. Of course, this difference is entirely due to our accurate choice of the initial conditions.

5. Conclusions and perspectives

At the very beginning, our main motivation to start the investigations we have described in the present paper was essentially of mathematical character. Indeed, our aim was to select a set of initial conditions corresponding to an invariant KAM torus whose existence could have been proved rigorously. For such a purpose, the adoption of an approach based on a Computer-Assisted Proof (hereafter, CAP) is somehow unavoidable. In the last few years, the performances of CAPs have been improved so much that they are able to prove the existence of invariant tori for values of a small parameter (say, ε) that are amazingly close to the so called breakdown threshold, i.e., the critical value of ε beyond which the KAM manifold under study disappears (Figueras et al. 2017). For the time being, so-successful results have been obtained for benchmark systems (mappings with or without additional dissipative terms) that are quite interesting but intrinsically simple. On the other hand, the application of CAPs to realistic models of physical interest highlights that there is still a gap to fill in order to approach the numerical threshold (see, e.g., Calleja et al. 2022 and Valvo & Locatelli 2022; see also Caracciolo & Locatelli 2020 for the rigorous evaluation of an effective stability time, with a similar kind of CAP technique). This is the reason for which we were looking for initial conditions that were not only corresponding to an invariant torus (that could have been found by applying, e.g., the frequency analysis; see Laskar 2003), but also quite far from its breakdown threshold (which is somehow depending on the physical parameters characterising a planetary systems). This has been made with the hope that a rigorous proof of the existence of such a KAM manifold would have been so relatively easy to be completed even if the CAP technique we adopted needs further improvements, to be extensively applied to Hamiltonian models of physical interest. This strategy of ours has been successful: as it is discussed in Caracciolo et al. (2022), in the case of the secular dynamics of the υ Andromedæ planetary system we have been able to rigorously prove the existence of a KAM torus that is travelled by the motion law starting from the initial conditions we have selected and reported in the Table included on the right of Fig. 5.

In order to solve such a challenging problem, we have introduced a robustness criterion that we have named "of the minimal area". The practical implementation of this method of investigation is computationally inexpensive, making it suitable for extensive studies of extrasolar systems. Indeed, it just requires a few additional computations during the numerical integrations of the Hamilton equations, each of them starting from different initial conditions, that all together should give a reasonable coverage of a data range which is compatible with the observations. Our robustness criterion is also flexible enough to be applied jointly with numerical integrations of a complete planetary model or a secular one without any need of additional efforts for the adaptation. Moreover, the comparisons reported in the right panel of Fig. 6 shows that in the case of the selected initial conditions there is a good agreement between the Poincaré sections for the secular model at order two in the masses and those related to the complete planetary system. Since the fraction of the chaotic motions is expected to be much more relevant in the latter case than in the former one (according to the discussions and figures widely commented in Sections 2–3), this result is not a priori obvious and enforces our confidence in the accuracy of the secular model, at least in the region where the invariant tori are more robust.

In our opinion, the possible applications of our approach are not limited to problems which are interesting for reasons that are mainly mathematical. From an astronomical point of view, we think that the most interesting result described in this work of ours concerns with the masses of the planets in the υ Andromedæ system. Our analysis allow to conclude that configurations with a large mass of υ And c have to be considered as more probable, because they are more robust In other words, one can expect that configurations with larger values of the mass of υ And c are within a region that is extremely stable because it is filled by tori so robust that they can eventually persist also when other perturbing terms are considered. For instance, additional gravitational effects could be taken into account, because of the eventual reintroduction of υ And b and υ And e in the planetary model.

The conclusion we have commented just above could be thought as counter-intuitive, because one might expect that stability is always gained by decreasing the values of the planetary masses. On the other hand, the following easy remark could explain such a situation which appears in contradiction: for fixed values of the semi-major axes and the eccentricities, in the case of υ Andromedæ system, the configuration that we identify as the most robust among the possible ones is that reducing as much as possible the imbalance between the angular momenta† of υ And c and υ And d. It is natural to argue about the real meaning of such a possible explanation: is this just by chance or is it quite general that planetary stability is gained by a better balance of the angular momenta? If the latter statement holds true, under which conditions? We think that there are also other natural questions about the generality of our approach, which are mainly due to the fact that our robustness criterion has been devised by studying the secular dynamics of a planetary three-body problem in an apsidal locking regime. Could it be extended to systems where the difference of the arguments of the pericentres is in rotation? Could our approach be significantly adapted to systems hosting more than two exoplanets? In our opinion, all these questions deserve to be further investigated.

Acknowledgements

This work was partially supported by the MIUR-PRIN project 20178CJA2B – 'New Frontiers of Celestial Mechanics: theory and Applications'. The authors acknowledge also INdAM-GNFM and the MIUR Excellence Department Project awarded to the Department of Mathematics of the University of Rome 'Tor Vergata' (CUP E83C18000100006).

References

Arnold, V.I., 1963, *Russ. Math. Surv.*, 18, 9

Arnold, V.I., 1989, *Mathematical methods of classical mechanics*, 2nd edition, Springer-Verlag

Butler, R.P., et al., 1999, *Astroph. Jour.*, 526, 916

Calleja, R.C., Celletti, A., Gimeno, J., & de la Llave, R., 2022, *Commun. Nonlinear Sc. Numer. Simulat.*, 106, 106099

Caracciolo, C., 2021, *Math. in Engineering*, 4, 1

Caracciolo, C., & Locatelli, U., 2020, *Jour. of Comput. Dynamics*, 7, 425

Caracciolo, C., & Locatelli, U., 2021, *Commun. Nonlinear Sc. Numer. Simulat.*, 97, 105759

Caracciolo, C., Locatelli, U., Sansottera, M., & Volpi, M., 2022, *Mon. Not. Royal Astron. Soc.*, 510, 2147

Chiang, E.I., Tabachnik, S., & Tremaine, S., 2001, *Astron. Jour.*, 122, 1607

† The angular momenta of υ And c and υ And d are such that $\mathcal{G}_j = \Lambda_j(1 - \sqrt{1-e_j^2})$, with $\Lambda_j = m_0 m_j \sqrt{G(m_0+m_j)a_j}/(m_0+m_j)\ \forall\, j=1,2$. Looking at the data about semi-major axes, eccentricities and minimal masses that are reported in Table 2, one can easily check that the minimum difference between the angular momenta (i.e., $\mathcal{G}_2 - \mathcal{G}_1$) corresponds to the maximum possible value of the mass of the inner planet and the minimum of that of the outer one, which are $m_1 = 1.91/\sin(\iota_1(0))$ and $m_2 = 4.22/\sin(\iota_2(0))$, respectively, where the ranges of values of the initial inclinations $\iota_1(0)$ and $\iota_2(0)$ are reported in Table 1.

Cloutier, R., et al., 2019, *Astron. & Astroph.*, 629, A111
Deitrick, R., et al., 2015, *Astroph. Jour.*, 798, 46
Figueras, J.-Ll., Haro, A., & Luque, A., 2017, *Found. Comput. Math.*, 17, 1123
Giorgilli, A., Locatelli, U., & Sansottera, M., 2009, *Cel. Mech. & Dyn. Astr.*, 104, 159
Giorgilli, A., Locatelli, U., & Sansottera, M., 2017, *Reg. & Chaot. Dyn.*, 22, 54
Giorgilli, A., & Sansottera, M., 2012, in P.M. Cincotta, C.M. Giordano & C. Efthymiopoulos (eds.), *Chaos, Diffusion and Non-integrability in Hamiltonian Systems*, Universidad Nacional de La Plata and Asociación Argentina de Astronomía Publishers
Kolmogorov, A.N., 1954, Engl. transl. in *Lecture Notes in Physics*, 1979, 93, 51
Laskar, J., *Astron. & Astroph.*, 1988, 198, 341
Laskar, J., 1989, *Notes scientifiques et techniques du Bureau des Longitudes* S026, available at `https://www.imcce.fr/content/medias/publications/publications-recherche/nst/docs/S026.pdf`
Laskar, J., 2003, in D. Benest, C. Froeschlé, & E. Lega E. (eds.), *Hamiltonian systems and Fourier analysis*, Taylor and Francis
Laskar, J., & Petit, A.C., 2017, *Astron. & Astroph.*, 605, A72
Laskar, J., & Robutel, P., 2001, *Cel. Mech. & Dyn. Astr.*, 80, 39
Libert, A.-S., Sansottera, M. 2013, *Cel. Mech. & Dyn. Astr.*, 117, 149
Locatelli, U., Caracciolo, C., Sansottera, M., & Volpi, M., 2022, in: G. Baù, S. Di Ruzza, R.I. Páez, T. Penati & M. Sansottera (eds.), *I-CELMECH Training School – New frontiers of Celestial Mechanics: theory and applications*, Springer PROMS (in press)
Locatelli, U., & Giorgilli, A., 2000, *Cel. Mech. & Dyn. Astr.*, 78, 47
Mayor, M., & Queloz, D., 1995, *Nature*, 378, 355
McArthur, B.E., et al., 2010, *Astroph. J.*, 715, 1203
McArthur, B.E., et al., 2014, *Astroph. J.*, 795, 41
Michtchenko, T.A., & Malhotra, R., 2004, *Icarus*, 168, 237
Morbidelli, A., & Giorgilli, A., 1995, *J. Stat. Phys.*, 78, 1607
Moser, J., 1962, *Nachr. Akad. Wiss. Gött., Math. Phys.*, 1, 1
Perryman, M., 2018, *The Exoplanet Handbook*, Cambridge Univ. Press, ISBN 9781108419772
Petit, A.C., Laskar, J., & Boué, G., 2017, *Astron. & Astroph.*, 607, A35
Sansottera, M., & Libert, A.-S., 2019, *Cel. Mech. & Dyn. Astr.*, 131, 38
Valvo, L., & Locatelli, U., 2022, *Jour. of Comput. Dynamics*, in press
Volpi, M., Locatelli, U., & Sansottera, M., 2018, *Cel. Mech. & Dyn. Astr.*, 130, 36

Multi-scale (time and mass) dynamics of space objects
Proceedings IAU Symposium No. 364, 2022
A. Celletti, C. Galeş, C. Beaugé, A. Lemaitre, eds.
doi:10.1017/S1743921321001411

New results on orbital resonances

Renu Malhotra

Lunar and Planetary Laboratory, The University of Arizona
Tucson, Arizona 85721, USA
email: malhotra@arizona.edu

Abstract. Perturbative analyses of planetary resonances commonly predict singularities and/or divergences of resonance widths at very low and very high eccentricities. We have recently re-examined the nature of these divergences using non-perturbative numerical analyses, making use of Poincaré sections but from a different perspective relative to previous implementations of this method. This perspective reveals fine structure of resonances which otherwise remains hidden in conventional approaches, including analytical, semi-analytical and numerical-averaging approaches based on the critical resonant angle. At low eccentricity, first order resonances do not have diverging widths but have two asymmetric branches leading away from the nominal resonance location. A sequence of structures called "low-eccentricity resonant bridges" connecting neighboring resonances is revealed. At planet-grazing eccentricity, the true resonance width is non-divergent. At higher eccentricities, the new results reveal hitherto unknown resonant structures and show that these parameter regions have a loss of some – though not necessarily entire – resonance libration zones to chaos. The chaos at high eccentricities was previously attributed to the overlap of neighboring resonances. The new results reveal the additional role of bifurcations and co-existence of phase-shifted resonance zones at higher eccentricities. By employing a geometric point of view, we relate the high eccentricity phase space structures and their transitions to the shapes of resonant orbits in the rotating frame. We outline some directions for future research to advance understanding of the dynamics of mean motion resonances.

Keywords. celestial mechanics, (stars:) planetary systems, planets and satellites: general, solar system: general, minor planets, asteroids, Kuiper Belt, methods: numerical.

---

1. Introduction

Orbital resonances have influenced the properties and distribution of planets and minor planets in the solar system as well as in exo-planetary systems. Their study has a long history in celestial mechanics, physics and mathematics. There are many examples of stable mean motion resonances (MMRs) in our solar system, such as the Hilda group and the Trojan group of asteroids in 3/2 and 1/1 MMRs with Jupiter, the 2/3 MMR of Pluto with Neptune, the 5/2 near-resonance between Jupiter and Saturn, and the chain of 2/1 MMRs amongst the Galilean moons Io, Europa and Ganymede. Unstable mean motion resonances are also of great significance; for example, the Kirkwood Gaps in the asteroid belt are linked to the chaotic MMRs of Jupiter (e.g. Moons 1996), and the long term stability of the solar system may be linked to the role of MMRs (e.g. Murray & Holman 1999; Guzzo 2006). There is also a growing body of literature on the importance of MMRs in the dynamics of exoplanetary systems (recent review by Zhu & Dong 2021).

Previous analytical as well as numerical treatments of MMRs have been based on or informed by perturbation theory. Commonly, these treatments have invoked the averaging principle to discard the short timescale terms to reduce the problem to one degree of freedom by identifying a slow timescale and a corresponding "critical resonant angle", and developing a pendulum-like model for the dynamics in that single degree of freedom. The common pendulum is the first approximation for describing the dynamics of MMRs. In this model, the phase space is divided into a zone of oscillations of the critical resonant angle (at sufficiently low relative energies), and rotation zones (both clockwise and counter-clockwise rotations, at sufficiently high energies); the boundaries between the oscillations and the rotations is a separatrix of zero frequency. For elementary expositions, the common pendulum model suffices (Dermott & Murray 1983; Winter & Murray 1997). The "second fundamental model of resonance" (SFMR) introduced by Henrard & Lemaitre (1983) provides a more accurate description and it has been a mainstay of such analytical treatments of MMRs. Similar to the common pendulum, the SFMR is also of one degree of freedom, but, unlike the common pendulum, its potential function is not independent of the canonical momentum. Specifically in the case of planetary mean motion resonances, the perturbative potential function possesses the d'Alembert characteristic in $(\sqrt{J}, \phi)$, where the canonical coordinate, ϕ, is identified with the critical resonant angle and J is its canonically conjugate momentum (obtained by means of canonical transformations starting from the Delaunay elements for the unperturbed Kepler problem).

While the above-mentioned approaches mitigated the problem of "small divisors" in historical perturbation theory treatment of mean motion resonances, they still retained certain features and ambiguities that were puzzling to this author; in particular, the widths of mean motion resonances exhibited singularities and/or divergences at low eccentricities. For example, the textbook by Murray & Dermott (1999) gives an analytical estimate that the resonance zone widths of Jupiter's first order MMRs in the asteroid belt diverge to infinity on both sides of the nominal resonance location as an asteroid's eccentricity approaches zero (their Eq. 8.76 and Fig. 8.7). Wisdom (1980) used an analytical perturbative approach and Winter & Murray (1997) used a non-perturbative numerical approach with Poincaré sections to measure the widths of Jupiter's resonances in the asteroid belt, and they also reported a similar divergence, albeit on only one side of the nominal resonance location. On the same topic, the textbook by Morbidelli (2002) also describes that the stable resonance center of Jupiter's 2/1 MMR diverges only on one side of the nominal resonance location as e approaches zero, and that a resonance separatrix "vanishes" for $e \lesssim 0.2$ making the resonance width undefined for smaller eccentricities (his Fig. 9.11.)

At higher eccentricities, previous studies either extrapolated the low-eccentricity analytical models or used a numerical approach to average the perturbation potential over the fast degrees of freedom. Such studies reported the divergence of resonance widths near planet-grazing eccentricities and a gradual decrease of resonance widths at higher eccentricities (Moons & Morbidelli 1993; Morbidelli et al. 1995; Nesvorný & Ferraz-Mello 1997; Deck et al. 2013; Hadden & Lithwick 2018; Gallardo 2019).

The high eccentricity regime of planetary mean motion resonances has gained increasing attention as discoveries of minor planets and dwarf planets in the distant solar system show that this regime of phase space does have a significant influence on the dynamics of these objects. For example, the phenomenon of long term "resonance sticking" is commonly observed, typically at high eccentricities, in numerical studies of a class of Kuiper belt objects called the "scattered disk" or the "scattering disk" (Duncan & Levison 1997; Lykawka & Mukai 2007; Gladman et al. 2008, 2012; Yu et al. 2018). Shorter term resonance sticking is reported amongst the Centaurs (Belbruno & Marsden 1997; Tiscareno & Malhotra 2003; Bailey & Malhotra 2009; Fernández et al. 2018; Roberts & Muñoz-Gutiérrez 2021). Recent interest in the possible observable effects of an unseen distant planet in the solar system have also stimulated interest in the dynamics of high eccentricity MMRs (e.g., Malhotra et al. 2016; Beust 2016; Hadden et al. 2018).

The desire to understand better the high eccentricity regime of MMRs as well as the puzzling features at low and moderate eccentricities was the motivation for a sequence of recent investigations in my research group (Wang & Malhotra 2017; Malhotra et al. 2018; Lan & Malhotra 2019; Malhotra & Zhang 2020). In these studies, we adopted a non-perturbative approach by computing Poincaré sections, but with some modifications relative to previous studies. The modifications have helped to significantly clarify some aspects of the dynamics of MMRs. Two additional recent publications, Lei & Li (2020) and Antoniadou & Libert (2021), have elaborated further on some of these results. The present article summarizes a part of these advances, specifically those related to interior mean motion resonances of the first order, that is, when the orbital periods of two planets orbiting a central host star are close to the ratio of two small integers differing only by unity, e.g., 2/1, 3/2, etc., and the interior planet is of negligible mass.

Results of our investigations of exterior MMRs can be found in Malhotra et al. (2018), Lan & Malhotra (2019), and are summarized in Malhotra (2019). We also refer the reader to Deck et al. (2013), Ramos et al. (2015), Hadden & Lithwick (2018), Hadden (2019), and Petit (2021) for additional new results on mean motion resonances, including in the non-restricted, non-circular co-planar three body model and on three-planet mean motion resonances; these are beyond the scope of the present review.

2. A new approach to Poincaré sections

We begin with the physical model known as the restricted three body problem, and its mathematically simplest version which has the massive planet ("the perturber"), of mass $m_2 = \mu$, in a circular orbit of unit radius about the host star whose mass is $m_1 = 1 - \mu$. This model admits an integral of the particle's motion, the Jacobi integral $C_J = 2(\Omega L - E)$, where Ω is the angular velocity of the massive bodies in their circular orbits and E and L are the specific energy and specific angular momentum of the particle (in the barycentric frame, with z-axis oriented along the total angular momentum of the massive bodies). A further simplification is that the zero-mass particle is restricted to move in the planet's orbital plane; this is the planar circular restricted three body problem (PCRTBP). Although it does not describe the full complexities of real planetary systems, this model remains a very useful approximation and provides insights into the phase space structure of MMRs. The model has two degrees of freedom, hence a phase space of four dimensions. The existence of the Jacobi integral implies that the motion of the test particle takes place on a three dimensional surface in the four dimensional phase space. This permits the visualization of the phase space structure in two dimensional surfaces called "surfaces of section" or Poincaré sections, devised by the mathematician Henri Poincaré (1854-1912). These are akin to stroboscopic plots with which we can track the motion of the particle in two variables to visualize on a flat plane.

Previous studies commonly followed the lead of Hénon (1966) on the choice of Poincaré sections for the PCRTBP by recording the test particle's state vector at every successive conjunction (or opposition) with the perturber (e.g. Duncan et al. 1989; Winter & Murray 1997)). With this choice, the Poincaré sections were presented as plots of $\dot{x}$ vs. x, where (x, y) and $(\dot{x}, \dot{y})$ are the cartesian position and velocity components in the synodic frame, that is, the barycenteric frame co-rotating with m_1 and m_2; the primaries' locations in this frame are fixed on the x-axis at $-\mu$ and $1 - \mu$, respectively. However this is not a unique choice. In our implementation, we made two modifications relative to previous studies:

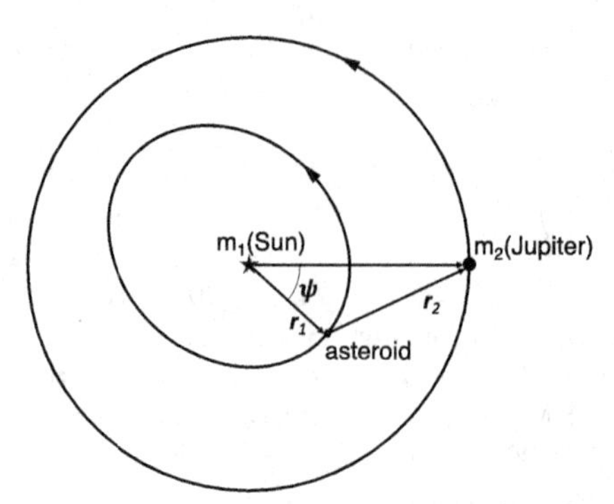

Figure 1. A schematic diagram to illustrate the definition of ψ, the angular separation of Jupiter from the test particle when the latter is at pericenter.

• First, we recorded the test particle's state vector at every successive perihelion passage. This choice is physically motivated so as to trace the behavior (libratory or not) of the test particle's perihelion longitude relative to the perturbing planet's position.

• Second, we recognized that there were other possible choices of dynamical variables for the Poincaré plots than the rotating frame configuration variables, $(x, \dot{x})$, most commonly adopted in previous studies. With the stroboscopic record of the state vector at successive perihelion passages, one could examine many different combinations of dynamical variables. Most useful are the two-dimensional plots in (x, y), $(e\cos\psi, e\sin\psi)$, and (ψ, a), where x, y are the cartesian position coordinates in the rotating frame, a and e are the osculating semimajor axis and eccentricity of the particle, and ψ measures the angular separation of the perturber from the test particle when the particle is at perihelion. The definition of ψ is illustrated in Figure 1 in which the perturber is Jupiter and the test particle is an asteroid.

These modifications enable a visualization of the phase space structure from a different point of view than in previous studies. In the previous implementations of Poincaré sections of the PCRTBP, only one point is recorded per synodic period, whereas in our implementation, typically there are multiple points recorded per synodic period. For example, in the neighborhood of an interior $(p+1)/p$ resonance, there are $(p+1)$ points recorded per synodic period. In this sense, our implementation gives a higher resolution view of the phase space structure than in the previous approach. The new perspective has helped to resolve the puzzles and ambiguities of resonance widths mentioned above (Section 1).

In addition, we examine the shape of the resonant orbit in the rotating frame and how it changes with increasing (or decreasing) value of the particle's eccentricity. And we correlate its shape with the evolving topology of the phase space as visible in the Poincaré sections (Section 4). The exercise of correlating the phase space structure with the geometry of the resonant orbit yields insights into the physical origins of the multiple resonance libration modes in the moderate-to-high eccentricity regime: it explains both their existence and their size as a function of the particle's eccentricity. It also reveals the physical connections between resonance zones and stable/unstable periodic orbits of the PCRTBP. Periodic orbits present as fixed points in the Poincaré sections and define the centers and boundaries of the resonance zones.

The relationship between the conventional approach taken in analytical and numerical-averaging approaches and this different perspective is actually rather simple. For a first order MMR, the usual "critical resonant angle" is defined as

$$\phi = (p+1)\lambda' - p\lambda - \varpi, \tag{2.1}$$

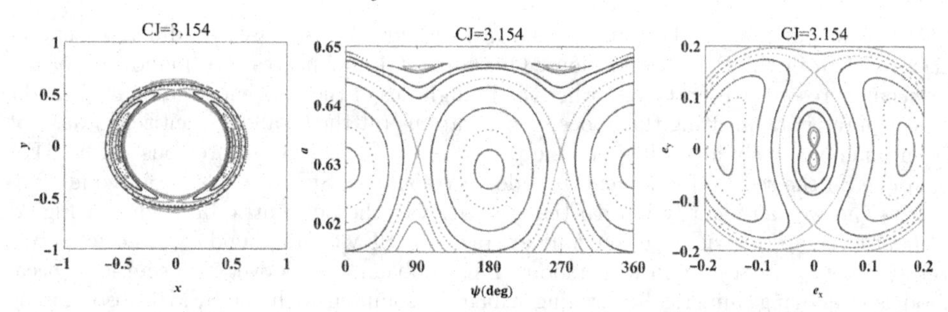

Figure 2. Poincaré sections near Jupiter's 2/1 interior resonance for one value of the Jacobi constant, as indicated by the legend at the top of each panel. The left panel displays the Poincaré section in the configuration space, (x, y), in the rotating frame; the locations of the Sun and Jupiter are fixed in this plane at $(x, y) = (-\mu, 0)$ and $(x, y) = (1 - \mu, 0)$, respectively. The middle panel plots the same Poincaré section in the (ψ, a) plane, and the right panel in the $(e_x, e_y) \equiv (e \cos \psi, e \sin \psi)$ plane.

where p is an integer, λ, λ' are the mean longitudes of the particle and of the perturbing planet, respectively, and ϖ is the particle's longitude of perihelion. In our definition of the Poincaré section, the particle is located at its perihelion, that is, when $\lambda = \varpi$. It then follows that ϕ and ψ are related:

$$\phi = (p + 1)\psi. \tag{2.2}$$

It must be understood that ϕ is a continuous function of time, whereas ψ is a stroboscopic coordinate, being defined at the point in time when the particle is at perihelion. The choice of ψ, together with our non-perturbative approach, reveals more details of the phase space structure than have been possible with the conventional use of ϕ in analytical, semi-analytical and numerical-averaging approaches in the previously existing literature. In particular, we will see below that the use of ψ proves to be the key to solving one of the puzzles in the previous literature, that of the apparent "vanishing of the separatrix" at low eccentricity.

Figure 2 shows an example set of Poincaré sections obtained with the above approach for one value of the Jacobi integral in the neighborhood of Jupiter's 2/1 interior MMR. In the (x, y) plane, we observe that there are two pairs of libration zones, one pair is centered on the x-axis and the other pair is centered (almost) on the y-axis. For the former, the test particle's alternate perihelion passages occur at conjunction and opposition with Jupiter; these can be called the pericentric resonant islands. For the latter, the alternate perihelion passages occur close to 90° leading and trailing Jupiter's longitudinal position; consequently, conjunctions and oppositions with Jupiter occur at the particle's aphelion, so these can be called the apocentric resonant islands. In the (ψ, a) plane as well as in the $(e \cos \psi, e \sin \psi)$ plane, the same two pairs of libration islands are visible, with the first pair (centered at $\psi = 0$ and $\psi = 180°$) being the larger size and at larger eccentricity, while the second pair (centered near $\psi = 90°$ and $\psi = 270°$) being the smaller size and at smaller eccentricity. It is striking that the larger pair of islands has a central value of a close to (slightly below) the nominal resonant value of $a = (1/2)^{\frac{2}{3}} = 0.630$, whereas the smaller pair is displaced to a higher value. (The keen reader will also notice that the centers of the smaller islands are also visibly displaced slightly from $\psi = 90°, 270°$; this is discussed further below.) The existence of two pairs of resonant islands illustrates the existence of two branches of the resonance: the pericentric branch in which the conjunctions of the particle and Jupiter librate about the particle's pericenter, and the apocentric branch in which the conjunctions librate about the particle's apocenter.

We point out that in the example shown in Figure 2, there are two different separatrices visible in the Poincaré sections. One separatrix delineates the boundaries of the pericentric resonant islands, the other the boundaries of the apocentric resonant islands. The separatrix delineating the apocentric librations (at the smaller eccentricity) has not been unambiguously identifiable in the perturbative treatments in previous studies. For example, in the second fundamental model of resonance, SFMR, there is only one separatrix, and we can identify it with the separatrix of the pericentric librations in Fig. 2; although the apocentric libration zone exists in the SFMR, its boundary is not clearly a zero-frequency separatrix in that model. This problem carries over into semi-analytical models employing numerical averaging which, in common with the SFMR, also employ the critical resonant angle, ϕ, as the single degree of freedom. This is the reason for the apparent "vanishing of the separatrix" at low eccentricity claimed in previous studies, such as in Morbidelli (2002) and in Ramos et al. (2015).

The exclusive use of the critical resonant angle in previous studies also suppresses the distinction amongst the different islands of the pericentric branches of the resonance. As illustration, we observe in Fig. 2 that the two pericentric islands in the Poincaré section for Jupiter's 2/1 interior MMR are not exactly symmetric: the value of a at the center of the island at $\psi = 0$ is slightly larger than that of the island at $\psi = 180°$. The asymmetry is also evident in the shapes of the boundaries of the two islands in the (ψ, a) plane. Consequently, when we measure the extent of the resonance zone in semi major axis (discussed in Section 3), we find that each of the two islands has slightly different values of $a_{\rm min}$ and $a_{\rm max}$, resulting in two different estimates of the resonance width.

In general, for interior first order $(p+1)/p$ MMRs, these Poincaré sections show a chain of $(p+1)$ islands for the pericentric branch and a chain of $(p+1)$ islands for the apocentric branch. Likewise, for exterior first order $p/(p+1)$ MMRs, these Poincaré sections will show a chain of p islands for the pericentric branch and a chain of p islands for the apocentric branch. For example, Jupiter's interior 3/2 MMR in the asteroid belt appears as a chain of three islands for the pericentric branch and another chain of three islands for the apocentric branch; Neptune's exterior 2/3 MMR in the Kuiper belt appears as a chain of two islands for the pericentric branch and another chain of two islands for the apocentric branch.

There are some ranges of the Jacobi integral for which one of the branches vanishes. For example, in the low eccentricity regime for interior MMRs, it is the apocentric branch that does the vanishing, and for exterior MMRs it is the pericentric branch that does the vanishing trick. These branches reappear at higher eccentricities, and, in some cases, vanish and reappear multiple times at high eccentricities. This is discussed further in the following sections.

3. Resonance Widths

3.1. *Measuring resonance width*

To measure the resonance width in semi major axis as a function of eccentricity, we first identify the center of the resonance by the values a^* and e^* at the center of each resonant island. For a^*, we read this central value in the (ψ, a) plane, and for e^*, we read it in the $(e\cos\psi, e\sin\psi)$ plane. Then we measure the minimum and maximum values of a in the libration zone of each resonant island in the (ψ, a) plane. For some ranges of the Jacobi integral, the resonant islands may not possess a regular separatrix but present a fuzzy boundary due to a chaotic zone. In these cases we estimate values of $a_{\rm min}$ and $a_{\rm max}$ for the largest visible regular librating trajectory in the (ψ, a) plane. This procedure is rather labor-intensive, and the resolution in a, e that has been achieved in the results reported thus far is certainly limited by the available time and capacity of the researchers.

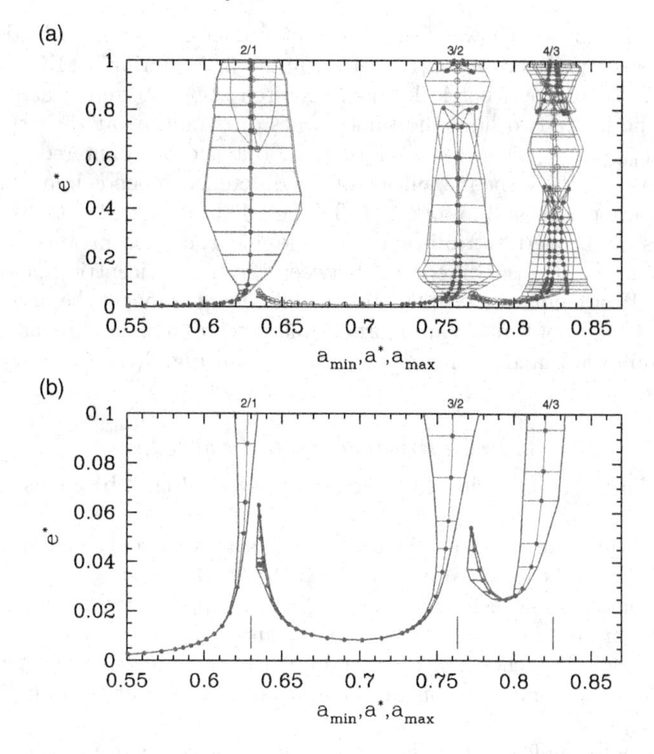

Figure 3. (a) The libration centers and widths of Jupiter's 2/1, 3/2 and 4/3 interior resonances for the full range of eccentricities, $0 < e < 1$, for the pericentric and apocentric libration islands, in blue and red, respectively. The center (a^*, e^*) of the pericentric libration islands is indicated by the filled circles, that of the apocentric libration islands is indicated by the open circles. The maximum libration range of a in each case is indicated by the horizontal bars. (b) Expanded view of one of each of the pericentric and apocentric branches at low eccentricity. The pericentric island shown is centered near $360°/(p+1)$ while the apocentric island is centered near $180°/(p+1)$ where $p = 1, 2, 3$ for the 2/1, 3/2, 4/3 MMRs, respectively. The apocentric branch of the 4/3 is not shown because it has dissolved into a chaotic zone. The short vertical lines indicate the nominal location, $a_{\rm res} = (p/(p+1))^{\frac{2}{3}}$.

Using this procedure, we recently investigated several first order interior MMRs of Jupiter (as well as lower-mass perturbers) and many exterior MMRs of Neptune (Wang & Malhotra 2017; Lan & Malhotra 2019; Malhotra & Zhang 2020). The correlated measure, Δe, of the range of eccentricity libration accompanying the maximal range, Δa, is additionally reported in Lei & Li (2020). Here we illustrate the main new results with a few examples. The numerically determined resonance centers and widths of the 2/1, 3/2 and 4/3 interior MMRs of Jupiter are shown in the (a, e) parameter space in Figure 3(a) for the entire eccentricity range, $0 < e < 1$; in Fig. 3(b), the low eccentricity regime, $0 < e < 0.1$, is shown in an expanded scale. There are several interesting features in this figure, as we discuss below.

As mentioned above, a $(p+1)/p$ interior MMR presents in the Poincaré sections as a chain of $(p+1)$ islands each for the apocentric and pericentric branches. (In the terminology of period orbit theory, the centers of a chain of p resonant islands represent a single stable periodic orbit of period p.) Only those islands located symmetrically relative to the x-axis (equivalently, $\psi = 0$) have identical central values, a^* and e^*, and of the resonance width (measured with $a_{\rm min}$ and $a_{\rm max}$); the other islands have slightly different resonance centers and widths. This means that in the (a, e) parameter space, the number of distinct resonance boundaries is not always equal to the number of resonant islands that appear in the Poincaré sections. For example, for Jupiter's 3/2 interior MMR, there are three resonant islands of the pericentric branch, centered at $\psi = 0$ and near $\psi = 120°, 240°$, respectively. The latter two have the same values of a^* and e^* at their centers and the same width ($a_{\rm min}$ to $a_{\rm max}$) in semi major axis, but the one centered at $\psi = 0$ differs slightly in both respects. (The magnitude of the difference depends upon the perturber's mass.) In Fig. 3, for each of Jupiter's 2/1, 3/2 and 4/3 interior MMRs, we plot the center and widths of the pericentric islands in blue and the apocentric islands in red. At the resolution of Fig. 3(a) the differences between the two pericentric islands of Jupiter's interior 2/1 MMR are not very visible, but the differences amongst the pericentric islands of the interior 3/2 and of the 4/3 are more visible. Such differences are also more evident amongst the multiple islands at higher eccentricities in Fig. 3(a).

3.2. *Fine structure at low eccentricities*

For small values of eccentricity, the features of note in Fig. 3(b) are as follows:

(i) the pericentric and apocentric branches both exist continuously at significant distances from the nominal location of each MMR;
(ii) the pericentric branches increase in width as a^* approaches the nominal resonance location from the left (from lower values); and
(iii) the apocentric branches first increase in width, then decrease and terminate as a^* approaches the nominal resonance location from the right (from higher values).

With these results we learn that the divergence of the resonance boundaries described in the previous literature (e.g. Wisdom 1980; Murray & Dermott 1999) is actually resolved into two branches of the resonance, each of which diverges away from the nominal resonance location, but the width of each branch does not diverge, rather it decreases as eccentricity decreases. Moreover, the Poincaré sections near very small eccentricity show that the apocentric branch is bounded by a bonafide separatrix which passes through $e = 0$. This separatrix is not visible in previous analytical treatments nor in the semi-analytical and numerical-averaging treatments because those approaches employed the critical resonant angle as the dynamical coordinate (Henrard 1983; Nesvorný & Ferraz-Mello 1997; Morbidelli 2002). The choice of this coordinate hides the existence of the separatrix passing through $e = 0$. This is the reason why the fine structure of first order resonances at low eccentricities has remained unresolved and unremarked in the previous literature on resonance widths.

Our change of coordinate, from the critical resonant angle, ϕ, to its sub-multiple, $\psi = \phi/(p+1)$ (cf. Eq. 2.2), is essential to reveal the fine structure at low eccentricity. This has also been demonstrated by Lei & Li (2020) who recover most of the low-eccentricity fine structure of a first order MMR described in Malhotra & Zhang (2020) within a semi-analytical approach parallel to the SFMR, but using $\sigma = \phi/(p+1)$ as the canonical coordinate. These authors further show that the SFMR, with its truncation of the resonant potential to only one leading-order term, is insufficient to describe accurately the dynamics near a first order MMR at low eccentricities (see also Beaugé (1994)); at least two additional terms of the resonant perturbation potential are necessary to recover the topology of the phase space approximately consistent with the non-perturbative results of Malhotra & Zhang (2020). Some features revealed with the non-perturbative approach nevertheless remain hidden in the semi-analytical results of Lei & Li (2020). This is due to the fundamental approximation in any analytical or semi-analytical approach that all but the specific resonant terms of the perturbation potential are discarded or numerically averaged out. One significant feature revealed uniquely in the non-perturbative approach is the existence of low eccentricity bridges between neighboring first order MMRs; this is described in the next section.

3.3. *Low eccentricity resonant bridges*

We point out one of the most interesting features of first order MMRs at low eccentricity: that the pericentric branch of the 4/3 MMR smoothly connects with the apocentric branch of the 3/2, and, similarly, the pericentric branch of the 3/2 smoothly connects with the apocentric branch of the 2/1. This can be seen in Fig. 3(a), and at higher resolution in Fig. 3(b). These connections between neighboring first order MMRs occur with a gradual transformation of the phase space structure as we tune the Jacobi integral across a range of values encompassing neighboring MMRs. For example, the transformation of the phase space structure from the chain of four resonant islands of the pericentric branch of the 4/3 MMR to the chain of three resonant islands of the apocentric branch of the 3/2 occurs as follows: as the Jacobi integral increases, the four pericentric islands decrease in size; two of these islands – those centered near 90° and 270° – gradually move their centers towards 60° and 300°, respectively; one of the four islands, the one centered at $\psi = 0$, migrates to smaller values of e^*, and its size decreases faster than that of the other three, until it vanishes. This leaves only three islands, of similar size, which then form the three apocentric islands of the 3/2 resonance. Similarly, the transformation of the chain of three resonant islands of the pericentric branch of the 3/2 MMR to the two-island chain of the apocentric branch of the 2/1 occurs by the gradual shrinking and disappearance of one of the three pericentric libration islands of the 3/2, the one centered at $\psi = 0$, and the gradual migration of the other two islands – those centered near $\psi = 120°$ and $\psi = 240°$ – to positions near $\psi = 90°$ and $\psi = 270°$ of the 2/1 apocentric branch. These features have remained unremarked in the previous literature.

We christened these features as "resonant bridges" between first order MMRs, and conjectured that these structures could serve as long range transport conduits under weak dissipative forces, so that a particle could adiabatically move long radial distances along these resonant paths, including transferring amongst different MMRs. The study by Antoniadou & Libert (2021) confirms this conjecture, with the caveat that the dissipative forces be strong enough to avoid the traps of higher order MMRs in that path.

Orbital migration and adiabatic evolution near individual MMRs has been a subject of many previous investigations. Applications include the phenomena of capture into resonance during convergent migration and eccentricity excitation during either convergent or divergent migration (e.g. Dermott et al. 1988; Peale 1999; Chiang & Jordan 2002; Mustill & Wyatt 2011). The phenomenon of resonant repulsion has also been investigated (Lithwick & Wu 2012; Petrovich et al. 2013; Terquem & Papaloizou 2019). However, smooth transfer between neighboring first order MMRs is a novel possibility revealed by the low eccentricity resonant bridges. We suggest that it may have interesting applications in the dynamical transport and mixing of small bodies in planetary systems.

3.4. *High eccentricity resonances*

For moderate to high values of e^*, we observe the following in Fig. 3(a):

(i) The pericentric branch achieves a maximum width near but slightly below the planet-grazing value, $e_{\rm cross} = (\frac{p+1}{p})^{\frac{2}{3}} - 1$, and then decreases with increasing eccentricity. Our non-perturbative approach finds that the maximum width is finite,

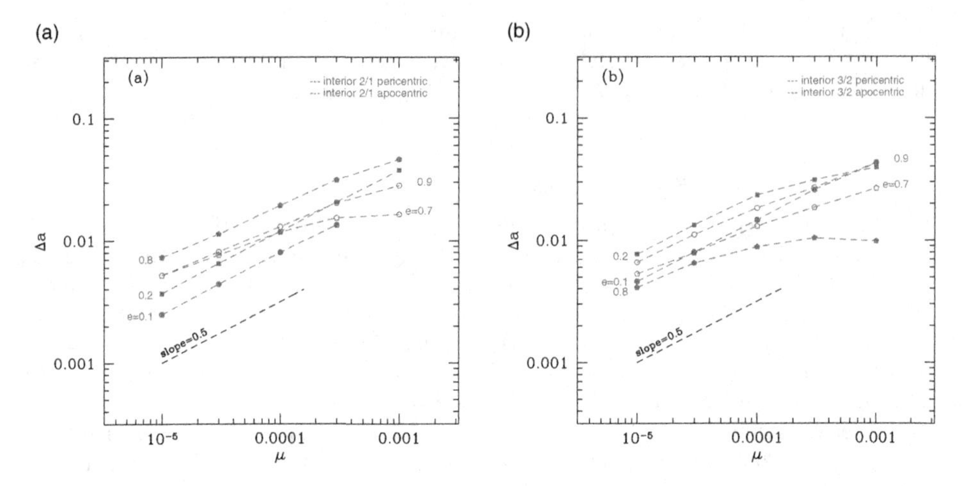

Figure 4. Resonance widths, Δa, as a function of perturber's mass, μ, (a) for the 2/1 interior MMR, and (b) for the 3/2 interior MMR. The blue curves are for the pericentric branch and the red curves are for the apocentric branch; each curve is for a specific eccentricity of the test particle, as marked on each plot. Based on results reported in Wang & Malhotra (2017).

in contrast with some previous studies reporting divergent widths near the planet-grazing eccentricity.

(ii) The apocentric branch re-emerges at eccentricities exceeding $e_{\rm cross}$, and grows wider with increasing e^*. Both the pericentric and apocentric branches co-exist in the high eccentricity regime.

(iii) In the case of the MMRs with larger p ($p > 1$), there are multiple terminations and re-emergences of pericentric and/or apocentric branches at higher values of e^*.

The co-existence of the pericentric and apocentric branches and their terminations/re-emergences at high eccentricities have remained largely unremarked in previous studies. It is rather remarkable that the widths of resonances at high eccentricities are not much smaller than their maximum width achieved in the lower eccentricity regime. In Section 4, we discuss the properties and physical origin of these high-eccentricity features from a geometric point of view. These features have applications in the high eccentricity populations of minor planets throughout the solar system, and potentially also in exo-planetary systems and multiple star systems.

Wang & Malhotra (2017) referred to the pericentric and apocentric branches as the "first resonance zone" and the "second resonance zone", respectively; later we adopted the terminology of "pericentric" and "apocentric" branches, more descriptive of their physical properties.

3.5. *Dependence on μ*

A selection of results for the dependence of the resonance widths, Δa, on the perturber's mass, μ, is shown in Fig. 4 for the interior 2/1 and 3/2 MMRs, at a few different eccentricities. Fitting the measured widths to power laws, $\Delta a \propto \mu^\beta$, finds that the pericentric branches at the low eccentricity regime have $\beta \simeq 0.5$. In the high eccentricity regime, $\beta \simeq 0.4$ for the 2/1 MMR while $\beta \simeq 0.33$ for the 3/2 MMR, respectively. The apocentric branches in the high eccentricity regime have β in the range 0.3–0.4.

In general, guidance from analytical theory leads to the expectation of a sub-linear dependence of resonance widths on the perturber's mass. For example, from the pendulum model one would expect resonance widths to scale as the square root of the magnitude of the resonant potential. At higher accuracy, the second fundamental model for resonance, SFMR, predicts first order resonance widths to scale as $\sim \mu^{\frac{2}{3}}$ in the low eccentricity regime (Wisdom 1980; Henrard & Lemaitre 1983). In the high eccentricity regime, there is limited analytical guidance available for the μ–dependence of resonance widths. Extrapolation of the analytical SFMR to higher eccentricities predicts the scaling $\sim \mu^{\frac{1}{2}}$ (Malhotra 1998). These perturbation theory estimates of the scalings are somewhat different from the results of our non-perturbative approach. The reasons for the differences are not yet known.

It is also noteworthy that, in the case of the 2/1 interior MMR, the power laws $\Delta a \propto \mu^{\beta}$ hold well for μ values up to 0.003 whereas for the 3/2 interior MMR, significant deviations (toward smaller power law index) are observed for $\mu \gtrsim 10^{-4}$. This shallowing of the μ–dependence is likely owed to the crowding of neighboring MMRs, crowding that becomes increasingly significant for larger μ. The surprise is that this crowding appears already for the 3/2 MMR at $\mu \sim 10^{-4}$, a value that would not usually be considered large.

4. Connection of Phase Space Topology to the Geometry of the Resonant Orbit

The geometry of resonant orbits has a fundamental physical connection to the phase space structures and the eccentricity dependence of the resonance widths described in the previous sections. We illustrate this for the interior 2/1 MMR with the help of Figure 5. In the frame co-rotating with the constant angular velocity of the primaries, and with origin at their center-of-mass, the position of the sun and the planet are fixed at distance $-\mu$ and $1-\mu$ from the origin; these are indicated with the brown open circles in Fig. 5(top panels). In this rotating frame, an exact resonant orbit is a periodic orbit: its trace is a closed curve and its shape has $(p+1)$-fold symmetry. For a specified eccentricity and resonant ratio, there exist multiple periodic orbits; all have the same shape, but are distinguished from each other by their different (but specific) orientations relative to the fixed locations of the sun and planet in the rotating frame. Each chain of resonant islands in the Poincaré section corresponds to a particular orientation of the resonant orbit relative to the (fixed) location of the sun and planet in the rotating frame.

It is worth stating that non-resonant orbits in proximity to the resonance also trace curves whose shape resembles that of an exact resonant orbit, but their trace does not close; rather, the whole trace gradually rotates all the way around without closing on itself. Librating orbits also trace curves that are not closed, but they oscillate (librate) around the orientation of a stable exact resonant orbit. The exact resonant orbit is a periodic orbit whereas the librating resonant orbits and the nearby non-resonant orbits are quasi-periodic. A visualization (with animations) of a Pluto-like librating resonant orbit and a non-resonant orbit near Neptune's 2/3 MMR can be found in Zaveri & Malhotra (2021).

The closed blue curve in Figure 5(top) shows the shape of an eccentric 2/1 interior resonant (periodic) orbit in the rotating frame, for one stable orientation; the left panel shows the case of eccentricity 0.3, the right panel is the case of eccentricity 0.8. The corresponding phase space portraits in the vicinity of these cases are shown in the bottom two panels. We ask the reader to imagine rotating the two-fold symmetric shape of the resonant orbit to different orientations to understand the stability or instability of different orientations that we describe now. This description uses Fig. 5 as the main aid, but also will refer to the structures of the resonances in the (a, e) plane in Fig. 3.

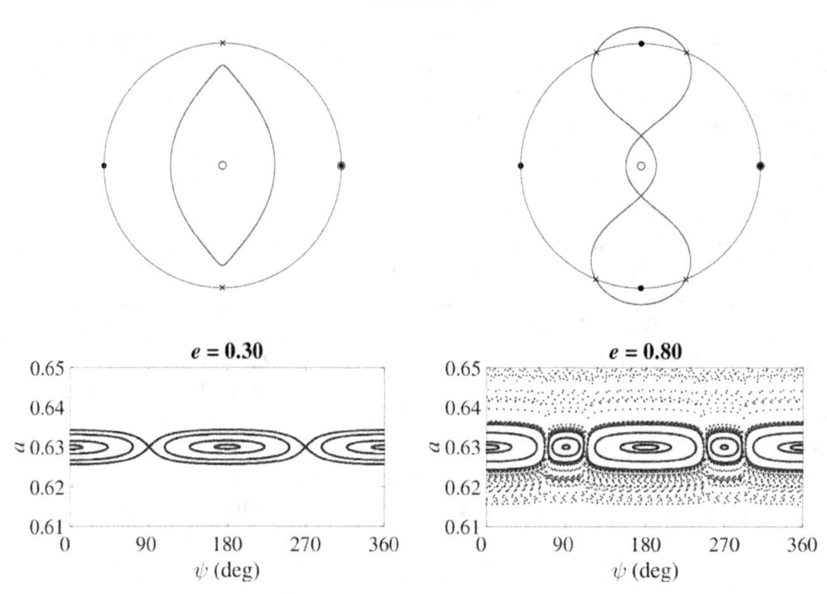

Figure 5. (top) The geometry of the exact resonant orbit in the rotating frame is traced by the closed curve of two-fold symmetry (shown in blue), for eccentricity 0.3 (left panel) and 0.8 (right panel). The test particle has two perihelion passages in this closed shape, one on the positive abscissa and one on the negative abscissa. The green circle is of radius $1-\mu$; this would be the path of the planet in the barycentric inertial frame, but in the rotating frame the Sun and the planet's positions are fixed on the abscissa at $-\mu$ and $1-\mu$, respectively, as indicated by the open brown circles. The black dots on the green circle denote the stable orientations of the test particle's perihelion, while 'x's denote its unstable orientations.
(bottom) Poincaré sections in (ψ, a) near the interior 2/1 MMR, for mass ratio $\mu = 3 \times 10^{-5}$, for eccentricity near $e^* = 0.3$ (left panel) and near $e^* = 0.8$ (right panel). The chains of resonant islands indicate the stable libration regions surrounding the stable periodic orbit in exact 2/1 resonance with the planet; each periodic orbit appears as a pair of points in the Poincaré section. Note that the stable and unstable orientations of the resonant orbit (top plots) correspond, respectively, to the centers of the libration islands and to the saddle points in the Poincaré sections (bottom plots). The lengths of arcs in-between the stable and unstable orientations correspond to the maximal possible extent in ψ of the libration islands.

On the green circle of radius $1-\mu$, we denote with a black dot the particle's longitude of perihelion corresponding to each of the stable centers of libration visible in the Poincaré section; it is intuitively evident that these orientations are stable because they maximally avoid close encounters with the planet. Indeed, integrated over the trace of the closed blue curve, these orientations minimize the planet's perturbation on the particle. Similarly, on the green circle, we denote with a black 'x' the particle's longitude of perihelion corresponding to the geometry of each of the unstable fixed points visible in the Poincaré section ((which are saddle points of the potential function). Again, it is intuitively evident that these orientations are unstable because the particle approaches the perturber most closely in these paths. And, integrated over the trace of the closed blue curve, these orientations maximize the planet's perturbation on the particle.

When the eccentricity is not too large, the particle's resonant orbit is completely inside the orbit of the planet and its aphelion is far from the planet. For the case of $\mu = 3 \times 10^{-5}$ and $e = 0.3$ illustrated in our example, we see that there is a chain of two libration islands in the Poincaré section, and two saddle points at $\psi = 90°, 270°$ (Figure 5, bottom left). A separatrix passing though the saddle points delineates the extent of the resonance libration zone. The two-fold symmetry of the trace of the test particle's resonant orbit in the rotating frame leads directly to the two centers of the stable islands at $\psi = 0, 180°$ in the corresponding Poincaré section. The stable geometry is one in which the test particle's longitude of perihelion is oriented near one of the two locations denoted by the black dots, i.e., with $\psi = 0$, or $\psi = 180°$. In this geometry, the planet's perturbation on the particle is minimized.

For larger eccentricity, the aphelion of the test particle gets closer to the orbit of the planet, so the perturbation of the planet on the test particle also becomes larger. Chaotic regions appear in the Poincaré section, bounding the stable libration islands. When the eccentricity exceeds $e_{\rm cross} = (1/2)^{\frac{2}{3}} - 1 = 0.59$, the test particle's orbit is planet-crossing and new stable islands appear in the Poincaré section (Figure 5, bottom right). When the aphelion distance of the test particle's orbit exceeds the orbit radius of the planet, the two lobes of the resonant orbit intersect the green circle; the green circle is cut into four arcs. The intersection points, denoted by 'x' in the figure, are the collision points. They also represent the new unstable orientations of the test particle's perihelion, and are visible as the new unstable fixed points in the Poincaré section. The ranges of longitudes of the two arcs in-between the unstable orientations are where new stable islands appear in the Poincaré section. The length of arc of the green circle which is enclosed by each of the two lobes delineates the maximum possible range of ψ for the new resonant islands in the Poincaré section, whereas the lengths of arc outside the lobes correspond to the maximum possible range of ψ in the old resonant islands. The chaotic regions in the Poincaré section are the perihelion orientations near the 'x' points; in these orientations, the particle would have close approaches to the planet. As the eccentricity becomes larger, the length of arc enclosed by each lobe also becomes larger, whereas the length of arc outside the lobes shrinks. So the new stable islands grow and expand at the expense of the range of the old stable islands (Fig. 3(a)). In other words, the apocentric islands grow at the expense of the pericentric islands.

As the eccentricity approaches 0.90, the two lobes intersect each other, and their intersection points get close to the green circle. At this high eccentricity, the trace of the resonant orbit in the rotating frame again defines only two arcs, but these two are the ones enclosed by the lobes. This means that the original two islands have disappeared and only the new stable islands exist, marking the termination of the pericentric branch in Fig 3. At even higher eccentricity, when the self-intersections of the lobes occur outside the green circle, the green circle is again cut into four arcs, so the "old" stable islands centered at $\psi = 0, 180°$ reappear, albeit with smaller sizes. This marks the reappearance of the pericentric branch at high eccentricity in Fig. 3(a).

Why are the widths of the different islands in the pericentric branch (respectively, apocentric branch) not all the same (Fig. 3)? The answer lies in the fact that the planet's perturbation on the particle is not invariant under the transformation $x \to -x$, but is only invariant under the transformation $y \to -y$. (Recall that x, y are the coordinates in the rotating frame.) Although the closed shape of the resonant orbit (periodic orbit) nominally appears to be symmetric about the y-axis, it is easy to see that this is not an exact symmetry. For example, in the cases of the 2/1 resonant orbits illustrated in Fig. 5, the perihelion passage that occurs at $\psi = 0$ has smaller distance to the planet than the perihelion passage that occurs at $\psi = 180°$. This difference translates into slightly different velocities of the particle at alternate perihelion passages, hence slightly different orbital elements a^*, e^* at the center of the corresponding two pericentric islands visible in the Poincaré sections, as well as different values of $a_{\rm min}$ and $a_{\rm max}$ at the extrema of the libration zones of each resonant island. This is also the reason for the deviation of the centers of the apocentric islands from exactly 90° and 270°, as noted in Section 2.

There is an extensive body of literature in periodic orbit theory on families of periodic orbits in the restricted three body problem. From the perspective of periodic orbit theory, the phenomenon of the doubling of the stable islands at some of the transitions in the phase space structure is described as arising from the bifurcation of a periodic orbit from a collision orbit (e.g., Hadjidemetriou & Voyatzis 2000; Voyatzis & Kotoulas 2005). Here we have given a physically-intuitive description from a geometric point of view, by reference to the shape and orientation of the resonant orbits in the rotating frame.

We note that Wang & Malhotra (2017) described the physical connection between the phase space structures and the shape and orientation of the resonant orbit in the rotating frame from a different point of view: by considering different possible locations of the planet on the green circle, rather than fixing its location on the positive x-axis. That explanation as well as the one given here are both correct and indeed equivalent to each other, but one or the other is more useful depending upon the application. For example, allowing the planet's location on the green circle to 'float' is of value in predicting the location (within its orbit) of a hypothetical unseen Planet Nine in the distant solar system (Malhotra et al. 2016).

5. Future Directions

We end this review with a non-exhaustive list of potential future directions for research on this topic.

i. The connection between the geometry of a resonant orbit and the phase space structure (as revealed in the Poincaré sections) is most clear for the regime of moderate-to-high eccentricities; it enables an understanding of resonance dynamics from a geometric and physical point of view. However, in the low eccentricity regime, this geometric point of view and corresponding physical understanding is not so clear. Particularly vexing is the behavior of the apocentric branch of first order interior MMRs (and pericentric branch, in the case of exterior MMRs). A physical explanation for the low-eccentricity phase space topology awaits future work.

ii. There is much room to investigate further the low eccentricity resonant bridges and their applications to natural systems. These bridges appear continuously over a large range of semi major axis in-between adjacent first order MMRs, but we can imagine that higher order MMRs in-between may play a role as "off-ramps" from these bridges. This requires investigation with higher numerical resolution in the Jacobi constant than in the studies published thus far.

It is also interesting to examine the relationship between the low eccentricity resonant bridges and the near-resonant and secular approximations. The concepts of "forced" and "free eccentricity" in proximity to a mean motion resonance and in secular perturbation theory are usually a good approximation for the regions in-between low order MMRs (e.g., Murray & Dermott 1999). A quantitative comparison between the non-perturbative results and the analytical estimates of forced and free eccentricity awaits future work.

iii. The locations of the centers, (a^*, e^*), of the pericentric and apocentric resonance branches computed in our non-perturbative analysis with Poincaré sections have some notable differences compared to those of the corresponding periodic orbits computed by Antoniadou & Libert (2021). This is particularly visible in the locations of high-eccentricity apocentric branches. A study to reconcile these differences would be potentially useful to clarify the quantitative details of the phase space structure and possibly also to improve numerical algorithms.

iv. For first order interior resonances, the scaling of resonance widths with the perturber's mass found with our non-perturbative approach has some fundamental discrepancies with the analytical guidance from the second fundamental model for resonance (Sec. 3.5). It would be of value to understand the reasons for these discrepancies. It would also be of value to investigate these scalings for other MMRs, including exterior MMRs and higher order MMRs.

v. The concept of "overlapping resonances" as the origin of deterministic chaos has been well accepted for the past few decades. This has provided fairly good quantitative results in applications to the solar system, such as the Kirkwood gaps in the asteroid belt. Some of the new results reported here highlight the role of a distinctly different mechanism, namely bifurcations of resonance zones (equivalently, bifurcations of periodic orbits) near the planet-crossing eccentricity, in generating chaos and reducing or even eliminating stable libration zones. Investigation of the complementary roles of bifurcations and overlapping resonances could significantly advance our understanding of the origin of dynamical chaos near orbital resonances.
Furthermore, investigation of the transport pathways within the chaotic zones near MMRs has potentially very interesting and useful applications to the migration and mixing of small bodies in planetary systems.

vi. We have seen that for some ranges of parameters, first order resonances have fuzzy boundaries. Dynamical systems theory explains such fuzzy boundaries as homoclinic tangles arising from perturbations of a separatrix that exists in the one-degree-of-freedom idealization of the system. In the planetary dynamics literature of the past few decades it is usually attributed to overlapping higher order resonances that accumulate near the separatrix. Is there a physical explanation, even an approximate one, as an alternative to the hand-wavy explanation embodied in "overlapping resonances"? For example, it is conceivable that the stable libration zone boundaries are related to the closest approach distance (and relative velocity at close approach) to the perturber. The scaling of the stable resonance widths with the perturber mass would suggest this to be the case. A nice start in this direction has been made by Pousse & Alessi (2021) who have obtained estimates of the limits of the averaged solutions for mean motion resonances related to the close approach distance to the perturber; this approach merits further investigation.

vii. Our specific approach to Poincaré sections of the PCRTBP has the potential to be extendable to more realistic models for the dynamics of mean motion resonances. The stroboscopic record of state vectors at successive perihelion passages can enable analysis of the dynamics and properties of the phase space neighborhood generated in the non-restricted three body problem, or in non-coplanar models or by multiple planetary perturbers. In these more complex and more-than-two-degrees-of-freedom models, it is still possible to capture the remnants or "shadows" of the resonant structures with this stroboscopic record in a useful way. By employing so-called *pseudo*-Poincaré sections inspired by the approach reviewed here, a new study (Volk et al. 2021) reports such an attempt in an investigation of Neptune's exterior MMRs in a three-dimensional spatial setting with a realistic numerical model including the perturbations of all four giant planets, Jupiter, Saturn, Uranus and Neptune.

viii. Several recent analyses of planetary mean motion resonances in the context of exo-planetary systems have been based on the conventional critical resonant angle (e.g., Deck et al. 2013; Ramos et al. 2015; Hadden & Lithwick 2018; Hadden 2019; Petit 2021). These could potentially also benefit from validation with the lens of the sub-multiple of the critical resonant angle used in our work, and/or employing physically-motivated stroboscopic records to generate *pseudo*-Poincaré sections for comparison with the semi-analytic approaches.

ix. The generation of many Poincaré sections and the task of measuring resonance centers and widths by visual examination is labor intensive and fatiguing for a researcher. It would be useful to develop methods to carry out these tasks with machine-learning algorithms.

Acknowledgements

Figures 1–3 and Figure 5 in this paper are adapted from Malhotra & Zhang (2020); Figure 4 is based on results reported in Wang & Malhotra (2017). This work was partially supported by research funding from NSF (grant AST-1824869), NASA (grant 80NSSC18K0397), and the Marshall Foundation of Tucson, AZ.

References

Antoniadou, K. I., & Libert, A.-S. 2021, MNRAS, 506, 3010, doi: 10.1093/mnras/stab1900

Bailey, B. L., & Malhotra, R. 2009, Icarus, 203, 155, doi: 10.1016/j.icarus.2009.03.044

Beaugé, C. 1994, Celestial Mechanics and Dynamical Astronomy, 60, 225, doi: 10.1007/BF00693323

Belbruno, E., & Marsden, B. G. 1997, AJ, 113, 1433, doi: 10.1086/118359

Beust, H. 2016, A&A, 590, L2, doi: 10.1051/0004-6361/201628638

Chiang, E. I., & Jordan, A. B. 2002, AJ, 124, 3430, doi: 10.1086/344605

Deck, K. M., Payne, M., & Holman, M. J. 2013, ApJ, 774, 129, doi: 10.1088/0004-637X/774/2/129

Dermott, S. F., Malhotra, R., & Murray, C. D. 1988, Icarus, 76, 295, doi: 10.1016/0019-1035(88)90074-7

Dermott, S. F., & Murray, C. D. 1983, Nature, 301, 201, doi: 10.1038/301201a0

Duncan, M., Quinn, T., & Tremaine, S. 1989, Icarus, 82, 402, doi: 10.1016/0019-1035(89)90047-X

Duncan, M. J., & Levison, H. F. 1997, Science, 276, 1670, doi: 10.1126/science.276.5319.1670

Fernández, J. A., Helal, M., & Gallardo, T. 2018, Planet. Space Sci., 158, 6, doi: 10.1016/j.pss.2018.05.013

Gallardo, T. 2019, Icarus, 317, 121, doi: 10.1016/j.icarus.2018.07.002

Gladman, B., Marsden, B. G., & Vanlaerhoven, C. 2008, Nomenclature in the Outer Solar System (in The Solar System Beyond Neptune, Barucci, M. A. and Boehnhardt, H. and Cruikshank, D. P. and Morbidelli, A. and Dotson, R. (eds.), University of Arizona Press, Tucson), 43–57

Gladman, B., Lawler, S. M., Petit, J.-M., et al. 2012, AJ, 144, 23, doi: 10.1088/0004-6256/144/1/23

Guzzo, M. 2006, Icarus, 181, 475, doi: 10.1016/j.icarus.2005.11.019

Hadden, S. 2019, AJ, 158, 238, doi: 10.3847/1538-3881/ab5287

Hadden, S., Li, G., Payne, M. J., & Holman, M. J. 2018, AJ, 155, 249, doi: 10.3847/1538-3881/aab88c

Hadden, S., & Lithwick, Y. 2018, AJ, 156, 95, doi: 10.3847/1538-3881/aad32c

Hadjidemetriou, J., & Voyatzis, G. 2000, Celestial Mechanics and Dynamical Astronomy, 78, 137

Hénon, M. 1966, in IAU Symposium, Vol. 25, The Theory of Orbits in the Solar System and in Stellar Systems, ed. G. I. Kontopoulos, 157

Henrard, J. 1983, Celestial Mechanics, 31, 115, doi: 10.1007/BF01686813

Henrard, J., & Lemaitre, A. 1983, Celestial Mechanics, 30, 197, doi: 10.1007/BF01234306

Lan, L., & Malhotra, R. 2019, Celestial Mechanics and Dynamical Astronomy, 131, 39, doi: 10.1007/s10569-019-9917-1

Lei, H., & Li, J. 2020, MNRAS, 499, 4887, doi: 10.1093/mnras/staa3115

Lithwick, Y., & Wu, Y. 2012, ApJL, 756, L11, doi: 10.1088/2041-8205/756/1/L11

Lykawka, P. S., & Mukai, T. 2007, Icarus, 192, 238, doi: 10.1016/j.icarus.2007.06.007
Malhotra, R. 1998, Solar System Formation and Evolution: ASP Conference Series, 149, 37
Malhotra, R. 2019, Geoscience Letters, 6, 12, doi: 10.1186/s40562-019-0142-2
Malhotra, R., Lan, L., Volk, K., & Wang, X. 2018, AJ, 156, 55, doi: 10.3847/1538-3881/aac9c3
Malhotra, R., Volk, K., & Wang, X. 2016, ApJL, 824, L22, doi: 10.3847/2041-8205/824/2/L22
Malhotra, R., & Zhang, N. 2020, MNRAS, 496, 3152, doi: 10.1093/mnras/staa1751
Moons, M. 1996, Celestial Mechanics and Dynamical Astronomy, 65, 175
Moons, M., & Morbidelli, A. 1993, Celestial Mechanics and Dynamical Astronomy, 56, 273, doi: 10.1007/BF00699737
Morbidelli, A. 2002, Modern celestial mechanics: aspects of solar system dynamics (Taylor & Francis London)
Morbidelli, A., Thomas, F., & Moons, M. 1995, Icarus, 118, 322, doi: 10.1006/icar.1995.1194
Murray, C. D., & Dermott, S. F. 1999, Solar system dynamics, 1st edn. (New York, New York: Cambridge University Press)
Murray, N., & Holman, M. 1999, Science, 283, 1877, doi: 10.1126/science.283.5409.1877
Mustill, A. J., & Wyatt, M. C. 2011, MNRAS, 413, 554, doi: 10.1111/j.1365-2966.2011.18201.x
Nesvorný, D., & Ferraz-Mello, S. 1997, Icarus, 130, 247, doi: 10.1006/icar.1997.5807
Peale, S. J. 1999, ARA&A, 37, 533, doi: 10.1146/annurev.astro.37.1.533
Petit, A. C. 2021, Celestial Mechanics and Dynamical Astronomy, 133, 39, doi: 10.1007/s10569-021-10035-7
Petrovich, C., Malhotra, R., & Tremaine, S. 2013, ApJ, 770, 24, doi: 10.1088/0004-637X/770/1/24
Pousse, A., & Alessi, E. M. 2021, arXiv e-prints, arXiv:2106.14810. https://arxiv.org/abs/2106.14810
Ramos, X. S., Correa-Otto, J. A., & Beaugé, C. 2015, Celestial Mechanics and Dynamical Astronomy, 123, 453, doi: 10.1007/s10569-015-9646-z
Roberts, A. C., & Muñoz-Gutiérrez, M. A. 2021, Icarus, 358, 114201, doi: 10.1016/j.icarus.2020.114201
Terquem, C., & Papaloizou, J. C. B. 2019, MNRAS, 482, 530, doi: 10.1093/mnras/sty2693
Tiscareno, M. S., & Malhotra, R. 2003, AJ, 126, 3122, doi: 10.1086/379554
Volk, K., Malhotra, R., & Graham, S. 2021, in AAS/Division of Dynamical Astronomy Meeting, Vol. 53, AAS/Division of Dynamical Astronomy Meeting, 305.01
Voyatzis, G., & Kotoulas, T. 2005, Planet. Space Sci., 53, 1189, doi: 10.1016/j.pss.2005.05.001
Wang, X., & Malhotra, R. 2017, AJ, 154, 20, doi: 10.3847/1538-3881/aa762b
Winter, O. C., & Murray, C. D. 1997, A&A, 319, 290
Wisdom, J. 1980, AJ, 85, 1122, doi: 10.1086/112778
Yu, T. Y. M., Murray-Clay, R., & Volk, K. 2018, AJ, 156, 33, doi: 10.3847/1538-3881/aac6cd
Zaveri, N., & Malhotra, R. 2021, Research Notes of the AAS, 5, 235
Zhu, W., & Dong, S. 2021, Annual Review of Astronomy and Astrophysics, 59, 291, doi: 10.1146/annurev-astro-112420-020055

Multi-scale (time and mass) dynamics of space objects
Proceedings IAU Symposium No. 364, 2022
A. Celletti, C. Galeş, C. Beaugé, A. Lemaitre, eds.
doi:10.1017/S1743921322000758

Latitudinal variations of charged dust in co-orbital resonance with Jupiter

Stefanie Reiter[1] and Christoph Lhotka[1,2]

[1]Department of Astrophysics, University of Vienna,
Türkenschanzstrasse 17,1180 Vienna, Austria
email: stefanie.reiter@univie.ac.at

[2]Department of Mathematics, University of Rome Tor Vergata,
Via della Ricerca Scientifica 1, 00133 Rome, Italy
email: lhotka@mat.uniroma2.it

Abstract. The interplanetary magnetic field may cause large amplitude changes in the orbital inclinations of charged dust particles. In order to study this effect in the case of dust grains moving in 1:1 mean motion resonance with planet Jupiter, a simplified semi-analytical model is developed to reduce the full dynamics of the system to the terms containing the information of the secular evolution dominated by the Lorentz force. It was found that while the planet causes variations in all orbital elements, the influence of the magnetic field most heavily impacts the long-term evolution of the inclination and the longitude of the ascending node. The simplified secular-resonant model recreates the oscillations in these parameters very well in comparison to the full solution, despite neglecting the influence of the magnetic field on the other orbital parameters.

Keywords. Dust dynamics, Lagrange problem, interplanetary magnetic field

1. Introduction & Background

Interplanetary space is not empty but rather filled with dust particles of various sizes. Dust grains are produced in several physical processes, i.e. the activity of the comets and collisions of asteroids, see e.g. Koschny *et al.* (2019). Additionally, dust may also be released into space via plumes (e.g. Enceladus) or by the interaction of celestial bodies with the solar wind (by sputtering and erosion). The actual origin and resulting distribution of interplanetary dust still remain an unsolved problem today. One reason for this is found in the lack of in-situ observations in large areas of interplanetary space. While observations of the zodiacal light indicate the presence of dust bands near the ecliptic plane, the presence of dust at high ecliptic latitudes can generally not be ruled out. In Jorgensen *et al.* (2020), the authors propose that a secondary population of dust at higher inclination wrt. the ecliptic is caused by scattering of a primary population at low latitudes by orbital resonances with planet Jupiter via the Kozai-Lidov (KL) effect. Alternatively, these features may be explained by dust bands that originate from a few asteroid families which deliver the dust to the inner solar system at high enough orbital inclinations – Dermott *et al.* (1984). In the present paper, we provide an alternative explanation for high latitudinal motions of small dust grains, namely via oscillations induced by the heliospheric magnetic field. Dust in space gets positively charged due to the photo-electric effect, see Lhotka *et al.* (2020). The orbital motion of charged dust results in a Lorentz force term in the equations of motion which acts normal to the orbital plane of the dust grains, leading to perturbations of the orbit. The present study

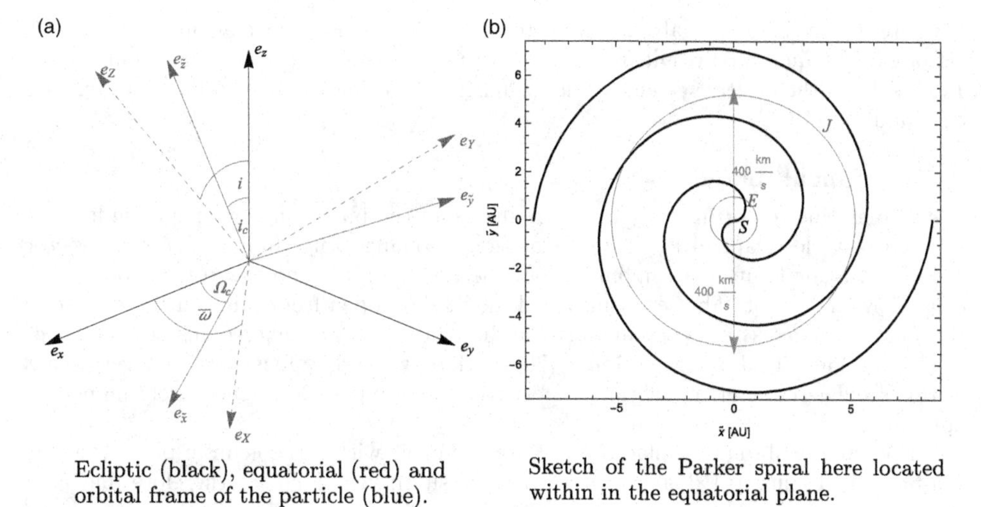

Ecliptic (black), equatorial (red) and orbital frame of the particle (blue).

Sketch of the Parker spiral here located within in the equatorial plane.

Figure 1. Overview of the coordinate systems and the shape of the magnetic field.

is motivated by recent findings in Liu & Schmidt (2018a), Liu & Schmidt (2018b), and can be seen as a continuation of previous studies: While the focus in Lhotka *et al.* (2016) was the role of the normal component of the interplanetary magnetic field on the radial drift of particle motions, the role of outer mean motion resonances has been investigated in Lhotka & Galeş (2019). In Zhou *et al.* (2021); Lhotka & Zhou (2021) extensive numerical studies are used to investigate the role of the interplanetary magnetic field on the location and extent of the 1:1 mean motion resonance (MMR) for co-orbital dust with a planet (Jupiter and Venus), yielding interesting results on the asymmetry between the Lagrange points L_4 and L_5. While these former studies heavily relied on numerical simulations, in the present work we report on the progress of deriving a semi-analytical model for co-orbital motion of dust with planet Jupiter.

Let the position vector of a micron-sized dust particle be $\vec{r} = (x, y, z)$ in the reference frame of the ecliptic coordinate system. The rotation of the orbital frame into the ecliptic frame (a comparison between the two is given in Figure 1a) allows it to define all orbital elements, i.e. the semi-major axis a, the eccentricity e, the orbital inclination i, the argument of perihelion ω, the longitude of the ascending node Ω, and the mean anomaly M. Assuming that the particle is only influenced by the Sun, it would move on a Keplerian orbit with constant values for the orbital elements. The angular momentum h of the particle in this case is conserved and given by $h = \sqrt{(1-e^2)a^2 n}$, where n gives the mean motion of the particle. Any force (per unit mass) term beside the solar gravity is introduced into the body's equation of motion in the following way:

$$\ddot{\vec{r}} + \mu \frac{\vec{r}}{r^3} = \vec{F}, \tag{1.1}$$

with $\ddot{\vec{r}}$ representing the acceleration vector of the particle, μ the term Gm_0 (i.e. the gravitational constant G and the solar mass m_0), and $\vec{F}$ an arbitrary external force (which is already assumed to be divided by particle mass in this case). Due to the influence of this perturbing force $\vec{F}$, h is no longer a conserved quantity, which relates to variations in the orbital elements. It is thereby possible to derive the so-called Gauss equations – for all Kepler elements see e.g. Fitzpatrick (2016) – in the form of

$$\frac{\mathrm{d}i}{\mathrm{d}t} = \frac{F_z r \cos(\vartheta + \omega)}{h} \qquad \frac{\mathrm{d}\Omega}{\mathrm{d}t} = \frac{F_z r \sin(\vartheta + \omega)}{h \sin i} \tag{1.2}$$

where ϑ is the true anomaly. The force terms F_r, F_ϑ and F_z (which all appear in the full set of Gauss' equations) result from a transformation of the Cartesian force components F_x, F_y, F_z in the orbital system to the cylindrical coordinates in the ecliptic system, see e.g. Moulton (1914).

2. Dynamical model

As we are interested in studying the influence of the interplanetary magnetic field on charged particles trapped in 1:1 mean motion resonance with Jupiter, the force vector $\vec{F}$ in (1.1) is made up of a combination of the Lorentz force and the gravitational force caused by the planet. The semi-analytical model developed from this is used to analyse the long-term effects of the Lorentz force on the orbital plane of charged dust grains (in co-orbital motion with Jupiter) in Reiter (2021). Here, we mainly focus on the development of this secular-resonant model, and its accuracy in comparison to the purely numerical approach.

For simplicity, Jupiter is placed on a circular orbit within the ecliptic plane (i.e. the x-y-plane in Figure 1a). For the perturber's position vector $\vec{r}_1$, we therefore find $\vec{r}_1 = (x_1, y_1, z_1) = (a_1 \cos M_1, a_1 \sin M_1, 0)$, with a_1 as the semi-major axis and M_1 as the mean anomaly of the planet. For the implementation of the Lorentz force, we assume the form of the modified Parker spiral model in Webb *et al.* (2010), taking into account the dipole structure of the magnetic field. According to Parker (1958), the solar wind carries the field lines out into the solar system, and due to the solar rotation, they curl up in the shape of a spiral. This coiled-up structure of the field lines is sketched in Figure 1b. The dipole structure of the interplanetary magnetic field leads to a polarity reversal at the transition region, the so-called heliospheric current sheet. Hence, the configuration of the magnetic field depends on both the orientation of the dipole axis and the solar rotation axis, which is given by the $\vec{e}_{\tilde{z}}$-axis of the equatorial system in Figure 1a. For simplicity, we assume that these two axes align, placing the current sheet within the equatorial plane. The equatorial coordinate system results from a rotation of the ecliptic system around the angles i_c and Ω_c, as indicated in Figure 1a. In this work, they are fixed to $\Omega_c = 73.5$deg and $i_c = 7.15$deg, as in Beck & Giles (2005). This results in $\vec{e}_{\tilde{z}} = (x_0, y_0, z_0)$ for the expression in ecliptic coordinates. Finally, we note that we neglect the time-dependency of the magnetic field (i.e. the solar cycle).

First, we look at the implementation of the planetary perturbations into (1.1). The gravitational influence of the perturbing body on the dust grain can be defined by the potential energy U – compare e.g. Lhotka & Celletti (2015) – according to

$$U = -Gmm_1 \left(\frac{1}{\|\vec{r} - \vec{r}_1\|} - \frac{\vec{r}\vec{r}_1}{r_1^3} - \frac{1}{r} \right), \tag{2.1}$$

where m_1 represents the mass of the planet. We notice that in order to obtain $n = n_1$ at $a = a_1$ the gravitational mass parameter $G(m_0 + m_1)$ results in the additional term $-1/r$ in U stemming from the potential part of the Kepler problem $G(m_0 + m_1)/r$, that we collect with respect to m_1 in U. As for any conservative force, the perturbing force term $\vec{F}_1$ can be derived from the gradient of the potential, i.e. via $\vec{F}_1 = -\nabla U/m$, for which we find:

$$\vec{F}_1 = -Gm_1 \left(-\frac{\vec{r}}{r^3} + \frac{\vec{r}_1}{r_1^3} + \frac{\vec{r} - \vec{r}_1}{\|\vec{r} - \vec{r}_1\|^3} \right), \tag{2.2}$$

Plugging (2.2) into (1.1) results in the full dynamical (Newton) solution of the influence of Jupiter. The process is similar for the implementation of the Lorentz force. Generally, the effect of the interplanetary magnetic field is expressed, see e.g. Gruen *et al.* (1994), as $\vec{F}_L = \frac{q}{m}(\vec{v} - \vec{u}_{SW}) \times \vec{B}$, where q is the charge of the particle, m its mass, $\vec{v}$ its velocity, $\vec{u}_{SW} = u_{SW}\vec{e}_r$ the velocity of the solar wind moving radially away from the

Sun†, and $\vec{B}$ the magnetic field vector. The full expression of the Lorentz force used in the computations is adopted from Eq. 16 in Lhotka & Galeş (2019) in the form of

$$\vec{F}_L = -\frac{qB_0 r_0^2}{mr^2}\left(\frac{\vec{r}\times\dot{\vec{r}}}{r} + \frac{\Omega_s}{r}\vec{r}\times(\vec{r}\times\vec{e}_{\tilde{z}}) + \frac{\Omega_s}{u_{SW}}(\vec{r}\times\vec{e}_{\tilde{z}})\times\dot{\vec{r}}\right)\tanh\left(\frac{\alpha\vec{r}\vec{e}_{\tilde{z}}}{r}\right), \tag{2.3}$$

where B_0 gives the background magnetic field strength at a reference distance of r_0 (typically 1AU), and Ω_s represents the solar rotation rate. The unit vector $\vec{e}_{\tilde{z}}$ yields the $\tilde{z}$-direction in the equatorial system, so it represents both the dipole and the rotation axis of the Sun, as already explained above. Finally, the parameter α is used to model the sign change of the magnetic field at the solar equator (i.e. it represents the effect at the heliospheric current sheet). Combining (2.3) and (2.2) into (1.1) as $\vec{F} = \vec{F}_1 + \vec{F}_L$ describes the numerical approach for the particle dynamics.

As we are primarily interested in studying the secular evolution of the particle analytically rather than numerically, it is necessary to express the force terms (2.2) and (2.3) in cylindrical coordinates and plug the resulting values for F_r, F_ϑ and F_z into all Gauss equations exemplified in (1.2). Starting again with the influence of the perturbing body, it is beneficial to expand the potential U, i.e. (2.1), into small parameters (e and $\rho = r/r_1 - 1$) and the angle $\cos\psi = \frac{\vec{r}\vec{r}_1}{rr_1}$ between the two position vectors. The gradient of this expression is then used to compute the cylindrical vector components. For more details on this series expansion, see Lhotka & Celletti (2015). This computation is advantageous, as it facilitates isolating the secular terms from the full solution and hence from the short-period oscillations induced by M. With the example of a, we note that the terms for the long-term evolution of each orbital element take the form of a Fourier series according to

$$\frac{\mathrm{d}a}{\mathrm{d}t} = \sum_{k=0}^{K}\sum_{l=0}^{L}\sum_{m}^{M'} c_{klm}^{(a)}(e)\cos(ki)\sin(l\Phi + m\omega), \tag{2.4}$$

where $\Phi = M + \omega + \Omega - M_1$ is the resonant angle – see e.g. Beauge (1994) – and c denotes polynomials in e. The three orders of expansion (k, l, m) occur due to the expression of U in terms of e, ρ and $cos(\psi)$. A comparison of the results from the full solution in terms of the planetary perturbations using the Newton equations – i.e. (2.2) in (1.1) – and of the secular evolution from the Gauss equations is given in Figure 2.

As for the influence of the interplanetary magnetic field, (2.3) also needs to be implemented into all Gauss equations, from e.g. Fitzpatrick (2016). In Lhotka & Galeş (2019), the authors have been able to isolate those terms of the equations, which do not average out on longer timescales, from the full solution. As it was solely the normal component of the magnetic field causing long-term oscillations, only i and Ω are the dynamically relevant terms for our semi-analytical model. The equations of Lhotka & Galeş (2019) are implemented here in the form of

$$\frac{di}{dt} = -\alpha\frac{qB_0}{2m}\left(\frac{r_0}{a}\right)^2\left(\left(1 - z_0\cos i\frac{\Omega_s}{n}\right)(x_0\cos\Omega + y_0\sin\Omega)\right) \tag{2.5}$$

$$\frac{\mathrm{d}\Omega}{\mathrm{d}t} = -\alpha\frac{qB_0}{2m}\left(\frac{r_0}{a}\right)^2\left(\left(\cot i - z_0\frac{\Omega_s\cos(2i)}{n\sin i}\right)(y_0\cos\Omega - x_0\sin\Omega) + \cos i(1-2z_0)\frac{\Omega_s}{n} + z_0\right), \tag{2.6}$$

† Due to the combination of a so-called frozen-in magnetic field and no background current resistivity (Lhotka *et al.* (2016)), the solar wind vector appears due to the form of the electric field vector $\vec{E} = -\vec{u}_{SW}\times\vec{B}$.

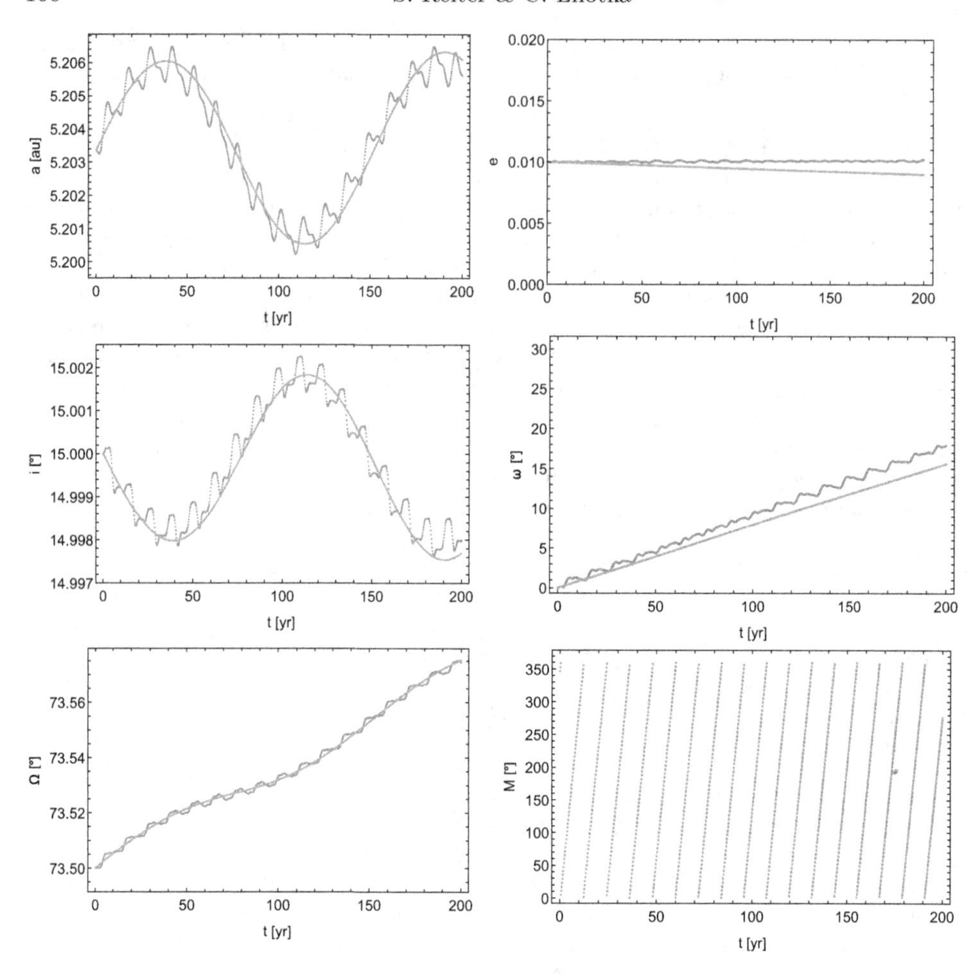

Figure 2. Comparison of the Newton solution (blue) and the resonant secular model (yellow) for uncharged particles in 1:1 mean motion resonance with Jupiter.

where $\vec{e}_{\tilde{z}} = (x_0, y_0, z_0)$. The combination of these two expressions with the resonant terms resulting from the Gauss equations in the form of (2.4) yields the secular dynamics of a charged dust particle in co-orbital motion with Jupiter, as it is affected by the magnetic field. As we only implement the effect of the Lorentz force on the inclination and the longitude of the ascending node into the model, we limit the analysis of the dynamics to the parameter space of i and Ω. Figure 3 shows that in the case of these two orbital elements, the agreement between the full Newton solution and the approximated Gauss model is very good. A comparison of the results in Figure 3 and the middle and lower left graphs in Figure 2 indicates that the magnetic field, rather than the influence of Jupiter, governs the secular dynamics in the i-Ω-space. The results demonstrate that the semi-analytical model comprised of (2.5) and (2.6) in combination with (2.4) for all orbital elements is very useful for analysing the long-term evolution of the orbital plane of charged dust grains in 1:1 mean motion resonance with Jupiter, as is done in detail in Reiter (2021). Most notably, we find that the latitudinal motion of micron-sized dust can easily be explained by the Lorentz force stemming from the interaction of micron-sized charged dust with the heliospheric magnetic field. A study on the dynamics in the full

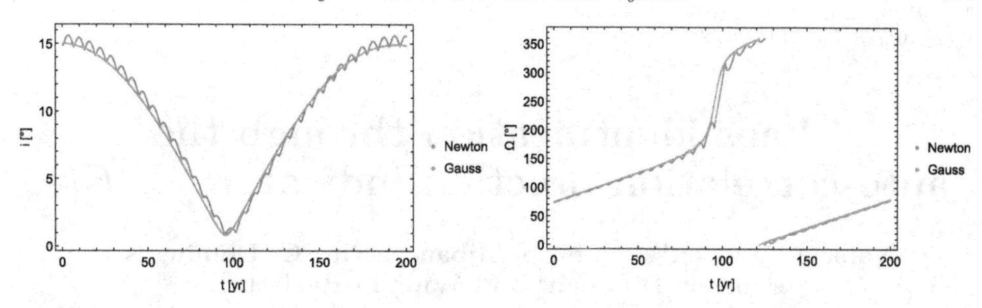

Figure 3. As Figure 2 for the inclination i and the longitude of the ascending node Ω, but in the case of a charged dust grain affected by both Jupiter and the interplanetary magnetic field.

(Ω, i)-plane for charged dust of varying particle size, and located at different regions in the solar system is currently in progress.

Acknowledgments

This work is fully supported by the Austrian Science Fund FWF with project number P-30542.

References

Beauge, C. 1994, *Celestial Mechanics and Dynamical Astronomy*, 60, 225
Beck, J. G. & Giles, P. 2005, *ApJ*, 621, L153
Dermott, S.F., Nicholson, P.D., Burns, J.A., Houck, J.R. 1984, *Nature*, 312, 505
Fitzpatrick, R. 2016, *An Introduction to Celestial Mechanics*
Gruen, E., Gustafson, B., Mann, I., et al. 1994, *AAP*, 286, 915
Jorgensen, J.L., Benn, M., Connereny, J.E.P., Denver, T., Jorgensen, P.S., Andersen, A.C., and Bolton, S.J 2020, *JGR Planets*, 126, e2020JE006509
Koschny, D., Soja, R. H., Engrand, C., et al. 2019, *Space Sci. Revs*, 215, 34
Lhotka, C., Zhou, L. 2021, *CNSNS*, https://doi.org/10.1016/j.cnsns.2021.106024 (accepted)
Lhotka, C. & Celletti, A. 2015, *Icarus*, 250, 249
Lhotka, C., Bourdin, P., Narita, Y. 2016 *ApJ*, 828, 10
Lhotka, C. & Galeş, C. 2019, *Celestial Mechanics and Dynamical Astronomy*, 131, 49
Lhotka, C., Rubab, N., Roberts, W.W., Holmes, J., Torkar, K., Nakamura, R. 2020 *PoP*, 27, 103704
Liu, X. & Schmidt, J. 2018, *A&A*, 609, A57
Liu, X. & Schmidt, J. 2018, *A&A*, 614, A97
Moulton, F. R. 1914,*An introduction to celestial mechanics* (New York, The Macmillan company)
Parker, E. N. 1958, *ApJ*, 128, 664
Reiter, S. 2021, *The spatial distribution of charged dust particles in the outer solar system*, (University of Vienna, in preparation)
Webb, G. M., Hu, Q., Dasgupta, B., et al. 2010, *Journal of Geophysical Research (Space Physics)*, 115, A10112
Zhou, L., Lhotka, C., Gales, C., Narita, Y., Zhou, L.-Y. 2021, *A&A*, 645, A63

Multi-scale (time and mass) dynamics of space objects
Proceedings IAU Symposium No. 364, 2022
A. Celletti, C. Galeş, C. Beaugé, A. Lemaitre, eds.
doi:10.1017/S1743921321001307

Chaos identification through the auto-correlation function indicator ($ACFI$)

Valerio Carruba[1], Safwan Aljbaae[2], Rita C. Domingos[3], Mariela Huaman[4] and William Barletta[1]

[1]São Paulo State University (UNESP), Guaratinguetá, SP, 12516-410, Brazil
email: valerio.carruba@unesp.br

[2]National Space Research Institute (INPE), C.P. 515, 12227-310, São José dos Campos, SP, Brazil

[3]São Paulo State University (UNESP), São João da Boa Vista, SP, 13876-750, Brazil

[4]Universidad tecnológica del Perú (UTP), Cercado de Lima, 15046, Perú

Abstract. Close encounters or resonances overlaps can create chaotic motion in small bodies in the Solar System. Approaches that measure the separation rate of trajectories that start infinitesimally near, or changes in the frequency power spectrum of time series, among others, can discover chaotic motion. In this paper, we introduce the ACF index ($ACFI$), which is based on the auto-correlation function of time series. Auto-correlation coefficients measure the correlation of a time-series with a lagged duplicate of itself. By counting the number of auto-correlation coefficients that are larger than 5% after a certain amount of time has passed, we can assess how the time series auto-correlates with each other. This allows for the detection of chaotic time-series characterized by low $ACFI$ values.

Keywords. Celestial mechanics; Asteroid Belt; Chaotic Motions; Statistical Methods.

1. Introduction

Fidelity, which measures the degree of similarity between two quantum states, has been employed to detect chaotic behavior in quantum computing Frahm *et al.* (2004), Pellegrini & Montangero (2007), Lewis-Swan *et al.* (2019). In the saw-tooth map, Pellegrini & Montangero (2007) determined fidelity values for quantum pure states in chaotic and integrable dynamics. In general, fidelity began at one, decreased until it reached a saturation point, and then oscillated about it. The saturation value for chaotic dynamics was substantially nearer to zero than for integrable systems, permitting the two forms of behavior to be distinguished.

In this research, we present a study about detecting chaotic behavior using the auto-correlation function. The correlation coefficient R between two time-series, as described by Pearson (1895), indicates how strong the association is. R near to 1 indicates strongly correlated series, R close to -1 indicates anti-correlated series, and $R \simeq 0$ indicates uncorrelated series †.

The auto-correlation function (ACF) is obtained by computing values of R for the series with a lagged copy of itself. In essence, R values are computed for the series at

† Contrary to the case of quantum fidelity, anti-correlated series can have negative R values.

lag 0 in relation to the series at lag 1, 2, and so on. ACF depicts a range of R values connected with various time-lags. Most R values will be close to 1 for substantially auto-correlated time-series. Unpredictable series, such as white noise, would display the majority of R values near to zero once a sufficient amount of time has passed. After some time delay, which is a free parameter of the method, we can count the fraction of auto-correlation coefficients that are larger than the 5% value, which is commonly used for auto-regressive functions to set the null hypothesis level of negligible correlation. This new approach is called $ACFI$, which stands for auto-correlation function index. To test $ACFI$, in this work we will apply it to the well-understood Hénon-Heiles dynamical system Skokos *et al.* (2016), and compare its outcome to those of other chaos indicators, like the Smaller Alignment Index ($SALI$) approach Skokos *et al.* (2004).

2. Methods

Correlation coefficients can be defined in a variety of ways, but Pearson's approach is the most frequently used (Pearson (1895)). If the i-th term of the series in x and y is defined as x_i and y_i, then:

$$R = \frac{cov(X,Y)}{\sigma_X \sigma_Y}, \tag{2.1}$$

where $cov(X,Y)$ denotes the covariance between the two series, which is defined as:

$$cov(X,Y) = \frac{1}{N^2} \sum_{i=1}^{N} \sum_{j=1}^{N} \frac{1}{2}(x_i - x_j)(y_i - y_j). \tag{2.2}$$

The number of terms in the two series is N, and the standard deviation of the x_i series is σ_X, which is defined as:

$$\sigma_X = \sqrt{\frac{1}{N} \sum_{i=1}^{N} (x_i - \mu_x)^2}. \tag{2.3}$$

Here $\mu_x = \frac{1}{N} \sum_{i=1}^{N} x_i$ is the series mean value, and an analogous expression exists for σ_Y. The correlation function of a time series with a lagged copy of itself is the auto-correlation coefficient of the series. Assume we have built a time series with a lag of one, $y_i = x_{i-1}$. Equation 2.1 will be used to calculate the auto-correlation coefficient for this y_i. For lags of 2 ($y_i = x_{i-2}$), 3 ($y_i = x_{i-3}$), and so on, analogous auto-correlation coefficients can be found. The spectrum of auto-correlation coefficients for several values of the time lag is the auto-correlation function (ACF) of x_i. The auto-correlation function is useful for determining the predictability of a series temporal behavior.

Figure 1 depicts two ACF for time series of semi-major axis a of a regular (left panel) and chaotic (right panel) particle in the Veritas orbital region. The regular particle exhibits a significantly higher proportion of auto-correlation coefficients outside the null hypothesis levels of ± 0.05 than the chaotic one, especially for time lags greater than 200. On short time scales, it becomes impossible to predict the time behavior of the a series for the chaotic particle. We can create a chaos indicator built on the ACF, the auto-correlation function index ($ACFI$), based on these considerations:

$$ACFI = \frac{1}{i_{fin} - i_{in}} \sum_{i=i_{in}}^{i=i_{fin}} n_i(|R| > 0.05), \tag{2.4}$$

where $n_i(|R| > 0.05)$ denotes the number of auto-correlation coefficients greater than 5% in absolute value. In our instance $i_{fin} - i_{in} = 500 - 200 = 300$, and we only consider

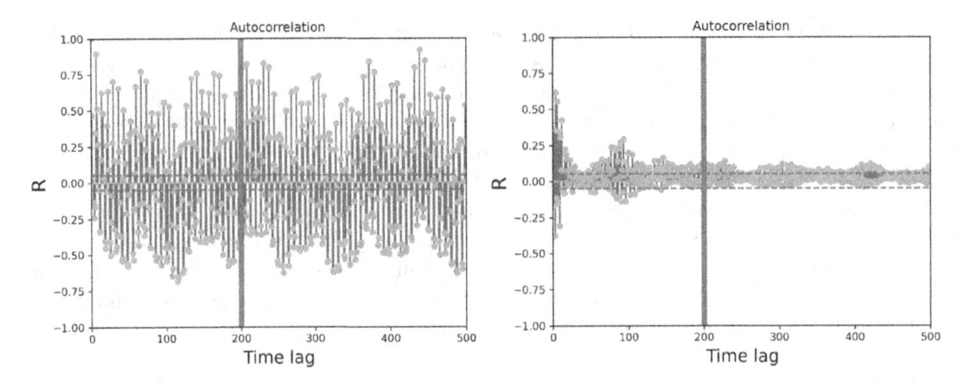

Figure 1. The ACF of a normal asteroid in the Veritas family orbital region is shown in the left panel. The ACF of a somewhat chaotic orbit is shown in the right panel. The vertical line shows a 200 time-step lag. The area between dashed horizontal lines represents the region where auto-correlation coefficients are less than 5% and represents negligible auto-correlation.

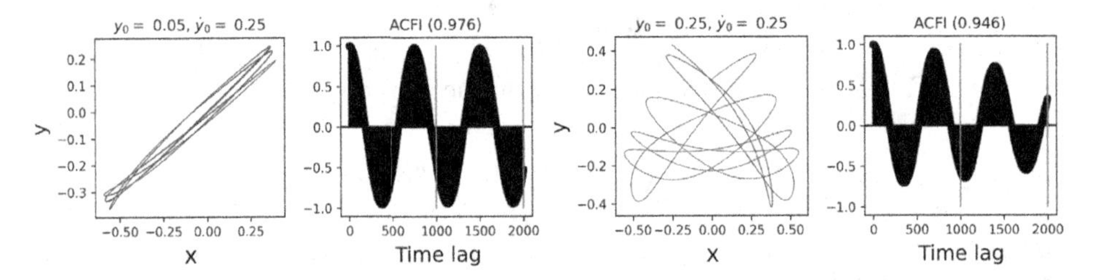

Figure 2. A (x, y) projection of a normal orbit (left panel) and its ACF, shown in the panel in the center left position. The values of the i_{in} and i_{fin} parameters are represented by the vertical red lines. The panels on the center right and right positions display the same quantities, but for a chaotic orbit.

coefficients between 200 and 500 to avoid include auto-correlation at short time-frames. The values of i_{in} and i_{fin} were chosen after experimenting with lower and upper bounds in the ranges of 100 to 300 and 300 to 1000, respectively. Changing either of the given parameters has a maximum effect on $ACFI$ values of 0.04 for regular particles and of 0.01 for chaotic ones. By adjusting either parameter, the mean values of $ACFI$ are modified by less than 1%.

We will apply $ACFI$ to the Hénon-Heiles system in the next section to see how well this method may be utilized to detect chaotic behavior in a well-known system.

3. The Hénon-Heiles Hamiltonian system case

The Hamiltonian of the Hénon-Heiles system (HH) is:

$$H(x, y, \dot{x}, \dot{y}) = \frac{1}{2}(\dot{x}^2 + \dot{y}^2) + \frac{1}{2}(x^2 + y^2) + x^2 y - \frac{1}{3}y^3, \tag{3.1}$$

The Hamiltonian was integrated in such a way that the trajectory passes across the x-axis 1000 times. A (x, y) projection of the orbit and the ACF of a regular (left side) and chaotic (right side) orbits for this system are shown in figure (2). Here, we set the i_{in} and i_{fin} free parameters of $ACFI$ to 1000 and 2000, respectively. The HH system's regular orbits have higher $ACFI$ values.

We first utilize the Smaller Alignment Index ($SALI$) approach, which uses the area of a parallelogram generated by two deviation vectors in the tangent space of the orbit to

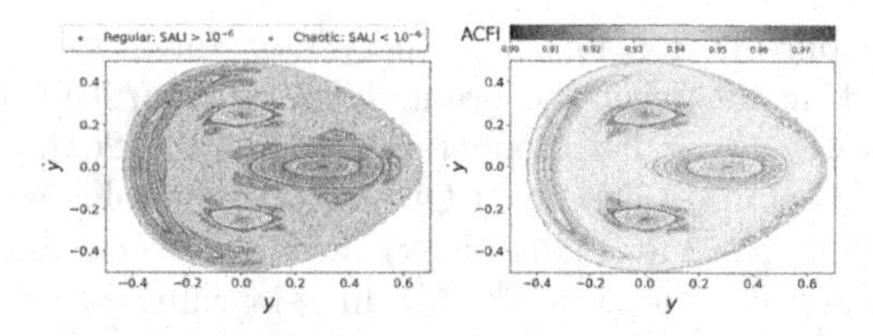

Figure 3. The 2D Hénon-Heiles system with $H = 0.125PSS$. The $SALI$ algorithm is used to classify orbits in the left panel. The PSS values of $ACFI$ are shown in the right panel.

assess the effect of applying $ACFI$ to this system. The $SALI$ method's definition is as follows:

$$\begin{aligned} SALI(t) &= min(d_-, d_+) \\ d_- &= \parallel \frac{w_1(t)}{\parallel w_1(t) \parallel} - \frac{w_2(t)}{\parallel w_2(t) \parallel} \parallel \\ d_+ &= \parallel \frac{w_1(t)}{\parallel w_1(t) \parallel} + \frac{w_2(t)}{\parallel w_2(t) \parallel} \parallel \end{aligned} \tag{3.2}$$

$w1$ and $w2$ are two deviation vectors that start off orthogonal and point in two random directions. We refer interested readers to Skokos (2001), Skokos *et al.* (2003), and Skokos *et al.* (2004) for more details on this method.

The Poincaré Surface of Section (PSS) in the $(y, \dot{y})$ plane for the Hénon-Heiles system with Hamiltonian $H = 0.125$ is shown in figure (3). For all of the orbits evaluated, we set x_0 to 0 and varied y_0 from -0.43 to 0.65 with a 0.001 step. We then changed the value of $\dot{y}_0$ from -0.5 to 0.5 with a 0.005 step and estimated the value of $\dot{x}_0$ using eq. (3.1). The PSS is made up of the spots on the x-axis where the trajectories cross. On each revolution, each orbit passes over the Poincare section twice, but only the one with positive y velocity is considered. In the PSS, a quasi-periodic orbit appears as a set of points on a smooth closed curve. Chaos, on the other hand, will result in scattered locations on the map. The $SALI$ approach clearly differentiates between regular and chaotic behavior. In our example, a value of $SALI = 10^{-6}$ is used to distinguish between the two types of motion (see left panel of Fig. (3)).

After that, we compute $ACFI$ for each orbit in the system. In the right panel of Fig. 3, we show our findings. A greater value of $ACFI$ is associated with most regular behavior, and the two techniques appear to give qualitatively identical outcomes. The advantage of $SALI$ for this system is that it can distinguish between chaotic and regular behavior with a single value, whereas the $ACFI$ distribution is more subtle. We can see that, unlike the SALI method, there is no distinct value distinguishing the regular and chaotic motions; however, a high value of this indicator can characterize the majority of the regular motion in the system. With a very low $ACFI$ value, certain small islands of regular orbits appear on the map, and the $SALI$ indicator detects these as regular zones. This might alter if integration times were extended. However, exploring these areas in greater depth is not the main focus of our research. Despite this, $ACFI$ appears to be able to recognize all of the chaotic regions in the $SALI$ PSS.

More detailed results on the theory and applications to small bodies dynamics of $ACFI$ can be found at Carruba *et al.* (2021).

Acknowledgements

We are grateful to the reviewer of this work, Dr. Cătălin Galeş, for helpful comments and suggestions. We would like to thank the Brazilian National Research Council (CNPq), that supported VC with the grant 301577/2017-0, and WB with the PIBIC grant

121889/2020-3, The Coordination for the Improvement of Higher Education Personnel (CAPES), for supporting SA with the grant 88887.374148/2019-00. RD is supported by the São Paulo Research Foundation (FAPESP, Grant 2016/024561-0). VC, RD, and WB are part of "Grupo de Dinâmica Orbital & Planetologia (GDOP)" (Research Group in Orbital Dynamics and Planetology) at UNESP, campus of Guaratinguetá. This is a publication from the MASB (Machine-learning applied to small bodies, https://valeriocarruba.github.io/Site-MASB/) research group.

4. Code availability

The Python code for identifying chaotic behavior is available at the GitHub repository: https://github.com/valeriocarruba/ACFI-Chaos-identification-through-the-autocorrelation-function-indicator.

References

Carruba V., Aljbaae S., Domingos R. C., Huaman M., Barletta W., (2021) *CMDA*, 133, A38.

Frahm KM., Fleckinger R., Shepelyansky DL. (2004) *Eu- ropean Physical Journal D* 29(1), 139.

Lewis-Swan RJ., Safavi-Naini A., Bollinger JJ., Rey AM. (2019) *Nature Communications 10*, 1581.

Pearson K. (1895) *Proceed. Royal Society of London*, Series I, 58, 240.

Pellegrini F., Montagero S. (2007) *Physical Review A* 76(5):052327.

Skokos CH. (2001) *Journal of Physics A Mathematical Gen- eral 34*, 10029.

Skokos C., Antonopoulos C., Bountis T., Vrahatis M. (2003) *Progress of Theoretical Physics Supplement* 150, 439.

Skokos C., Antonopoulos C., Bountis TC., Vrahatis MN. (2004) *Journal of Physics A Mathematical General 37*, 6269.

Skokos CH., Gottwald GA., Laskar J. (2016) *Lecture Notes in Physics, Berlin Springer Verlag* 915:E1.

Multi-scale (time and mass) dynamics of space objects
Proceedings IAU Symposium No. 364, 2022
A. Celletti, C. Galeş, C. Beaugé, A. Lemaitre, eds.
doi:10.1017/S1743921321001356

Closed-form perturbation theory in the Sun-Jupiter restricted three body problem without relegation

Irene Cavallari[1] and **Christos Efthymiopoulos**[2]

[1]Dept. of Mathematics , University of Pisa,
Pisa, 56127, Italy
email: irene.cavallari@dm.unipi.it

[2]Dept. of Mathematics Tullio Levi-Civita, University of Padua,
Padua, 35121, Italy
email: cefthym@math.unipd.it

Abstract. We present a closed-form normalization method suitable for the study of the secular dynamics of small bodies inside the trajectory of Jupiter. The method is based on a convenient use of a book-keeping parameter introduced not only in the Lie series organization but also in the Poisson bracket structure employed in all perturbative steps. In particular, we show how the above scheme leads to a redefinition of the remainder of the normal form at every step of the formal solution of the homological equation. An application is given for the semi-analytical representation of the orbits of main belt asteroids.

Keywords. celestial mechanics, methods: analytical, solar system

---

1. Introduction

In the present short presentation we summarize our work concerning the development of a normalization method in the framework of the Sun-Jupiter restricted three-body problem (R3BP) in order to represent semi-analytically the secular dynamics of a massless particle inside the planet trajectory. We search for a normal form not depending on the fast angles of the problem; using modified Delaunay variables, the latter are the mean longitudes of the particle and the planet, λ and λ_P. We will briefly describe below the main steps for the elimination of these angles in the Hamiltonian by a normalization procedure in closed form. A more detailed presentation will be given elsewhere (Cavallari & Efthymiopoulos (2021)).

In our problem, the initial Hamiltonian has two components, a leading term Z_0 not depending on λ and λ_P and a disturbing function R. The dependence of the Hamiltonian on modified Delaunay variables $(\Lambda, \Gamma, \Theta, \lambda, \gamma, \vartheta)$ and on λ_P is through the orbital elements of the particle $(a, e, i, \Omega, \omega, u)$ (with u the eccentric anomaly) and of the planet (a_P, e_P, f_P) (with f_P the planet true anomaly). We have:

$$Z_0 = -\frac{\mathcal{G}\mathcal{M}}{2a} + n_P I_P, \quad R = \mu \sum_{s=s_0}^{+\infty} \varepsilon^s R_s^{(0)}(a, e, i, \Omega, \omega, u, f_P; e_P, a_P), \quad s_0 \in \mathbb{Z}^+,$$

where n_P is the planet mean motion and I_P is the dummy action variable conjugated to λ_P; $\mathcal{M}$ is the mass of the Sun and $\mu = \mathcal{G} m_P$, with m_P the mass of Jupiter; ε is a so-called *book-keeping* parameter, i.e. a formal parameter with numerical value $\varepsilon = 1$ whose powers keep track of the relative size of each perturbing term in R.

In perturbation theory the normal form is typically computed in an iterative way through a composition of Lie transformations (see Deprit (1969)). At the j-th iteration the term $R^{(j-1)}_{s_0+j-1}$ (of book-keeping order s_0+j-1) of the Hamiltonian $\mathcal{H}^{(j-1)}$, computed at the previous step, is normalized by means of a Lie generating function $\chi^{(j)}_{s_0+j-1}$, determined by solving the homological equation

$$\mathcal{L}_{\chi^{(j)}_{s_0+j-1}}(Z_0)+\varepsilon^{s_0+j-1}R^{(j-1)}_{s_0+j-1}=\varepsilon^{s_0+j-1}Z_{s_0+j-1}, \tag{1.1}$$

with

$$\mathcal{L}_{\chi^{(j)}_{s_0+j-1}}(Z_0)=-\left(n\frac{\partial}{\partial\lambda}+n_P\frac{\partial}{\partial\lambda_P}\right)\chi^{(j)}_{s_0+j-1},$$

where $n=\sqrt{\mathcal{GM}/a^3}$ and $\mathcal{L}_{\chi^{(j)}_{s_0+j-1}}=\{\cdot,\chi^{(j)}_{s_0+j-1}\}$ is the Poisson bracket operator. Solving (1.1) in our problem can be complicated since the initial disturbing function R depends on λ and λ_P through u and f_P, which implies solving Kepler equation in series form to obtain the required trigonometric expansions. Two techniques are typically used to overcome this difficulty. One consists of approximating the original Hamiltonian by means of a Taylor expansion in the eccentricities e, e_P, to make explicit the trigonometric dependence on the fast angles (see Brouwer & Clemence (1961) and Kaula (1966)). The drawback of this technique is that it can be used only for low values of the eccentricity. A second technique, introduced in Palacián (1992) and formalized in Segerman & Coffey (2000); Deprit & Palacián (2001), is the so-called relegation method: since $n_P<n$ for our problem, the term $n_P I_P$ of the leading term is neglected in equation (1.1), so that this can be solved in closed form. We refer to Sansottera & Ceccaroni (2017) for a discussion about the algorithm convergence, and to Lara (2021) for more general references on closed-form perturbative techniques.

Here, we propose a closed-form normalization method alternative to relegation, which is suitable for orbits with relatively high values of the eccentricity. A method similar to ours was introduced in Lara et al. (2013), referring to the motion of a test particle under a multiple expansion of the geopotential (e.g. with J_2 and C_{22} terms). In summary, the method works as follows. We perform a multipolar (Legendre) expansion of the initial disturbing function and we expand the semi-major axis as $a=a^*+\delta a$, where a^* is a reference value characteristic of each considered individual trajectory. This last step aims at having constant frequencies in the leading term Z_0, which turns out to be useful for algorithmic convenience purposes (see below). The starting Hamiltonian takes the form

$$\mathcal{H}^{(0)}=Z_0+R,\quad Z_0=n^*\delta\Lambda+n_P I_P,\quad R=\sum_{s=s_0}^{+\infty}\varepsilon^s R^{(0)}_s. \tag{1.2}$$

The book-keeping parameter ε separates terms in groups of different order of smallness, depending on four small quantities: e, e_P, $\delta\Lambda$ and the ratio between the planet and the Sun masses. To overcome the difficulty of solving (1.1), the main idea is, now, to accept a remainder generated by the homological equation (to be normalized at successive steps):

$$\mathcal{L}_{\chi^{(j)}_{s_0+j-1}}(Z_0)+\varepsilon^{s0+j-1}R^{(j-1)}_{s_0+j-1}=\varepsilon^{s_0+j-1}Z_{s_0+j-1}+\mathcal{O}(\varepsilon^{s_0+j}) \tag{1.3}$$

where Z_{s_0+j-1} contains the terms of $R^{(j-1)}_{s_0+j-1}$ not depending on λ and λ_P.

The steps to perform in order to apply the normalization method are described in Section 2. In Section 3, we discuss the applicability of the method in the case of the planar elliptic R3BP.

2. Normalization Method

2.1. *Hamiltonian preparation*

We consider a heliocentric inertial reference frame with the $\widehat{x}$ axis and $\widehat{z}$ axis parallel to the planet orbital eccentricity vector and the angular momentum respectively. Let $\mathbf{r}$ be the heliocentric position vector, $r=|\mathbf{r}|$, $\mathbf{p}$ the vector of conjugated momenta, $p=|\mathbf{p}|$ and $\mathbf{r}_P$ the planet position vector, $r_P=|\mathbf{r}_P|$; the Hamiltonian of the R3BP is

$$H=\frac{p^2}{2}-\frac{\mathcal{GM}}{r}+\mathcal{R},\qquad \text{with}\quad \mathcal{R}=-\mu\left(\frac{1}{\sqrt{r^2+r_P^2-2\mathbf{r}\cdot\mathbf{r}_P}}-\frac{\mathbf{r}\cdot\mathbf{r}_P}{r_P^3}\right)$$

The following operations must be performed:

(*a*) *Multipolar Expansion:*

$$\mathcal{R}\sim\mathsf{R}=-\frac{\mu}{r_P}\sum_{j=2}^{o}\frac{r^j}{r_P^j}P_j(\cos\alpha),\quad \text{with}\quad \cos\alpha=\frac{\mathbf{r}\cdot\mathbf{r_P}}{rr_P}$$

where $P_j(\cdot)$ are Legendre polynomials;

(*b*) *Extended Hamiltonian:* the Hamiltonian is expressed as an implicit function of the modified Delaunay variables by means of the orbital elements of the particle and of the planet. A dummy action variable I_P, conjugated to λ_P, is introduced. We get:

$$\mathsf{H}=-\frac{\mathcal{GM}}{2a}+n_PI_P+\mathsf{R},\quad \text{with}\quad \mathsf{R}=\mathsf{R}(a,e,i,\omega,\Omega,u,f_P;a_P,e_P).$$

(*c*) *Expansion of the semi-major axis:* Considering that $a=\frac{\Lambda^2}{\mu}$, we perform the expansion of the semi-major axis as

$$a=a^*+\frac{2\delta\Lambda}{a^*n^*}+\dots,\qquad n^*=\sqrt{\frac{\mathcal{GM}}{a^{*3}}}.$$

Thus, we obtain

$$\mathcal{H}=n_PI_P+n^*\delta\Lambda-\frac{3}{2}\frac{\delta\Lambda^2}{a^{*2}}+\dots+\mathsf{R}(a^*+\frac{2\delta\Lambda}{a^*n^*}+\dots,e,i,\omega,\Omega,u,f_P;a_P,e_P).$$

(*d*) *RM-reduction:* using the identity $r=a(1-e\cos u)$, we re-write $\mathcal{H}$ as

$$\mathcal{H}=n_PI_P+n^*\delta\Lambda+\Bigg(-\frac{3}{2}\frac{\delta\Lambda^2}{a^{*2}}+\dots$$
$$+\mathsf{R}\left(a^*+\frac{2\delta\Lambda}{a^*n^*}+\dots,e,i,\omega,\Omega,u,f_P;a_P,e_P\right)\Bigg)Q$$

$$\text{with}\quad Q=\frac{a(1-e\cos u)}{r}=\frac{a^*(1-e\cos u)}{r}+2\frac{(1-e\cos u)}{a^*n^*r}\delta\Lambda+\dots=1.\tag{2.1}$$

After performing RM-reduction, the Hamiltonian $\mathcal{H}$ is expressed as a sum of trigonometric monomials $\cos(k_1u+k_2f_P+k_3\omega+k_4\Omega)$, all divided by r. As discussed below, such a form allows for a straightforward calculation of the solution of the homological equation yielding the generating function at each normalization step.

2.2. *Book Keeping*

To write the initial Hamiltonian in the form (1.2), we use the book-keeping parameter ε to keep track of the relative size of the several terms of $\mathcal{H}$. There are four different small parameters to consider in the problem: μ, $\delta\Lambda$, e and e_P. We define $s_0 = \left\lceil \frac{\log\left(\frac{m_P}{M}\right)}{\log(e)} \right\rceil$ and we adopt the following book-keeping rules:

- terms depending on $e^j e_P^k$ $(j, k \in \mathbb{Z})$ are multiplied by $\varepsilon^{(j+k)}$;
- terms depending on $(1+\eta)^j$ and $(1-\eta)^k$, with $\eta = \sqrt{1-e^2}$, $(j, k \in \mathbb{N})$ are multiplied by ε^0 and ε^{2k} respectively;
- terms depending on $\mu^j \delta\Lambda^k$ $(j, k \in \mathbb{N})$ are multiplied by $\varepsilon^{(j+k)s_0}$;
- terms depending on $\delta\Lambda^k$ $(k \in \mathbb{N})$ are multiplied by $\varepsilon^{(k-1)s_0}$.

Finally, the value of s_0 is specified, for any particular trajectory, by the initial value $e(0)$.

2.3. *Poisson bracket structure*

During the normalization process, we need to compute Poisson brackets of the form $\{A_1, A_2\}$, where A_1 and A_2 are implicit functions of $(\delta\Lambda, \Gamma, \Theta, I_P, \lambda, \gamma, \vartheta, \lambda_P)$ through the variables $(e, i, \omega, \Omega, u, f_P)$, r, $\eta = \sqrt{1-e^2}$, $\phi = u - M$ (with M the mean anomaly). To compute $\{A_1, A_2\}$, we use the formula:

$$\begin{aligned}\{A_1, A_2\} = {} & \frac{\partial A_1}{\partial \lambda}\frac{\partial A_2}{\partial \delta\Lambda} - \frac{\partial A_1}{\partial \delta\Lambda}\frac{\partial A_2}{\partial \lambda} + \frac{\partial A_1}{\partial \gamma}\frac{\partial A_2}{\partial \Gamma} - \frac{\partial A_1}{\partial \Gamma}\frac{\partial A_2}{\partial \gamma} + \frac{\partial A_1}{\partial \vartheta}\frac{\partial A_2}{\partial \Theta} - \frac{\partial A_1}{\partial \Theta}\frac{\partial A_2}{\partial \vartheta} \\ & + \left(\frac{\partial A_1}{\partial \lambda_P}\frac{\partial A_2}{\partial I_P} - \frac{\partial A_1}{\partial I_P}\frac{\partial A_2}{\partial \lambda_P}\right) Q.\end{aligned} \tag{2.2}$$

To evaluate the partial derivates with respect to the modified Delaunay variables in the formula, we must perform a composition of partial derivates. The last term is multiplied by Q defined in (2.1) to allow significant simplifications to be carried out automatically during the normalization process. For the same reason, the following relations must be used for any $A = A_{1,2}$ in the computation of the partial derivatives:

$$\frac{\partial e}{\partial \delta\Lambda} = -\frac{\eta e}{(1+\eta)n^* {a^*}^2}\varepsilon + \mathcal{O}(\varepsilon^{s_0}\delta\Lambda), \quad \frac{\partial e}{\partial \Gamma} = -\frac{\eta}{{a^*}^2 n^* e}\varepsilon^{-1} + \mathcal{O}(\varepsilon^{s_0}\delta\Lambda),$$
$$\frac{\partial \eta}{\partial \delta\Lambda} = \frac{1-\eta}{n^* {a^*}^2}\varepsilon^2 + \mathcal{O}(\varepsilon^{s_0}\delta\Lambda),$$

$$\frac{\partial \eta}{\partial \Gamma} = \frac{1}{{a^*}^2 n^*} + \mathcal{O}(\varepsilon^{s_0}\delta\Lambda), \quad \frac{\partial \cos i}{\partial \delta\Lambda} = \frac{1-\cos i}{{a^*}^2 n^* \eta} + \mathcal{O}(\varepsilon^{s_0}\delta), \quad \frac{\partial \cos i}{\partial \Gamma} = \frac{\cos i - 1}{{a^*}^2 n^* \eta} + \mathcal{O}(\varepsilon^{s_0}\delta\Lambda),$$

$$\frac{\partial \cos i}{\partial \Theta} = -\frac{1}{{a^*}^2 n^* \eta} + \mathcal{O}(\varepsilon^{s_0}\delta\Lambda), \quad \frac{\partial u}{\partial \lambda} = \frac{a^*}{r} + \mathcal{O}(\varepsilon^{s_0}\delta\Lambda), \quad \frac{\partial u}{\partial \gamma} = \frac{a^*}{r} + \mathcal{O}(\varepsilon^{s_0}\delta\Lambda),$$
$$\frac{\partial \phi}{\partial \lambda} = \frac{a^*}{r} - 1 + \mathcal{O}(\varepsilon^{s_0}\delta\Lambda),$$

$$\frac{\partial \phi}{\partial \gamma} = \frac{a^* e \cos u}{r}\varepsilon + \mathcal{O}(\varepsilon^{s_0}\delta\Lambda), \quad \frac{\partial u}{\partial \delta\Lambda} = \frac{\eta e \sin u}{r(1+\eta)n^* a^*}\varepsilon + \mathcal{O}(\varepsilon^{s_0}\delta\Lambda),$$
$$\frac{\partial u}{\partial \Gamma} = \frac{\eta \sin(u)}{a^* n^* e r}\varepsilon^{-1} + \mathcal{O}(\varepsilon^{s_0}\delta\Lambda),$$

$$\frac{\partial \phi}{\partial \delta\Lambda} = \frac{\eta e \sin u}{r(1+\eta)n^* a^*}\varepsilon + \mathcal{O}(\varepsilon^{s_0}\delta\Lambda), \quad \frac{\partial \phi}{\partial \Gamma} = \frac{\eta \sin(u)}{a^* n^* e r}\varepsilon^{-1} + \mathcal{O}(\varepsilon^{s_0}\delta\Lambda),$$

$$\frac{\partial \phi}{\partial \delta\Lambda} = \frac{\eta e \sin u}{r(1+\eta)n^* a^*}\varepsilon + \mathcal{O}(\varepsilon^{s_0}\delta\Lambda),$$

$$\frac{\partial \phi}{\partial \Gamma} = \frac{\eta \sin(u)}{a^* n^* e r}\varepsilon^{-1} + \mathcal{O}(\varepsilon^{s_0}\delta\Lambda), \quad \frac{\partial r}{\partial \lambda} = \frac{{a^*}^2 e \sin(u)}{r}\varepsilon + \mathcal{O}(\varepsilon^{s_0}\delta\Lambda),$$

$$\frac{\partial r}{\partial \gamma} = \frac{{a^*}^2 e \sin(u)}{r}\varepsilon + \mathcal{O}(\varepsilon^{s_0}\delta\Lambda),$$

$$\frac{\partial r}{\partial \delta\Lambda} = \frac{\eta e \cos u}{(1+\eta)n^* a^*}\varepsilon + \mathcal{O}(\varepsilon^{s_0}\delta\Lambda), \quad \frac{\partial r}{\partial \Gamma} = \frac{\eta(e\varepsilon - cos(u))}{n^* e r}\varepsilon^{-1} + \mathcal{O}(\varepsilon^{s_0}\delta\Lambda), \quad \frac{\partial \omega}{\partial \gamma} = -1, \quad \frac{\partial \omega}{\partial \vartheta} = 1,$$

$$\frac{\partial \Omega}{\partial \vartheta} = -1, \quad \frac{\partial f_P}{\partial \lambda_P} = 1 + \frac{2e_P \cos(f_P)}{\eta_P^3}\varepsilon + \left(\frac{1}{\eta_P^3} - 1 + \frac{e_P^2 \cos^2(f_P)}{\eta_P^3}\right)\varepsilon^2.$$

The fact that the book-keeping parameter ε is present in all partial derivatives of terms depending on e, e_P and $\delta\Lambda$ is an essential element of the method. We can readily see that, for every case in which $s_0 > 1$, the result of Poisson brackets between such terms contains terms of different powers of ε, which, however, are always larger than the current normalization order (for the case $s_0 = 1$, instead, see Cavallari & Efthymiopoulos (2021)).

2.4. *Normalization Process*

At the j-th iteration of the normalization process, we must determine the generating function $\chi^{(j)}_{s_0+j-1}$ satisfying the homological equation (1.3). The remainder term $R^{(j-1)}_{s_0+j-1}$ to normalize contains four different types of terms:

- type 1: $\frac{a^*}{r} f(e, i, \eta, \omega, \Omega)$,
- type 2: $\frac{a^*}{r} \widehat{f}_{\mathbf{k}}(e, i, \eta) \cos(k_1 u + k_2 f_P + k_3 \omega + k_4 \Omega)$,
- type 3: $\frac{a^*}{r^p} \bar{f}(e, i, \eta, \omega, \Omega)$, $p > 1$,
- type 4: $\frac{a^*}{r^p} \tilde{f}_{\mathbf{k}}(e, i, \eta) \cos(k_1 u + k_2 f_P + k_3 \omega + k_4 \Omega)$, $p > 1$.

The corresponding terms to be added in $\chi^{(j)}_{s_0+j-1}$ are

- for type 1: $\frac{1}{n^*} f(e, i, \eta, \omega, \Omega)\phi$,
- for type 2: $\frac{1}{k_1 n^* + k_2 n_P} \widehat{f}_{\mathbf{k}}(e, i, \eta) \cos(k_1 u + k_2 f_P + k_3 \omega + k_4 \Omega)$,
- for type 3: $\frac{1}{n^*} \phi \sum_{k=1}^{p} \frac{\bar{f}(e, \eta, i, \omega, \Omega)}{a^{k-1} r^{p-k}}$, $p > 1$,
- for type 4: $\frac{1}{k_1 n^* + k_2 n_P} \frac{1}{r^{p-1}} \tilde{f}_{\mathbf{k}}(e, \eta, i) \sin(k_1 u + k_2 f_P + k_3 \omega + k_4 \Omega)$, $p > 1$.

The new Hamiltonian is

$$\mathcal{H}^{(j)} = exp(\mathcal{L}_{\chi^{(j)}_{s_0+j-1}})\mathcal{H}^{(j-1)} = Z_0 + \sum_{s=s_0}^{s_0+j-1} \varepsilon^s Z_s + \sum_{s=s_0+j}^{r} \varepsilon^s R^{(j)}_s.$$

where $\exp(\mathcal{L}_\chi)$ denotes the operation $\exp(\mathcal{L}_\chi) = \sum_{k=0}^{\infty} \frac{1}{k!}\mathcal{L}^k_\chi$, with $\mathcal{L}^k_\chi = \underbrace{\{\dots\{\{\cdot, \chi\}, \chi\}\dots, \chi\}}_{k\,\text{times}}$. The normal form terms Z_s are independent of the fast angles λ, λ_P, while the remainder terms $R^{(j)}_s$ can be normalized at successive normalization steps.

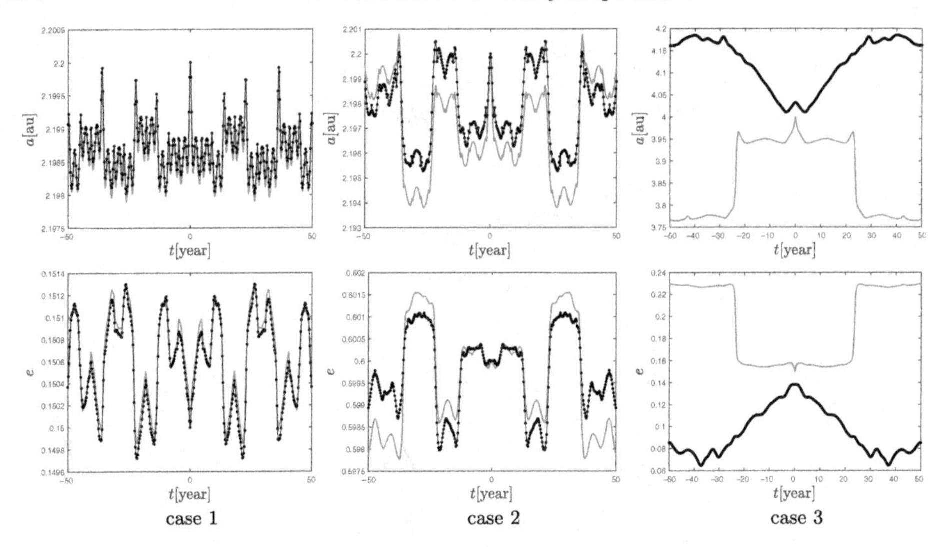

Figure 1. Evolution of the semi-major axis and of the eccentricity in the framework of the Sun-Jupiter restricted three-body problem. We select orbits with the following initial conditions: (case 1) $a(0) = 2.2$au, $e(0) = 0.15$; (case 2) $a(0) = 2.2$au, $e(0) = 0.5$; (case 3) $a(0) = 4$au, $e(0) = 0.15$. For all of them we set $a^* = a(0)$ and $\omega(0) = 90°$, $M(0) = 90°$, $i(0) = 20°$, $\Omega(0) = 0°$, $\lambda_P(0) = 0°$. We show the evolution computed by means of a numerical propagation of the trajectory (grey line) and semi-analytically through the normal form computed through the algorithm of section 2 (black line).

3. Results

The error of the method increases with an increase of the semi-major axis or the eccentricity. The trend with respect to the semi-major axis has two main causes: first of all, increasing a, we obtain trajectories which arrive closer to the planet and to the region of Hill-unstable orbits. The other reason is related to the multipolar expansion: as a increases, the heliocentric distances of the particle and of the planet become comparable. Dealing with this problem requires performing a multipolar expansion of high order, which, however, implies a substantial increase in the number of terms to normalize. Similarly, reducing the error for highly eccentric orbits requires performing a large number of normalization steps. However, the consequent increase of the computational time can become significant. In the case of the Sun-Jupiter planar elliptic R3BP, performing a multipolar expansion of order equal to 5 and doing from 4 to 7 normalization steps allows to obtain accurate results up to an initial semi-major axis $a \leqslant 0.6a_P$ and for relatively high values of e, up to almost 0.7.

Figure 1 shows the evolution of the semi-major axis and of the eccentricity in three cases: a trajectory with a low value of the eccentricity, one with a relatively high value of eccentricity and one with a semi-major axis larger than $0.6a_P$. We compare the outcomes obtained semi-analytically through our normal form algorithm with those resulting from a numerical propagation. The method works well in the first two cases, while the errors increase in general with e. In fact, in case 1 the maximum relative error is $\sim 10^{-4.6}$ for the semi-major axis and $\sim 10^{-3.34}$ for the eccentricity, while in case 2 it is $\sim 10^{-3.5}$ for the semi-major axis and $\sim 10^{-2.75}$ for the eccentricity. Instead, in case 3 the method is not sufficiently precise: the maximum relative error is $\sim 10^{-0.97}$ and $\sim 10^{-0.2}$ for the semi-major axis and for the eccentricity respectively. More detailed examples and results are contained in Cavallari & Efthymiopoulos (2021).

Acknowledgements

I.C. has been supported by the MSCA-ITN Stardust-R, Grant Agreement n. 813644 under the H2020 research and innovation program. C.E. acknowledges the support of MIUR-PRIN 20178CJA2B "New frontiers of Celestial Mechanics: theory and applications".

References

Brouwer D., Clemence G. M. 1961, Methods of celestial mechanics, *Academic Press*
Cavallari I., Efthymiopoulos 2021, in preparation
Deprit, André 1969, *CM&DA*, 1(1), 12–30
Deprit A., Palacián J., Deprit E. 2001, *CM&DA*, 79(3), 157–182
Kaula W.M. 1966, Theory of Satellite Geodesy: Applications of Satellites to Geodesy, *Blaisdell Publishing Company*
Lara M., San-Juan J.F., López-Ochoa L.M. 2013, *Math. Probl. Eng.*, 2013, 1–11
Lara M. 2021, Hamiltonian Perturbation Solutions for Spacecraft Orbit Prediction, *De Gruyter*
Palacián J. F. 1992, *Ph.D. thesis, Universidad de Zaragoza*
Palacián J. F. 2002, *J. Differ. Equ.*, 180(2), 471–519
Sansottera M. and Ceccaroni M. 2017, *CM&DA*, 127(1), 1–18
Segerman A.M. and Coffey S.L. 2000, *CM&DA*, 76(3), 139–156

Multi-scale (time and mass) dynamics of space objects
Proceedings IAU Symposium No. 364, 2022
A. Celletti, C. Galeş, C. Beaugé, A. Lemaitre, eds.
doi:10.1017/S1743921321001319

The current orbit of Atlas (SXV)

Demétrio Tadeu Ceccatto[1], Nelson Callegari Jr.[1] and Adrián Rodríguez[2]

[1]São Paulo State University (UNESP), Institute of Geosciences and Exact Sciences, Av 24-A 1515, 13506-900, Rio Claro, Brazil
email: dt.ceccatto@unesp.br

[2]Observatório do Valongo, Universidade Federal do Rio de Janeiro, Ladeira do Pedro Antônio 43, 20080-090 Rio de Janeiro, Brazil

Abstract. With the success of the Cassini-Huygens mission, the dynamic complexity surrounding natural satellites of Saturn began to be elucidated. New ephemeris could be calculated with a higher level of precision, which made it possible to study in detail the resonant phenomena and, in particular, the 54:53 near mean-motion resonance between Prometheus and Atlas. For this task, we have mapped in details the domains of the resonance with dense sets of initial conditions and distinct ranges of parameters. Our initial goal was to identify possible regions in the phase space of Atlas for which some critical angles, associated with the 54:53 mean motion have a stable libration. Our investigations revealed that there is no possibility for the current Atlas orbital configuration to have any regular behavior since it is in a chaotic region located at the boundary of the 54:53 mean-motion resonance phase space. This result is in accordance with previous works (Cooper et al. 2015; Renner et al. 2016). In this work, we generalize such investigations by showing detailed aspects of the Atlas-Prometheus 54:53 mean-motion resonance, like the extension of the chaotic layers, the thin domain of the center of the 54:53 resonance, the proximity of other neighborhood resonances, among other secondary conclusions. In particular, we have also shown that even in the deep interior of the resonance, it is difficult to map periodic motion of the resonant pair for very long time spans.

Keywords. Resonant dynamics, mean-motion, dynamical maps, Atlas, Prometheus, Saturn.

---

1. Introduction

The natural satellites of Saturn display a variety of orbital configurations and unique topological features that have intrigued astronomers, physicists and mathematicians for several years (e.g. Peale 1999). Such dynamic environment is responsible for the most diverse gravitational disturbances such that tides, short and long periodic perturbations. In particular, small satellites with their mean radius of the order of ten kilometers or smaller suffer non-negligible quasi-periodic variations in their orbits due to mutual gravitational interactions and, this phenomena imply that many pairs of satellites have their orbits close to commensurability of the mean-motions. In this work, we study the dymanics of Atlas. According to Thomas and Helfenstein (2020), Atlas is a small satellite with radius of 14.9 km orbiting near the outer edge of Ring A, see Table 2.

Spitale et al. (2006) combining data previously obtained by the Voyager spacecraft, the Hubble Space Telescope and ground-based telescopes determined precise ephemeris for these small worlds, in particular for Atlas. In their studies, Spitale et al. (2006) suggested that Atlas orbit is perturbed due to mutual gravitational interaction with Prometheus generating a 54:53 resonant mean-motion and this can be associated with the presence of the chaotic motion.

Table 1. Physical constants for Saturn obtained from Cooper et al. (2015)[a], *Horizons*[b], J_2, J_4 and J_6 a dimensionless coefficients of Saturn potential expansion.

Constant	Numerical value
GM^a ($Km^3 s^{-2}$)	3.7931208×10^7
Equatorial radius[b] (Km)	60268 ± 4
J_2^a	16290.71×10^{-6}
J_4^a	-935.83×10^{-6}
J_6^a	86.14×10^{-6}

Cooper et al. (2015) investigating the existence of the chaotic motion of Atlas suggested by Spitale et al. (2006) by calculating the Fast Lyapunov Indicator (FLI) method to verify the presence of chaos in the Atlas orbit. They analyzed the critical angles associated with the 54:53 resonant mean-motion for a timespan of 20 years. Their integrations showed that the angle associated with the Corotation Eccentric Resonance (ϕ_{CER}) oscillates with temporay oscillation with period about 4.92 years and amplitude around 180° followed by circulation intervals. In contrast, the angle associated with Lindblad Eccentric Resonance (ϕ_{LER}) has temporary oscillation of about 3 years followed by a long period of circulation (see Figure 2 of Cooper et al. 2015).

In an attempt to determine the origin of this alternation between libration and circulation, Renner et al. (2016) studied the Atlas orbit using the CoraLin model proposed by (El Moutamid et al. (2014). Your results show that the region of the space occupied by the CER resonance is superimposed by the chaotic region.

Following the contributions of Cooper et al. (2015) and Renner et al. (2016), we will investigate the Atlas orbit looking for initial conditions that enable stability for the libration of the ϕ_{CER} and ϕ_{LER} angles. In additions, we identiffy possible causes for instability described in Cooper et al. (2015).

2. Methodology

In this work, we follow the methodology given in Callegari and Yokoyama (2010, 2020) and Callegari et al. (2021) and numerically integrate the exact equations of the motion for a system of N satellites mutually disturbed orbiting under the action of the main terms of the Saturn's potential expanded up to second order. In our simulations, for brevity we show the main results considering only $N = 2$, that is, a system formed by Atlas and Prometheus. The justification for this choice stems from Atlas orbital elements can well be determined over the influence of Prometheus, without considering the other effects arising from the perturbation caused by Pandora or other companion satellite (Spitale et al. 2006; Cooper et al. 2015; Renner et al. 2016). It is worth to note that we have considering more general models in this work, but for brevity only the result within the domains of the three-body problem are shown here.

Two different models were used: i) the system of exact differential equation (Eq. 1–5) present in Callegari and Yokoyama (2010), where the equations of motion are integrated under the influence of the terms J_2 and J_4 and; ii) direct application of the Mercury package (Chambers 1999) with the addition of the term J_6. In both cases i) and ii) we apply the Everhart code "RA15" to solve systems of ordinary differential equations (Everhart 1985).

Physical parameters for Saturn and initial conditions for Atlas and Prometheus provided by the *Horizons* system of ephemerides (`http://sdd.jpl.nasa.gov/horizons.cgi`) are listed in Tables 1 and 2, respectively. The masses were obtained from Thomas and Helfenstein (2020) and the details for the numerical simulations are described in the caption of the respective figure.

Table 2. Mass and osculating orbital elements at Epoch 2000 January 1 00: 00: 00.0000 UTC computed from Ephemeris system - *Horizons*, in December 13, 2019.

	Mass (Kg)	*a (Km)*	*e*	I (°)	ω (°)	Ω (°)	n (°/*day*)
Atlas	5.75×10^{15}	138325.32	0.00591	0.00419	200.7	235.45	592.63
Prometheus	1.5×10^{16}	140246.44	0.00252	0.00801	201.36	309.14	581.87

Often, some oscillating elements obtained through numerical integration can be highly influenced by the term J_2, causing a fast frequency component in the pericenter (ω), a fact that makes difficult its interpretation (Greenberg 1981). Thus, as an additional effort, we calculate the geometric elements through the direct application of the algorithm described by Renner and Sicardy (2006).

3. The current orbit of Atlas and the 54:53 Prometheus-Atlas mean-motion

Atlas finds itself orbiting a region of dynamic complexity. The understanding of the 54:53 resonant mean-motion make it necessary to determine the stability of its orbit.

3.1. *The critical angles ϕ_{CER} and ϕ_{LER}*

Resonant phenomena are common among satellites of Saturn, in particular the Corotation Eccentricity Resonance (CER) and Lindblad Eccentricity Resonance (LER). According to Callegari et al. (2021) the CER resonance occurs when the conjunction between disturber (larger body) and particle (smaller body) happens near the pericenter of the disturbed one, on the other hand, the LER resonance occurs when the conjunction occurs near the pericenter of the particle. The knowledge of the physical behavior of these two angles makes it possible to predict where the conjunctions between the two bodies occur, in addition to verifying the orbital stability of the system. For the 54:53 resonant mean-motion between Prometheus and Atlas, the angles ϕ_{CER} and ϕ_{LER} can be written as an angular combination of the form $\phi_{CER} = 54\lambda_P - 53\lambda_A - \varpi_P$ and $\phi_{LER} = 54\lambda_P - 53\lambda_A - \varpi_A$, where $\lambda = \varpi + l$ represents the mean longitude, ϖ the pericenter longitude, l the mean anomaly, whereas A and P represent Atlas and Prometheus, respectively. Physically, the libration of the ϕ_{CER} angle around 180° means that the conjunctions occur close to the apocenter of Prometheus. In the case of the current orbit of Atlas, Fig. 1 shows that there are periods of alternation between oscillation and circulation of ϕ_{CER}.

On the other hand, the ϕ_{LER} angle has a brief period of oscillation, about 5,000 days, followed by a long period of circulation, as shown in Fig. 1.

Cooper et al. (2015) state that Atlas could be added to the list of satellites that have both critical angles librating, however, for a satellite similar to Atlas, we cannot make such an affirmation, as we do not obtain stability for the dynamics of these angles as it occurs for Anthe (Callegari and Yokoyama 2020) and Methone (Callegari et al. 2021). See also El (El Moutamid et al. (2014).

Expanding the investigation done by Cooper et al. (2015) we will look for a possible orbital configuration that can guarantee the stability of the libration of ϕ_{CER} and ϕ_{LER}. For this task, we will use the phase space mapping through Fourier spectra, a methodology described in Callegari and Yokoyama (2010, 2020) and Callegari et al. (2021).

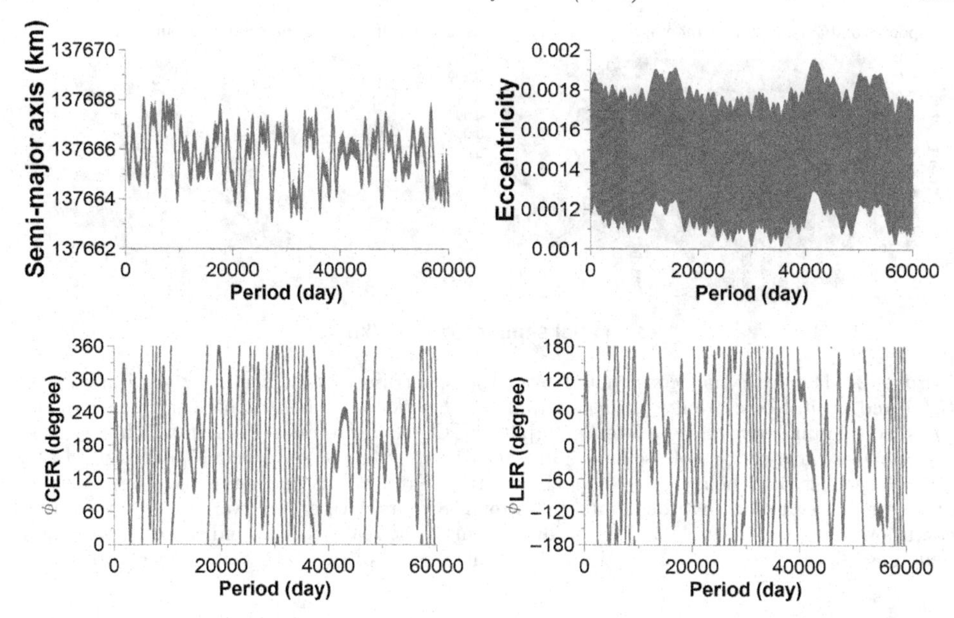

Figure 1. Simulation performed with the initial conditions given in Table 2 for an Atlas-like satellite. Semi-major axis and eccentricity geometric elements and critical angles in geometric elements associated with CER and LER type resonances. Note the lack of stability for the oscillation of ϕ_{CER} and ϕ_{LER}. The equations were integrated for 60,000 days with a 0.06 day step using the Mercury package.

3.2. *The mapping of the 54:53 Prometheus-Atlas resonant mean-motion*

We will now explore the resonant phase space domain. The mapping is carried out considering the domain of frequencies obtained with the Fourier spectrum for a set of numerically integrated orbits, whose initial conditions are close to the real orbit of Atlas.

Callegari and Yokoyama (2010, 2020) and Callegari et al. (2021) analyze the phase space around the orbital neighborhood of a satellite according to the value of the spectral number `N`. For the construction of this `N` it is considered an established value, usually 5% of the reference amplitude for the Fourier spectrum for each of the individual orbits of these test satellites. Then, `N` provides the number of peaks in the spectrum that are greater than or equal to the pre-set reference value.

We will now investigate the dependence of the resonant motion with the semi-major axis and the eccentricity for an Atlas-like satellite. Fig. 2 shows two dynamic maps built on a dense grid of initial conditions for (a_0,e_0). The red star represents the initial condition (a_0,e_0)=(138325.32 km, 0.0059) for Atlas on the initial date (see Table 2).

The map at the left in Fig. 2 represents the phase space of the spectral domain for the semi-major axis and the map on the right represents the spectral domain for the orbital eccentricity. We can identify four distinct regions:

a) A light "blue eye" region, with `N` less than 30, in the range [138321 km, 138326 km]x[0.004, 0.006] indicated by **C**. In this region all angles ϕ_{CER} are oscillating around 180° but the alternation occurs between the oscillation and circulation. However, for the initial conditions (138322 km, 0.004) blue disc and (138323 km, 0.005) magenta disc, in Fig. 2, the angle ϕ_{CER} is librating, see Fig. 3 items (b), (e) and (f). This would probably be the region of the corotation zone associated to the 54:53 Prometheus-Atlas mean-motion resonance, but, the absence of stability

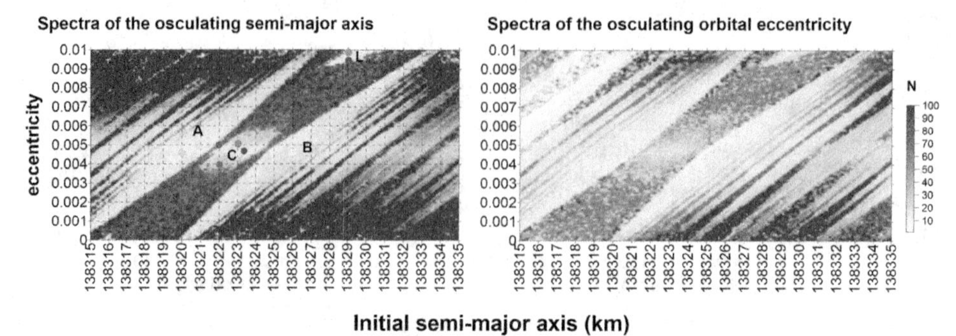

Figure 2. Dynamic map constructed from the spectrum of the osculating semi-major axis (left) and orbital eccentricity (right) of 15,000 Atlas-like satellites (numerical scheme i). `N` is the spectral number with 5% reference amplitude. **C** and **L** are the regions in which ϕ_{CER} and ϕ_{LER} angles have periods of oscillation of approximately 30,000 days and 15,000, respectively, but alternating their behavior. For this simulation, Prometheus and Atlas were used, the set of differential equations had been integrated for 258 years with an interval of 0.18 days. Initial conditions are given in Table 2. The initial element of the real Atlas are indicated by a red star. Different full discs correspond to initial conditions (a_0,e_0) of the orbits shown in Figure 4.

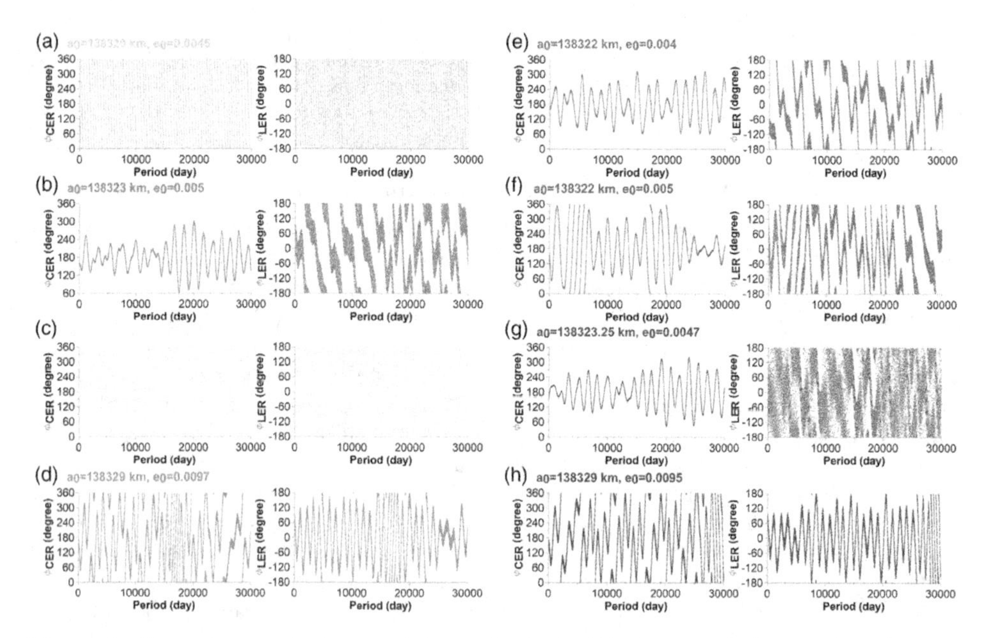

Figure 3. Geometrics angle ϕ_{CER} and ϕ_{LER} obtained with the numerical simulations (numerical scheme ii) corresponding to the colored discs in Fig. 2. In a) and c) both are circulating; b) e) and g) the angle ϕ_{CER} are librating and ϕ_{LER} circulating; and g) alternation between libration and circulation of angle ϕ_{CER} and circulation of angle ϕ_{LER}; d) and h) libration of angle ϕ_{LER} and circulation of angle ϕ_{CER}.

in the libration for the other initial conditions, Fig. 3 (f), does not allow this region to be considered as such.

b) The dark blue region, which fills practically the entire map, where the spectrum of the test satellites has a large value for `N`. In general, such behavior can be associated with chaotic orbits and strong or irregular disturbances in their motions (Callegari and Yokoyama 2020).
c) Region between [138327 km, 138330 km]x[0.009, 0.01], indicated by **L**. Location where the angle ϕ_{LER} shows oscillation around 0°. The absence of stability in the libration does not allow this region to be considered as such, see Fig. 3 items (d) and (g).
d) Bands A and B: in this region angles ϕ_{CER} and ϕ_{LER} show themselves circulating in a retrograde direction for a brief period of time, about 500 days (band A) and for region B, the angles are circulating in a prograde direction with a period of approximately 400 days. See Fig. 3 items (a) and (c).

We can verify, with the help of the dynamic map given in Figure 2, that there is no region with (a_0,e_0) in which there is stability for the critical angles ϕ_{CER} and ϕ_{LER} libration, neither for both critical angles.

We can observe that for the initial conditions given in Table 2, Atlas is found close to the edge of a possible chaotic region and this is probably responsible for the alternations observed in the angles ϕ_{CER} and ϕ_{LER} observed in Cooper et al. (2015) (See Figure 1).

(El Moutamid et al. (2014) have applied a general development of corotation and Lindblad resonaces (CoraLin model), ideal to be analyzed with the surface of section technique. The y-axes of the section are given by a variable χ, representing the resonance width (see Table 1 in (El Moutamid et al. (2014)) which is plotted against the corotation angle ϕ_{CER}.

In Fig. 4, we show several plots of χ x ϕ_{CER} for some orbits obtained numerically through the Mercury package taken in the regions (a), (c) and (d), where we consider the full values of χ obtained from numerical orbits, not the sections. Also we utilize $a' = 137665.519$ km as the reference geometric center of the corotation zone (Renner et al. 2016).

We have the following conclusions:

A) When values for the semi-major axis are between 138322 and 138323 km, χ assumes small values while ϕ_{CER} alternates between libration and circulation. See Figure 5 in Renner et al. (2016) and Figures 3 and 4 of this work.
B) For large values of eccentricity, χ becomes diffuse in the χ x ϕ_{CER} plane, while ϕ_{CER} is circulating.

4. Final considerations

The use of the dynamic map obtained through the spectrum of individual orbits revealed that:

i) Atlas is located on the edge of a possible chaotic region. Physically, this location implies the absence of stability for critical angles ϕ_{CER} and ϕ_{LER} found by Cooper et al. (2015);
ii) for the interval of semi-major axis and eccentricity values given by [138322 km, 138323 km] x [0.004, 0.005] the map reveal an initial conditions (a_0, e_0) in which we can find a stable libration for ϕ_{CER} and also, initial conditions in which the angle ϕ_{CER} presents episodes of libration oscillation followed by circulation;

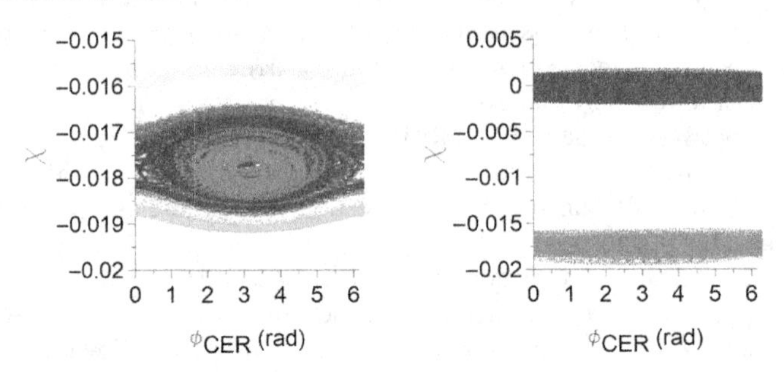

Figure 4. Plot for χ x ϕ_{CER} of several orbits with initial conditions gives in coloured full discs in Fig. 2.

iii) for eccentricity values greater than 0.009, and semi-major axis between 138328 and 138330 km we identified the region in which periods of libration occur for the angle ϕ_{LER}, as well as in ii) we did not find a region our intial condition in which there is stable libration for ϕ_{LER}.

iv) for regions A and B in Fig. 2, the orbits were presented as circulating according to (El Moutamid et al. (2014). Such feature indicates that the Atlas-like particle is far from a'.

Therefore, we conclude that Atlas is not deeply inserted in to the 54:53 resonance, due to the lack of stability in the ϕ_{CER} and ϕ_{LER} angle libration. This fact implies that Atlas cannot be added to Saturn's list of natural satellites that have both critical angles librating.

References

Callegari Jr., N., Yokoyama, T. Numerical exploration of resonant dynamics in the system of Saturnian inner Satellites. Planetary and Space Science 58, 1906–1921 (2010).

Callegari Jr., N., Yokoyama, T. Dynamics of the 11:10 Corotation and Lindblad Resonances with Mimas, and Application to Anthe, Icarus, 348, 113820 (2020).

Callegari Jr., N., Rodrigues, A., Ceccatto, D. T. The current orbit of Methone (S/2004). Celest Mech Dyn Astr, accepted for publication (2021).

Chambers, J. E. A hybrid sympletic integrator that permits close encounters betwwen massive bodies. Montly Notices of the Royal Astron. Society 304, 793–799 (1999).

Cooper, N. J. Renner, S. Murray C. D., Evans M. W. Saturn's inner satellites orbits, and the chaotic motion of Atlas from new Cassini imaging observations. The Astronomical Journal 149, 27–45, (2015).

El Moutamid, M. Renner, S. Sicardy, B. Coupling between corotation and Lindblad resonances in the elliptic planar three-body problem. Celest Mech Dyn Astr, 118, (2014).

Everhart, E. An efficient integrator that uses Gauss-Radau spacings. In: IAU Coloquium 83, 185–202 (1985).

Greenberg, R. Apsidal Precession of Orbits about an Oblate Planet. The Astronomical Journal 86, 912–914 (1981).

Murray, C. D, Dermott, S. F. Solar System Dynamics, Cambridge University Press (1999).

Peale, S. J. Origin and Evolution of the Natural Satellites. Annual Review of Astron. and Astrophys. 37, 533–602 (1999).

Renner, S., Sicardy, B. Use of the Geometric Elements in Numerical Simulations. Celestial Mechanics and Dynamical Astronomy 94, 237–248 (2006).

Renner S. Cooper J, N. El Moutmaid M.; Sicardy B. Vienne A.; Morray C. D. Sarlenfest M. Origin of the chaotic motion of the Saturnian satellite Atlas. The Astronomical Journal, 151 (2016).

Spitale, J. N., Jacobson, R. A., Porco, C. C., Owen, Jr, W. M. The Orbits of Saturn's Small Satellites Derived from Combined Historic and Cassini Imaging Observations. The Astronomical Journal 132, 792–810 (2006).

Thomas, P. C., Helfenstein, P. The small inner satellites of Saturn: Shapes, structures and some implications. Icarus 344, 113355 (2020).

Multi-scale (time and mass) dynamics of space objects
Proceedings IAU Symposium No. 364, 2022
A. Celletti, C. Galeş, C. Beaugé, A. Lemaitre, eds.
doi:10.1017/S1743921322000552

Evolution and stability of Laplace-like resonances under tidal dissipation

A. Celletti[1], E. Karampotsiou[1,2], C. Lhotka[2], G. Pucacco[3] and M. Volpi[1]

[1]Department of Mathematics, University of Roma Tor Vergata, Via della Ricerca Scientifica 1, 00133 Roma (Italy)
email: ekarampo@auth.gr

[2]Department of Physics, Aristotle University of Thessaloniki, 54124, Thessaloniki (Greece)

[3]Department of Physics, University of Roma Tor Vergata, Via della Ricerca Scientifica 1, 00133 Roma (Italy)

Abstract. The Laplace resonance is a configuration that involves the commensurability between the mean motions of three small bodies revolving around a massive central one. This resonance was first observed in the case of the three inner Galilean satellites, Io, Europa, and Ganymede. In this work the Laplace resonance is generalised by considering a system of three satellites orbiting a planet that are involved in mean motion resonances. These Laplace-like resonances are classified in three categories: first-order (2:1&2:1, 3:2&3:2, 2:1&3:2), second-order (3:1&3:1) and mixed-order resonances (2:1&3:1). In order to study the dynamics of the system we implement a model that includes the gravitational interaction with the central body, the mutual gravitational interactions of the satellites, the effects due to the oblateness of the central body and the secular interaction of a fourth satellite and a distant star. Along with these contributions we include the tidal interaction between the central body and the innermost satellite. We study the survival of the Laplace-like resonances and the evolution of the orbital elements of the satellites under the tidal effects. Moreover, we study the possibility of capture into resonance of the fourth satellite.

Keywords. Celestial mechanics; Planets and satellites; General methods; Numerical.

1. Introduction

The Laplace resonance is a notorious case of resonant configuration belonging to the class of three-body resonances, *i.e.*, resonances involving a commensurability between the mean motions of three small bodies revolving around a massive central body. A prominent example in our solar system is the configuration of the three Galilean satellites Io, Europa, and Ganymede, whose mean motions n_k satisfy the near-commensurability

$$n_I - 3n_E + 2n_G \approx 0\,. \tag{1.1}$$

The so-called Laplace angle is defined as

$$\phi_L = \lambda_1 - 3\lambda_2 + 2\lambda_3\,, \tag{1.2}$$

and it librates around π with a small amplitude.

In this work we generalize the relations that involve the mean longitudes of the three satellites, so that they describe a chain of resonances given by the form j:k&m:n. We will consider first-order resonances, when $j-k=1$ and $m-n=1$, second-order resonances, when $j-k=2$ and $m-n=2$, and, finally, mixed-order resonances, when $j-k=1$ and

$m-n=2$ or $j-k=2$ and $m-n=1$. In particular, we will study the resonant chains: 2:1&3:2, 3:2&3:2, 2:1&3:1, 3:1&3:1.

In order to perform numerical integrations of the equations of motion we use the parameters that correspond to the Galilean satellites. However this study can be easily applied to many examples of resonant chains of systems of satellites in the solar system, as well as of extrasolar planetary systems. For example, a resonant chain is observed in the system of satellites of Pluto. Specifically, Styx, Nix, Kerberos, and Hydra are near 3:1, 4:1, 5:1 and 6:1 resonance with Charon, respectively. There are also many examples of extrasolar planetary systems. Kepler-31 is a system of three planets in a 2:1&2:1 resonance, YZ Ceti is in a 3:2&3:2 resonance and Kepler-305 in a 3:2&2:1 resonance (Pichierri *et al.* (2019)). Moreover, three of the four planets of V1298 Tauri are in a 2:1&3:2 resonant chain (David *et al.* (2019)).

Following Celletti *et al.* (2021), we study the dynamical evolution of a system of three satellites orbiting around a central planet and they are involved in a first-, second- or mixed-order resonant chain. We implement a model which consists of the gravitational interaction among the central planet and the satellites, the mutual gravitational interaction of the satellites, the effects due to the oblateness of the planet and the secular interaction of a distant star and a fourth satellite. Finally we include the tidal interaction between the planet and the innermost satellite.

2. Hamiltonian model

In this section we present the Hamiltonian model adopted, which consists of the following contributions: (a) the gravitational interaction of the satellites due to the central planet, (b) the mutual gravitational interactons of the satellites, (c) the effects due to the oblateness of the planet and (d) the secular gravitational interaction due to a distant star and a fourth satellite. In the following, m_0 is the mass of the central planet, and m_j is the mass of the j-th satellite. The orbital elements of the j-th satellite are: the semi-major axis a_j, the eccentricity e_j, the inclination I_j with respect to the equatorial reference frame, the mean longitude λ_j, the longitude of the pericentre ϖ_j, the longitude of the ascending node Ω_j and the auxiliary variable $s_j = \sin(I_j/2)$.

2.1. *Keplerian part*

The Keplerian part of the Hamiltonian can be expressed using the semi-major axes of the satellites as

$$\mathcal{H} = -\frac{\mathcal{G}M_1\mu_1}{2a_1} - \frac{\mathcal{G}M_2\mu_2}{2a_2} - \frac{\mathcal{G}M_3\mu_3}{2a_3} - \frac{\mathcal{G}M_4\mu_4}{2a_4}, \tag{2.1}$$

where the following auxilliary variables are used:

$$\begin{aligned} \mu_1 &= \frac{m_0 m_1}{M_1}, & \mu_2 &= \frac{M_1 m_2}{M_2}, & \mu_3 &= \frac{M_2 m_3}{M_3}, & \mu_4 &= \frac{M_3 m_4}{M_4} \\ M_1 &= m_0 + m_1, & M_2 &= M_1 + m_2, & M_3 &= M_2 + m_3, & M_4 &= M_3 + m_4 . \end{aligned} \tag{2.2}$$

2.2. *Satellite interaction*

In order to express the Hamiltonian with respect to the orbital elements, the direct and the indirect part of the satellite-satellite interactions are expanded and truncated up to second order in the eccentricities and the inclinations (see for example Murray & Dermott (1999)). The Hamiltonian that corresponds to the satellite interaction is given by:

$$\mathcal{H} = \mathcal{H}_{(1,2)} + \mathcal{H}_{(2,3)} + \mathcal{H}_{(1,3)}, \tag{2.3}$$

where

$$
\begin{aligned}
\mathcal{H}_{(1,2)} = -\frac{\mathcal{G}m_1m_2}{a_2}\big(B_0(\alpha_{1,2}) &+ f_1^{1,2}({e_1}^2+{e_2}^2)\\
&+ f_2^{1,2}e_1\cos(j\lambda_2-k\lambda_1-\varpi_1)+f_3^{1,2}e_2\cos(j\lambda_2-k\lambda_1-\varpi_2)\\
&+ f_4^{1,2}{e_1}^2\cos(2j\lambda_2-2k\lambda_1-2\varpi_1)+f_5^{1,2}{e_2}^2\cos(2j\lambda_2-2k\lambda_1-2\varpi_2)\\
&+ f_6^{1,2}e_1e_2\cos(2j\lambda_2-2k\lambda_1-\varpi_1-\varpi_2)+f_7^{1,2}e_1e_2\cos(\varpi_2-\varpi_1)\\
&+ f_8^{1,2}{s_1}^2\cos(2j\lambda_2-2k\lambda_1-2\Omega_1)+f_9^{1,2}{s_2}^2\cos(2j\lambda_2-2k\lambda_1-2\Omega_2)\\
&+ f_{10}^{1,2}s_1s_2\cos(2j\lambda_2-2k\lambda_1-\Omega_1-\Omega_2)+f_{11}^{1,2}s_1s_2\cos(\Omega_1-\Omega_2)\big)
\end{aligned}
\tag{2.4}
$$

$$
\begin{aligned}
\mathcal{H}_{(2,3)} = -\frac{\mathcal{G}m_2m_3}{a_4}\big(B_0(\alpha_{2,3}) &+ f_1^{2,3}({e_2}^2+{e_3}^2)\\
&+ f_2^{2,3}e_2\cos(m\lambda_3-n\lambda_2-\varpi_2)+f_3^{2,3}e_3\cos(m\lambda_3-n\lambda_2-\varpi_3)\\
&+ f_4^{2,3}{e_2}^2\cos(2m\lambda_3-2n\lambda_2-2\varpi_2)+f_5^{2,3}{e_3}^2\cos(2m\lambda_3-2n\lambda_2-2\varpi_3)\\
&+ f_6^{2,3}e_2e_3\cos(2m\lambda_3-2n\lambda_2-\varpi_2-\varpi_3)+f_7^{2,3}e_2e_3\cos(\varpi_3-\varpi_2)\\
&+ f_8^{2,3}{s_2}^2\cos(2m\lambda_3-2n\lambda_2-2\Omega_2)+f_9^{2,3}{s_3}^2\cos(2m\lambda_3-2n\lambda_2-2\Omega_3)\\
&+ f_{10}^{2,3}s_2s_3\cos(2m\lambda_3-2n\lambda_2-\Omega_2-\Omega_3)+f_{11}^{2,3}s_2s_3\cos(\Omega_2-\Omega_3)\big)
\end{aligned}
\tag{2.5}
$$

$$
\begin{aligned}
\mathcal{H}_{1,3} = -\frac{\mathcal{G}m_1m_3}{a_3}\big(B_0(\alpha_{1,3}) &+ f_1^{1,3}(e_1^2+e_3^2)\\
&+ f_7^{1,3}e_1e_3\cos(\varpi_3-\varpi_1)+f_{11}^{1,3}s_1s_3\cos(\Omega_1-\Omega_3)\big)\ .
\end{aligned}
\tag{2.6}
$$

In the above expressions $\alpha_{ij}=a_i/a_j$ is the ratio of the semi-major axes of the i-th and j-th satellite, $B_0(\alpha_{i,j})=\frac{1}{2}{b_{1/2}}^{(0)}(\alpha_{i,j})-1$ and the functions $f^{i,j}$ are linear combinations of the Laplace coefficients ${b_s}^{(n)}$ and their derivatives (Murray & Dermott (1999), Ellis & Murray (2000)). Equations (2.4) and (2.5) are valid for a first-order resonant chain where $j-k=m-n=1$, however the expansion to the second- and mixed-order cases is straightforward.

2.3. *Secular interaction of the fourth satellite*

The secular gravitational attraction of the distant star ($\sigma=S$) or of a fourth satellite ($\sigma=4$) is given by:

$$
\begin{aligned}
\mathcal{H}_\sigma = -\sum_{i=1}^{\sigma}\Bigg[\frac{\mathcal{G}m_im_\sigma}{a_\sigma}\Bigg\{&\frac{1}{2}b_{1/2}^{(0)}\left(\frac{a_i}{a_\sigma}\right)-1+\frac{1}{8}\frac{a_i}{a_\sigma}b_{3/2}^{(1)}\left(\frac{a_i}{a_\sigma}\right)(e_i^2+e_\sigma^2)\\
&-\frac{1}{2}\frac{a_i}{a_\sigma}b_{3/2}^{(1)}\left(\frac{a_i}{a_\sigma}\right)(s_i^2+s_\sigma^2)\Bigg\}\Bigg]\ .
\end{aligned}
\tag{2.7}
$$

2.4. *Oblateness of the central planet*

The contribution due to the oblateness of the central planet, limited to the secular terms, is given by:

$$\begin{aligned}\mathcal{H}_{obl} = -\sum_{i=1}^{4}\frac{\mathcal{G}M_i\mu_i}{2a_i}\Bigg[J_2\left(\frac{R_P}{a_i}\right)^2\left(1+\frac{3}{2}e_i^2-\frac{3}{2}s_i^2\right)\\ -\frac{3}{4}J_4\left(\frac{R_P}{a_i}\right)^4\left(1+\frac{5}{2}e_i^2-\frac{5}{2}s_i^2\right)\Bigg],\end{aligned} \tag{2.8}$$

where R_P is the radius of the planet and J_2 is the spherical harmonic coefficient of degree two. To obtain the final formulation of the Hamiltonian, we express it in modified Delaunay variables, which are defined as $(j=1,2,3)$:

$$\begin{aligned} L_j &= \mu\sqrt{\mathcal{G}M_j\alpha_j}\,, & \lambda_j\\ P_j &= L_j(1-\sqrt{1-e_j^2})\,, & p_j=-\varpi_j\\ \Sigma_j &= L_j\sqrt{1-e_j^2}(1-\cos I_j)\,, & \sigma_j=\Omega_j\,. \end{aligned} \tag{2.9}$$

3. Tidal interaction

In order to take into account the tidal interaction between the central body (considered to be fast-rotating) and the closest satellite, we include in the model the following equations describing the evolution of semi-major axis, eccentricity, and inclination of the satellite (see Ferraz-Mello *et al.* (2008)):

$$\begin{aligned} \frac{\dot a}{a} &= \frac{2}{3}c\left(1+\frac{51}{4}e^2-D(7e^2+{S_B}^2)\right)\\ \frac{\dot e}{e} &= -\frac{1}{3}c\left(7D-\frac{19}{4}\right)\\ \dot I &= -\frac{3}{4}S_Bc(1+2D)\,. \end{aligned} \tag{3.1}$$

The parameters c and D are defined as:

$$c=\frac{9}{2}\frac{k_0}{Q_0}\frac{m_1}{m_0}\left(\frac{R_0}{a_1}\right)^5 n\,,\qquad D=\frac{Q_0}{Q_1}\frac{k_1}{k_0}\left(\frac{R_1}{R_0}\right)^5\left(\frac{m_0}{m_1}\right)^2 \tag{3.2}$$

following the notation of Malhotra (1991), where m_α are the masses, k_α/Q_α the tidal ratios, R_α the radii of the first satellite and the central planet, n is the mean motion of the satellite, and $S_B=\sin(I_1)$. As the tidal interaction is quite weaker than the gravitational forces, longer integrations are needed in order to appreciate its effects. We then multiply them by an enhancing parameter α before including the equations above in the model. This is equivalent to a rescaling of time (see Showman & Malhotra (1997), Lari *et al.* (2020), Celletti *et al.* (2021)). Equations (3.1) are translated in Delaunay variables and then added to the equations for the variables L_j, P_j, Σ_j that can be derived from the Hamiltonian.

4. Dynamical evolution

Having defined the model, we study the dynamical evolution of the system. The equations of motion are integrated numerically applying an 8-th order Runge-Kutta scheme implemented in a Mathematica program. The parameters used in this study are listed in Table 1 and Table 2.

Table 1. Values of the masses m_j, the eccentricities e_j and the inclinations I_j of the four satellites considered. These values correspond to those of the Galilean satellites.

Satellite	m_j	e_j	I_j
S_1	8.933×10^{22}	4.721×10^{-3}	3.758×10^{-2}
S_2	4.797×10^{22}	9.819×10^{-3}	4.622×10^{-1}
S_3	1.482×10^{23}	1.458×10^{-3}	2.069×10^{-1}
S_4	1.076×10^{23}	7.44×10^{-3}	1.996×10^{-1}

Table 2. Initial values of the semi-major axes of the four satellites, expressed in km.

	2:1&2:1	3:2&3:2	2:1&3:2	3:1&3:1	2:1&3:1
S_1	4.22×10^5	4.22×10^5	4.22×10^5	4.22×10^5	4.22×10^5
S_2	6.713×10^5	5.53×10^5	6.713×10^5	8.78×10^5	6.713×10^5
S_3	10.705×10^5	7.25×10^5	8.78×10^5	18.26×10^5	13.94×10^5
S_4	18.828×10^5	12.751×10^5	15.442×10^5	32.115×10^5	24.517×10^5

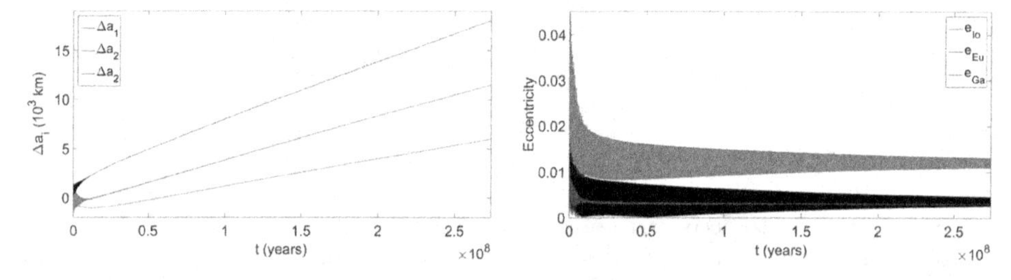

Figure 1. Evolution of the sem-major axes (left) and the eccentricities (right) of the three satellites in the case of the 2:1&3:2 first-order resonance. The tidal effects are multiplied by $\alpha = 10^4$.

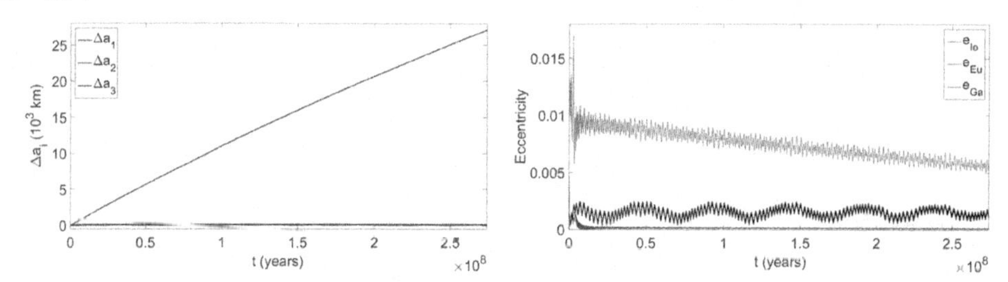

Figure 2. Evolution of the sem-major axes (left) and the eccentricities (right) of the three satellites in the case of the 2:1&3:1 mixed-order resonance. The tidal effects are multiplied by $\alpha = 10^5$.

In the case of the first-order resonances the three satellites move outwards due to the tidal interaction between the innermost satellite and the central planet (an example for the 2:1&3:2 resonance is given in the left panel of Fig. 1). The angular momentum is transferred through the tidal interaction from the central planet to the first satellite, from the first satellite to the second one, and so on. The resonant configuration is therefore maintained. However in the case of the 2:1&3:1 mixed-order resonance the two inner satellites move outwards, while the semi-major axis of the outer one remains constant, as shown in the left panel of Fig. 2. In this case the resonant configuration is destroyed.

The eccentricities of the satellites in the case of a first-order resonance converge to limiting values that are lower than the initial ones. An example is shown in the right panel of Fig. 1. In the mixed-order case the eccentricity of the innermost satellite converges to zero, the eccentricity of the second satellite is decreasing and the eccentricity of the third satellite oscillates around a value close to the initial one (see right panel of Fig. 2).

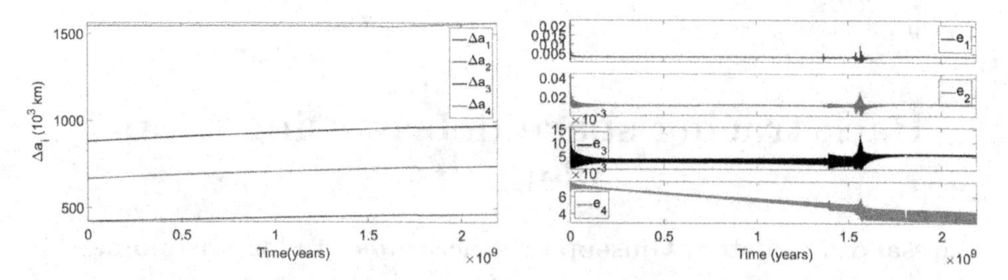

Figure 3. Evolution of the semi-major axes (left) and eccentricities (right) of the four satellites. The tidal effects are multiplied by $\alpha = 10^4$.

5. Capture into resonance of the fourth satellite

In this section we modify the Hamiltonian model in order to include the gravitational interaction of the fourth satellite (i.e., not limited to its secular part) in order to study the possibility of capture into resonance. We sample 100 different initial conditions for the mean longitude of the fourth satellite and integrate numerically the equations of motion. In the cases of the first-order resonances the fourth satellite is captured into resonance in all the different cases studied. In the case of the 2:1&3:2 resonance the fourth satellite is captured in a 2:1 resonance with the third satellite and the resonant argument $\phi_{outer} = \lambda_2 - 3\lambda_3 + 2\lambda_4$ librates around π. In the case of the 3:2&3:2 resonance the resonant angles $2\lambda_4 - \lambda_3 - \varpi_3$ and $2\lambda_4 - \lambda_3 - \varpi_4$ rotate, but the resonant argument ϕ_{outer} librates around π.

The semi-major axes of the satellites increase in a similar rate after the capture into resonance of the fourth satellite, which takes place after 1.5 Gyr (left panel of Fig. 3). The eccentricities of the first and second satellite remain the same after the capture of the fourth satellite, the eccentricity of the third satellite is larger after the capture than before and the eccentricity of the fourth satellite decreases (right panel of Fig. 3).

References

Celletti, A., Karampotsiou, E., Lhotka, C., Pucacco, G. & Volpi, M. 2021, *arXiv e-prints*, arXiv:2109.02694

David, T.J., Petigura, E.A., Luger, R. *et al.* 2019, *ApJ*, 885, L12

Ellis, K.M. & Murray, C.D. 2000, *Icarus*, 147 129

Ferraz-Mello, S., Rodríguez, A. & Hussmann, H. 2008, *Celestial Mechanics and Dynamical Astronomy*, 101, 171

Lari, G., Saillenfest, M. & Fenucci, M. 2020 *A&A*, 639, A40

Murray, C.D. & Dermott, S.F. 1999 *Cambridge university press*

Pichierri, G., Batygin, K., & Morbidelli, A., *A&A*, 625, A7

Showman, A.P., & Malhotra, R. 1997, *Icarus*, 127, 93

Multi-scale (time and mass) dynamics of space objects
Proceedings IAU Symposium No. 364, 2022
A. Celletti, C. Galeş, C. Beaugé, A. Lemaitre, eds.
doi:10.1017/S1743921322000588

Back-tracing space debris using proper elements

Alessandra Celletti[1], Giuseppe Pucacco[2] and Tudor Vartolomei[3]*

[1]Department of Mathematics, University of Rome Tor Vergata; Via della Ricerca Scientifica 1, 00133 Rome, Italy,
celletti@mat.uniroma2.it

[2]Department of Physics, University of Rome Tor Vergata; Via della Ricerca Scientifica 1, 00133 Rome, Italy,
Giuseppe.Pucacco@roma2.infn.it

[3]Department of Mathematics, University of Rome Tor Vergata; Via della Ricerca Scientifica 1, 00133 Rome, Italy,
vartolom@mat.uniroma2.it

*This extended abstract has been presented by Tudor Vartolomei at the IAU Symposium 364

Abstract. Normal form methods allow one to compute quasi-invariants of a Hamiltonian system, which are referred to as proper elements. The computation of the proper elements turns out to be useful to associate dynamical properties that lead to identify families of space debris, as it was done in the past for families of asteroids. In particular, through proper elements we are able to group fragments generated by the same break-up event and we possibly associate them to a parent body. A qualitative analysis of the results is given by the computation of the Pearson correlation coefficient and the probability of the Kolmogorov-Smirnov statistical test.

Keywords. Proper elements, Normal form, Space debris, Geopotential, Sun and Moon attractions, Solar radiation pressure, Statistical data analysis.

---

1. Introduction

The dynamics around the Earth can be classified taking into account the forces involved, which depend on the distance from the Earth's surface. To this end, we recall the three main regions above the Earth surface: Low-Earth-Orbits (LEO), Medium-Earth-Orbits (MEO) and Geosynchronous-Earth-Orbits (GEO). In all these regions, the motion is mainly governed by the gravitational field of the Earth, but other forces might influence the long term evolution of a spatial object, according to its altitude. Such forces include, in particular, the non-spherical shape of the Earth, the attraction of the Moon and the Sun, the solar radiation pressure. Once the Hamiltonian model is given, we implement a Lie series normalization procedure to compute quasi-invariants of the motion, namely the so-called proper elements, which are associated to semi-major axis, eccentricity and inclination. Using simulated data after a break-up event (a collision or an explosion), we analyzed the connection of the computation of the proper elements with the dynamics observed immediately after the catastrophic event. The results are corroborated by a statistical data analysis based on the check of the Kolmogorov-Smirnov test and the computation of the Pearson correlation coefficient. We refer to Celletti, Pucacco, & Vartolomei (2021a) for full details on the procedure and to Celletti, Pucacco, & Vartolomei (2021b) for an application to concrete cases.

2. The Model

A model of dynamics for space debris has been developed taking into account four main forces that act on a satellite or a space debris, namely the gravitational potential of the Earth, the attraction of the Moon and Sun, and the Solar radiation pressure. The entire model has been described in Celletti *et al.* (2017a) starting from the Cartesian equations of motion and using the expansion in the orbital elements. The Hamiltonian function has the following form

$$\mathcal{H} = \mathcal{H}_{Kep} + \mathcal{H}_E + \mathcal{H}_M + \mathcal{H}_S + \mathcal{H}_{SRP} \ ,$$

where $\mathcal{H}_{Kep}$ is given by

$$\mathcal{H}_{Kep} = -\frac{GM_E}{2a},$$

where G is the gravitational constant, M_E is the mass of the Earth and a denotes the semi-major axis.

Following Kaula (1966), the Hamiltonian part corresponding to the Earth's perturbation can be written as an expansion in orbital elements of the space object. In the quasi-inertial reference frame, the Hamiltonian function can be written as

$$\begin{aligned}\mathcal{H}_E = & \, GM_E \frac{R_E^2}{a^3} J_2 \left(\frac{3}{4}\sin^2 i - \frac{1}{2}\right) \frac{1}{(1-e^2)^{3/2}} \\ & + GM_E \frac{R_E^3}{a^4} J_3 \left(\frac{15}{8}\sin^3 i - \frac{3}{2}\sin i\right) e \sin\omega \frac{1}{(1-e^2)^{5/2}}.\end{aligned}$$

The perturbation of the space object due to the Moon and Sun attractions can be written as an expansion in orbital elements of the 3rd body perturber and the debris, using the following formula (see Kaula (1962)):

$$\begin{aligned}\mathcal{H}_M = & -Gm_M \sum_{l\geq 2}\sum_{m=0}^{l}\sum_{p=0}^{l}\sum_{s=0}^{l}\sum_{q=0}^{l}\sum_{j=-\infty}^{\infty}\sum_{r=-\infty}^{\infty} (-1)^{m+s}(-1)^{[m/2]}\frac{\varepsilon_m\varepsilon_s}{2a_M}\frac{(l-s)!}{(l+m)!}\left(\frac{a}{a_M}\right)^l \\ & F_{lmp}(i)F_{lsq}(i_M)H_{lpj}(e)G_{lqr}(e_M)\{(-1)^{t(m+s-1)+1}U_l^{m,-s}\cos(\phi_{lmpj}+\phi'_{lsqr}-y_s\pi) \\ & + (-1)^{t(m+s)}U_l^{m,-s}\cos(\phi_{lmpj}-\phi'_{lsqr}-y_s\pi)\} \ ,\end{aligned}$$

where m_M is the mass of the Moon, $y_s = 0$, if (s mod 2)=0, $y_s = \frac{1}{2}$, if (s mod 2)= 1, $t = (l-1)$ mod 2, and

$$\varepsilon_m = \begin{cases} 1, & m = 0 \\ 2, & m \in \mathbb{Z}\backslash\{0\} \end{cases}$$

$$\phi_{lmpj} = (l-2p)\omega + (l-2p+j)M + m\Omega$$

$$\phi'_{lsqr} = (l-2q)\omega_M + (l-2q+r)M_M + s\left(\Omega_M - \frac{\pi}{2}\right) \ .$$

The functions $F_{lmp}(i)$, $F_{lsq}(i_M)$ and $G_{lqr}(e_M)$ have been introduced in Kaula (1962), Celletti *et al.* (2017b); $H_{lpj}(e)$ are the Hansen coefficients, while the terms $U_l^{m,s}$ are given by

$$U_l^{m,s} = \sum_{r=\max(0,-(m+s))}^{\min(l-s,l-m)} (-1)^{l-m-r}\binom{l+m}{m+s+r}\binom{l-m}{r}\cos^{m+s+2r}\left(\frac{\varepsilon}{2}\right)\sin^{-m-s+2(l-r)}\left(\frac{\varepsilon}{2}\right),$$

where $\varepsilon = 23°26'21.406''$ is the Earth's obliquity. In applications, the expansion of the Moon's Hamiltonian will be truncated to $l = 2$ and averaged over the mean anomalies of the object M and of the Moon M_M.

The Hamiltonian due to the Sun depends on the orbital elements of the Sun and the debris. The expansion of $\mathcal{H}_S$ is given below and, again, we will consider the expansion to $l = 2$, averaging over the mean anomalies of the object and perturber body:

$$\mathcal{H}_S = -Gm_S \sum_{l\geq 2}\sum_{m=0}^{l}\sum_{p=0}^{l}\sum_{h=0}^{l}\sum_{q=-\infty}^{\infty}\sum_{j=-\infty}^{\infty} \frac{a^l}{a_S^{l+1}}\varepsilon_m \frac{(l-m)!}{(l+m)!} F_{lmp}(i)F_{lmh}(i_S)H_{lpq}(e)$$
$$G_{lhj}(e_S)\cos(\phi_{lmphqj}),$$

where m_S is the mass of the Sun and

$$\phi_{lmphqj} = (l-2p)\omega + (l-2p+q)M - (l-2h)\omega_S - (l-2h+j)M_S + m(\Omega - \Omega_S).$$

The contribution to the Hamiltonian due to Solar radiation pressure is given below:

$$\mathcal{H}_{SRP} = C_r P_r \frac{A}{m} a_S^2 \sum_{l=1}^{1}\sum_{s=0}^{l}\sum_{p=0}^{l}\sum_{h=0}^{l}\sum_{q=-\infty}^{\infty}\sum_{j=-\infty}^{\infty} \frac{a^l}{a_S^{l+1}}\varepsilon_s \frac{(l-s)!}{(l+s)!} F_{lsp}(i)F_{lsh}(i_S)H_{lpq}(e)$$
$$G_{lhj}(e_S)\cos(\phi_{lsphqj}) \ .$$

where C_r, P_r are, respectively, the reflectivity coefficient and the radiation pressure for an object located at $a_S = 1AU$, while $\frac{A}{m}$ denotes the area-to-mass ratio. This function is averaged over the mean anomaly of the object M, but it depends on the mean anomaly of the sun M_S.

3. Normal Forms

We construct a normal form, which consists of a procedure implementing iteratively canonical changes of coordinates, in such a way that the Hamiltonian function is transformed into a given form. In particular, we will require that the Hamiltonian is integrable up to a remainder term. Each time we implement the iterative procedure, we decrease the size of the norm of the remainder term; however, it is well known that such iteration is in general not converging Poincare (1892); besides, the complexity of the computation usually grows when increasing the normalization steps.

Proper elements are computed through normal form theory Efthymiopoulos (2011), which we shortly summarize as follows. Consider the following Hamiltonian function

$$\mathcal{H}(\underline{I}, \underline{\varphi}) = \mathcal{H}_0(\underline{I}) + \varepsilon \mathcal{H}_1(\underline{I}, \underline{\varphi}) \ , \tag{3.1}$$

where $(\underline{I}, \underline{\varphi})$ denote action-angle variables with $(\underline{I}, \underline{\varphi}) \in B \times \mathbb{T}^n$, where n is the number of degrees of freedom and $B \subset \mathbb{R}^n$ denotes an open set. The function $\mathcal{H}_0(\underline{I})$ appearing in (3.1) is called the integrable part, $\varepsilon \in \mathbb{R}$ is a small parameter, $\mathcal{H}_1(\underline{I}, \underline{\varphi})$ is called the perturbing function.

As we mentioned before, we implement a change of coordinates which transforms the Hamiltonian to remove the perturbation to orders of ε^2. Usually, such normalization procedure can only be iterated for some steps, after which it starts to diverge Poincare (1892).

Let the function $\mathcal{H}_1$ be expanded in Fourier series as

$$\mathcal{H}_1(\underline{I}, \underline{\varphi}) = \sum_{\underline{k} \in K} b_{\underline{k}}(\underline{I}) \exp(i\underline{k} \cdot \underline{\varphi}) \ ,$$

where $K \subseteq \mathbb{Z}^n$ and $b_{\underline{k}}$ denote functions with real coefficients. We call χ the generating function associated to the canonical transformation from the original coordinates $(\underline{I}, \underline{\varphi})$ to the new coordinates $(\underline{I}', \underline{\varphi}')$:

$$\underline{I} = S^{\varepsilon}_{\chi} \underline{I}' , \qquad \underline{\varphi} = S^{\varepsilon}_{\chi} \underline{\varphi}' ,$$

where S^{ε}_{χ} acts on a function $\mathscr{F}$ as

$$S^{\varepsilon}_{\chi} \mathscr{F} := \mathscr{F} + \sum_{i=1}^{\infty} \frac{\varepsilon^i}{i!} \{\{...\{\mathscr{F}, \chi\}, \dots\}, \chi\}$$

with $\{\cdot, \cdot\}$ representing the Poisson bracket operator. To compute S^{ε}_{χ}, we assume that the new Hamiltonian $\mathscr{H}^{(1)} = S^{\varepsilon}_{\chi} \mathscr{H}$ takes the form

$$\mathscr{H}^{(1)}(\underline{I}', \underline{\varphi}') = Z_1(\underline{I}') + \varepsilon^2 \mathscr{H}_2(\underline{I}', \underline{\varphi}') , \tag{3.2}$$

where $Z_1 = \mathscr{H}_0 + \varepsilon \overline{\mathscr{H}}_1$ is the new integrable part (the bar denotes the average over the angles), while $\mathscr{H}_2$ is the new remainder function which is of order ε^2. Inserting the transformation of coordinates in (3.1), the new Hamiltonian takes the desired form (3.2), if χ satisfies the following normal form equation:

$$\mathscr{H}_1(\underline{I}', \underline{\varphi}') + \{\mathscr{H}_0(\underline{I}'), \chi(\underline{I}', \underline{\varphi}')\} = \overline{\mathscr{H}}_1(\underline{I}').$$

Let us expand χ in Fourier series, let the frequency be $\underline{\omega}_0 = \frac{\partial \mathscr{H}_0}{\partial \underline{I}'}$, then the generating function is given by

$$\chi(\underline{I}', \underline{\varphi}') = -i \sum_{\underline{k} \in \mathbb{Z}^n} \frac{b_{\underline{k}}(\underline{I}')}{\underline{k} \cdot \underline{\omega}_0} \exp(i \underline{k} \cdot \underline{\varphi}') ,$$

provided the following non-resonance condition is satisfied: $\underline{k} \cdot \underline{\omega}_0 \neq 0$. Normal forms of higher order are obtained by iterating the procedure described above.

4. Results

4.1. *Simulator of break-up events*

Using a simulator of break-up events developed within the ongoing collaboration in Apetrii, Celletti, Efthymiopoulos, Galeş, Vartolomei (2021), we produce synthetic data in order to show the effectiveness of the proper elements. This simulator reproduces the break-up model Evolve 4.0 provided by NASA (see Johnson (2001), Klinkrad (2006)) and allows us to determine the cross-sections, masses, and imparted velocities of the fragments after an explosion or a collision. Our procedure consists in the following steps:

(*i*) simulate a break-up event and obtain the Cartesian coordinates for all generated fragments;

(*ii*) compute the orbital elements for each fragment;

(*iii*) using the orbital elements after the break-up, we propagate each fragment for a given period of time, typically up to 150 years;

(*iv*) we use the position of the fragments after the propagation up to a given interval of time (e.g., 150 years) to compute the proper elements of each fragment;

(*v*) the distribution of the proper elements is then compared to that of the elements at the initial time after the break-up.

4.2. *Proper Elements*

We take an example concerning an explosion that generates 465 fragments. The event occurs at a medium altitude of the parent body with $a = 25200$ km and with a relatively

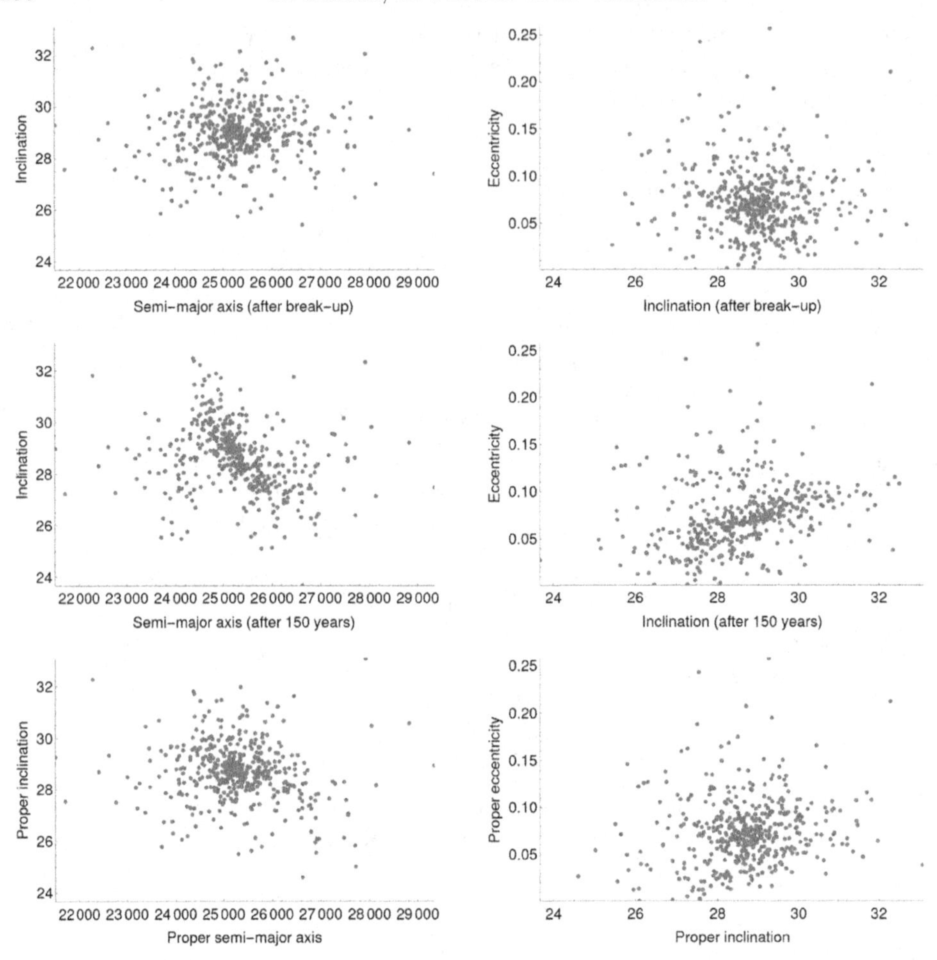

Figure 1. Distribution of initial osculating elements (first row), mean elements after 150 years (second row), and proper elements computed from mean elements after 150 years (third row) in the a-i plane (left), i-e plane (right). Initial osculating elements of the parent body: $a = 25200$ km, $e = 0.07$, $i = 29^o$, $\omega = 40^o$, $\Omega = 100^o$.

small inclination and eccentricity, $e = 0.07$, $i = 29^o$; the other elements are fixed as $\omega = 40^o$ and $\Omega = 100^o$.

Figure 1 shows the osculating elements after break-up (first row), mean elements after 150 years (second row), and proper elements computed after 150 years (third row) in the planes a-i and i-e. The scales have been fixed as the minimum and the maximum values of the evolution of the elements after 150 years.

While the distribution of the fragments in the mean elements is different from that in osculating elements, the first and third rows of Figure 1 show a connection between fragments at the break-up event and the distribution of the proper elements.

4.3. *Statistical analysis*

We compare the distributions of semimajor axis, eccentricity and inclination at break-up, after 150 years, and by computing the proper elements after 150 years, by implementing some statistical methods for data analysis Cowan (1997). Two of these methods are Kolmogorov-Smirnov (K-S) test that compares two distributions, and the Pearson correlation coefficient between the datasets.

As an example, we take the case of moderate orbits presented in Figure 1 and we implement the above methods to analyze the data. Since the semi-major axis is always constant, we are interested just in the analysis of eccentricity and inclination.

The K-S test for inclination gives a small probability, the so-called p-value, equal to 0.365534 when checking the similarity between the initial dataset and the mean elements after 150 years, while it gives a higher p-value equal to 0.968287 when looking at the initial data and the proper elements.

Pearson correlation coefficient provides evidence of the difference in inclination between the initial data and the data after 150 years, where its value turns out to be equal to 0.811243. Instead, a higher coefficient equal to 0.945672 is obtained when comparing the initial and proper elements.

5. Conclusions

In the present work we test the computation of the proper elements in the space debris problem. The Hamiltonian formulation of the model was used to describe the dynamics taking into account several perturbations: the potential of the Earth, the attraction of the Moon and the Sun, and the Solar radiation pressure. Using a break-up simulator, we analyzed the connection between proper elements and the initial osculating elements. In view of possible applications, we foresee several ways to improve the results, in primis the study of a more elaborated model including a larger number of spherical harmonics and a higher order expansion of the Hamiltonian.

Acknowledgements

All authors acknowledge EU H2020 MSCA ETN Stardust-R Grant Agreement 813644.

References

Celletti, A., Pucacco, G., and Vartolomei, T. *Proper elements for space debris.* Submitted to Celestial Mechanics and Dynamical Astronomy, (2021a).

Celletti, A., Pucacco, G., and Vartolomei, T. *Reconnecting groups of space debris to their parent body through proper elements.* Sci Rep 11, 22676 (2021). https://doi.org/10.1038/s41598-021-02010-x

Celletti, A., Gachet, F., Galeş, C., Pucacco, G., and Efthymiopoulos, C. Dynamical models and the onset of chaos in space debris. *Int. J. Nonlinear Mechanics* **90**, 47–163 March (2017a).

Kaula, W. M. *Theory of Satellite Geodesy.* Blaisdell Publ. Co., (1966).

Kaula, W. M. Development of the lunar and solar disturbing functions for a close satellite. *The Astronomical Journal* **67**, 300 June (1962).

Celletti, A., Galeş, C., Pucacco, G., and Rosengren, A. J. Analytical development of the lunisolar disturbing function and the critical inclination secular resonance. *Celestial Mechanics and Dynamical Astronomy* **127**(3), 259–283 March (2017b).

Poincaré, H. *Les méthodes nouvelles de la mécanique céleste.* Gauthier-Villars, Paris, (1892).

Efthymiopoulos, C. Canonical perturbation theory; stability and diffusion in Hamiltonian systems: applications in dynamical astronomy. *Workshop Series of the Asociacion Argentina de Astronomia* **3**, 3–146 January (2011).

Apetrii, M., Celletti, A., Efthymiopoulos, C., Galeş, C., and Vartolomei, T. *On a simulator of break-up events for space debris.* Work in progress, (2021).

Johnson, N. L., Krisko, P. H., Lieu, J.-C., and Am-Meador, P. *NASA's new break-up model of EVOLVE 4.0.* Adv. Space Res. 28, (2001).

Klinkrad, H. *Space Debris: Models and Risk Analysis.* (2006).

Cowan, G. *Statistical data analysis.* Clarendon Press, Oxford, (1997).

Multi-scale (time and mass) dynamics of space objects
Proceedings IAU Symposium No. 364, 2022
A. Celletti, C. Galeş, C. Beaugé, A. Lemaitre, eds.
doi:10.1017/S1743921321001265

Four- and five-body periodic Caledonian orbits

Valerie Chopovda† and Winston L. Sweatman

School of Mathematical and Computational Sciences, Massey University, Auckland, New Zealand. emails: valerie.chopovda@gmail.com, w.sweatman@massey.ac.nz

Abstract. We consider four- and five-body problems with symmetrical masses (Caledonian problems). Families of periodic orbits originate from the collinear Schubart orbits. We present and discuss some of these periodic orbits.

Keywords. Celestial mechanics; Stellar dynamics; Caledonian few-body problem.

---

1. Introduction

The Caledonian four-body and five-body problems are formed by point-mass systems with, respectively, four and five masses Roy & Steves (2000); Steves *et al.* (2020). Symmetry about the centre of mass makes these systems less complicated than the general N-body problem. Although simpler, Caledonian few-body problems can provide insight for the more general problem (Sweatman (2015)).

The present paper considers families of periodic orbits in the Caledonian few-body problem. Similar families for the three-body problem were found by Hénon (1976), who discovered a family of periodic orbits arising from Schubart's orbit (Schubart (1956)). We have explored various mass ratios and outline some of the results.

2. The Caledonian symmetrical four- and five-body problems

The Caledonian symmetrical four-body problem consists of four masses arranged in a system that has rotational symmetry about the centre of mass. The pair of masses m_1 and m_3 are sited opposite, across the centre of mass one another, with equal masses so that $m_3 = m_1$. Working within the rest frame of the centre of mass, their position vectors are equal and opposite, $\mathbf{r}_3 = -\mathbf{r}_1$, as are their velocities, $\mathbf{v}_3 = -\mathbf{v}_1$. Masses m_2 and m_4 are similarly related: $m_4 = m_2$, $\mathbf{r}_4 = -\mathbf{r}_2$ and $\mathbf{v}_4 = -\mathbf{v}_2$. If a system has such rotational symmetry at any time, then the rotational symmetry is preserved for all time.

The Caledonian five-body problem is created by including an additional mass m_0 at the centre of mass of a Caledonian four-body problem (see Figure 1). This fifth mass remains at rest for all time: $\mathbf{r}_0 = 0$, $\mathbf{v}_0 = 0$, $\forall t$. We can regard the Caledonian four-body problem as being the special case of the Caledonian five-body problem when this mass is zero: $m_0 = 0$.

To explore the Caledonian four- or five-body problem numerically, the equations of motion are regularised using an adaptation of the global regularisation method (Heggie (1974)). In the four-body case, all possible close two-body encounters are regularised by introducing Levi-Civita coordinates for each distinct inter-body distance and a corresponding time transformation (cf. Sivasankaran *et al.* (2010)). The same regularised

† Present address: Department of Engineering Science, University of Auckland, New Zealand.

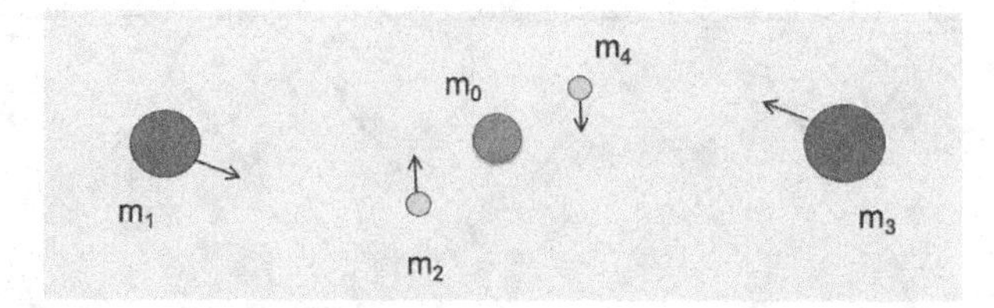

Figure 1. The Caledonian five-body problem: pairs $[m_1,m_3]$ and $[m_2,m_4]$ are symmetric about m_0. To obtain the Caledonian four-body problem m_0 is removed or, alternatively, set to zero ($m_0 = 0$).

coordinates also work for the five-body case, as the extra mass, m_0, will only come close to m_1 or m_3 when these masses approach one another. Similarly, m_0 is only close to m_2 or m_4 when these masses are themselves close. We cannot regularise close interactions involving all four or all five masses.

Families of periodic orbits can be found with the approach used previously by Hénon (1976) for three-body systems and by Chopovda & Sweatman (2018) for the equal-mass Caledonian four-body problem. Initial conditions from known periodic solutions are used to generate approximate initial conditions for a neighbouring orbit. These are integrated through a period and the difference between initial and final points is used to produce improved initial conditions by the process of differential correction (cf. Sweatman (2014) and papers cited therein).

3. Families of periodic orbits generated from the four-body Schubart orbits

The orbits within our families are considered periodic in the sense that the masses return to the same configuration at the end of a period as they had at the beginning. This is sometimes called a relatively periodic orbit. However, typically, the configuration is rotated by some angle during the orbital period. This angle of rotation and, also, the angular momentum vary continuously along each family of orbits.

The families of orbits are parameterised by the mass ratio $[m_1 : m_2]$. A range of mass ratios have been explored (cf. Chopovda (2019)) and from these some features can be identified that are common to all the families. For a particular mass ratio, the family begins and ends at the one-dimensional limit of motion, in which the four masses remain ordered and move along a fixed line. The initial periodic orbit of an individual family is a symmetric four-body Schubart orbit (Sweatman (2002, 2006); Sekiguchi & Tanikawa (2004)). In such orbits the masses perform an interplay motion in which close encounters alternate between the central and outer parts of the system. If we label the masses in their collinear order m_1, m_2, m_4, m_3, then, in the centre, the inner masses m_2 and m_4 have close interactions. On the outside m_1 interacts with m_2 whilst m_3 interacts with m_4. If a family begins at a particular Schubart orbit, then the final family member is also a Schubart orbit but with the masses reordered so that the inner and outer masses are exchanged, i.e., m_4, m_1, m_3, m_2 with the previous labelling. As the Schubart orbits are strictly one-dimensional, they have no angular momentum, and there is zero rotation during a period. Members of the families near either Schubart orbit similarly display an interplay-type motion.

Figure 2 presents three Schubart orbits. Each line on a chart represents the position of one of the masses, these values are plotted against time. The two orbits towards the top are for the case where the masses are unequal and one mass value is three times the size of the other. In the uppermost plot the smaller masses are on the inside and for the central plot they are on the outside. When the inner masses of the Schubart

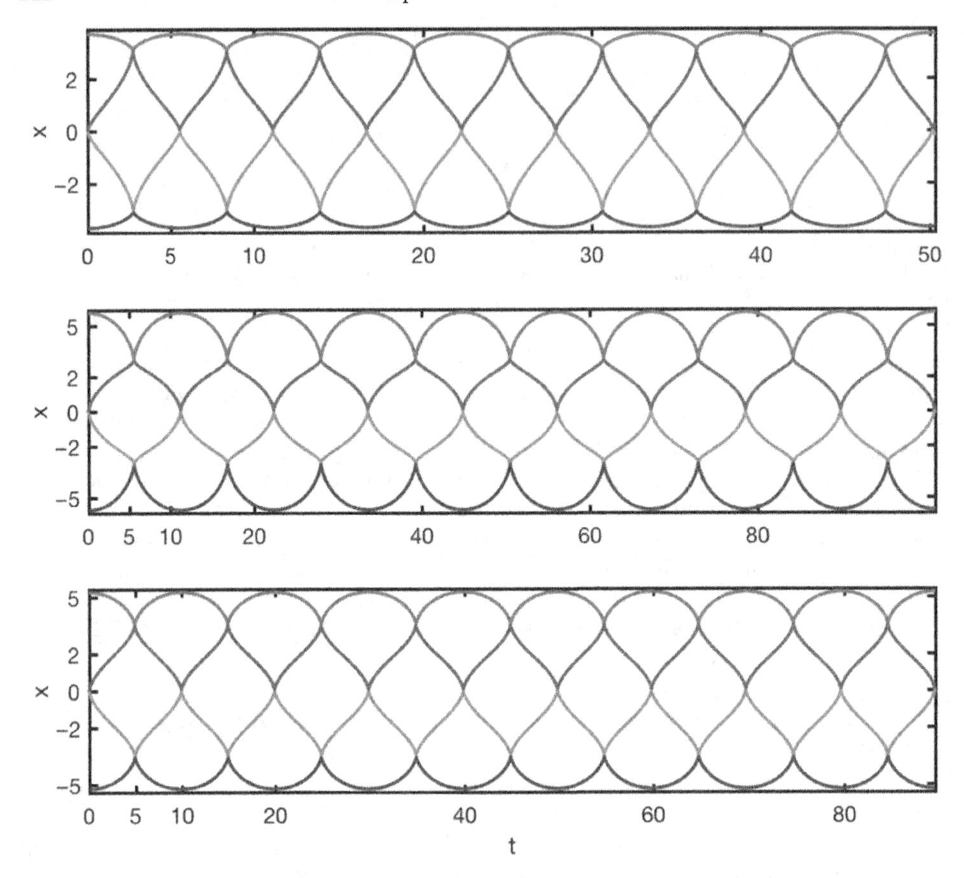

Figure 2. Four-body Schubart orbits for mass ratios $[m_1 : m_2] = [3 : 1]$ (top), $[m_1 : m_2] = [1 : 3]$ (middle) and $[m_1 : m_2] = [1 : 1]$ (bottom), where $m_1 = m_3$ is the mass of a body on the outside of the system and $m_2 = m_4$ is the mass of a body on the inside of the system.

orbit are relatively smaller, their motion is rapid compared with the outer masses so that sufficient momentum can be transferred to keep the outer masses apart (Figure 2 (top)). In contrast, when the inner masses are larger their relative motion approaches two-body motion, and the smaller outer masses have little effect on other masses unless passing very close (Figure 2 (middle)). The plot at the bottom of Figure 2 is for the equal masses case.

Another feature common to all the families of orbits is a double choreography orbit that occurs midway through the family. In the double choreography orbits each pair of symmetrical masses shares a common path and the masses return to their initial positions after two (relative) periods. Figure 3 shows double choreography orbits for three different mass ratios. In orbit D1, the larger mass is approximately seven times larger than the smaller. For orbit D2 the factor is close to three. The equal masses case, orbit D3, was described by Chen (2001) and has fourfold symmetry. This feature contrasts with what occurs for the three-body family, as shown by Hénon (1976), for which the size of the orbits approaches infinity (cf. Chopovda & Sweatman (2018)).

Figure 4 illustrates a specific family of orbits corresponding to the 3:1 mass ratio. Members of the family lie between the two Schubart orbits of this mass ratio (cf. Figure 2). Unlike the Schubart orbits, most of these orbits have a non-zero rotation. For example in

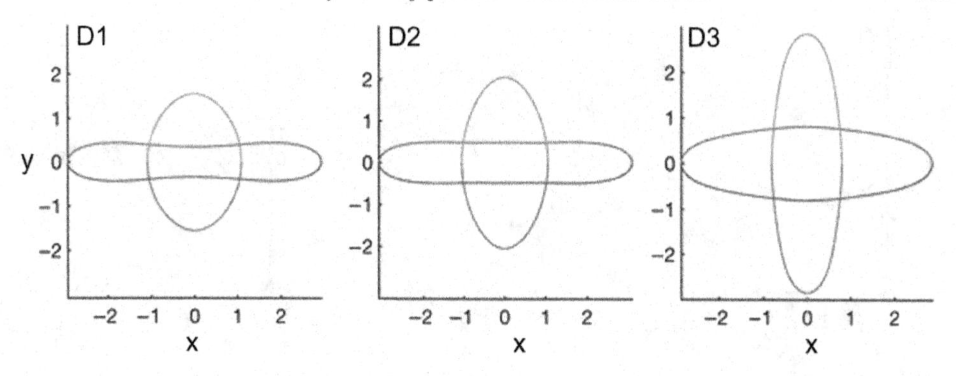

Figure 3. Double choreography orbits with mass ratios [254:1746], [496:1504] and [1:1]

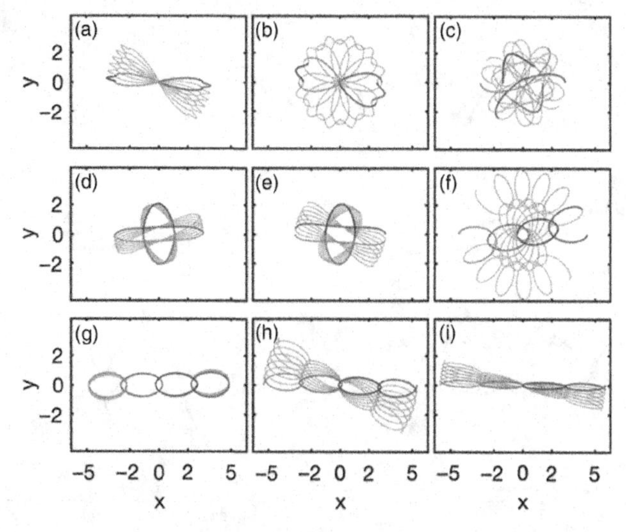

Figure 4. The family of orbits for the 3:1 mass ratio. The family begins from the Schubart orbit with the larger masses on the outside (Figure 2 (top)). It then progresses through the orbits shown, alphabetically, and finishes at the Schubart orbit with the smaller masses on the outside (Figure 2 (middle)).

orbit A, early in the family, a negative angle of rotation produces a clockwise precession of the orbit. Starting from the Schubart orbit with the larger masses on the outside (Figure 2 (top)), the angle of rotation decreases from zero into negative values. The angle of rotation continues to decrease through most of the family until near orbit H, at which stage it has changed by a magnitude greater than a complete rotation. From orbit H, the angle then increases until the end of the family at the final Schubart orbit (Figure 2 (middle)), at which stage the net change from the start is essentially a whole negative rotation which is equivalent to a zero rotation angle, modulo 360°. The interplay motion for family members close to a Schubart orbit is illustrated by orbits A and B near the start and orbits F, G, H and I towards the end. Orbit B completes about one twelfth of a rotation in each (relative) period and hence B is approximately absolutely periodic (i.e., returns to the initial conditions) over 12 (relative) periods. Such absolutely periodic orbits occur whenever the angle of rotation is a rational multiple of a degree. The double choreography orbit of the family will occur between orbits D and E. The interplay orbit

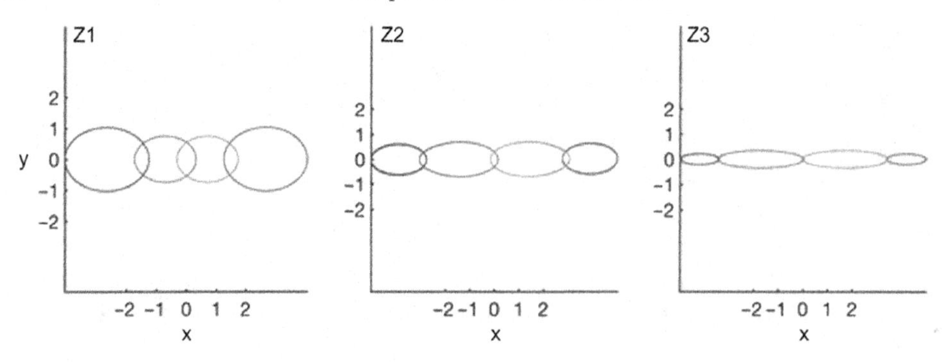

Figure 5. Absolutely periodic interplay orbits. The ratios of the outer masses to the inner masses are [3:197], [73:127] and [1056:944].

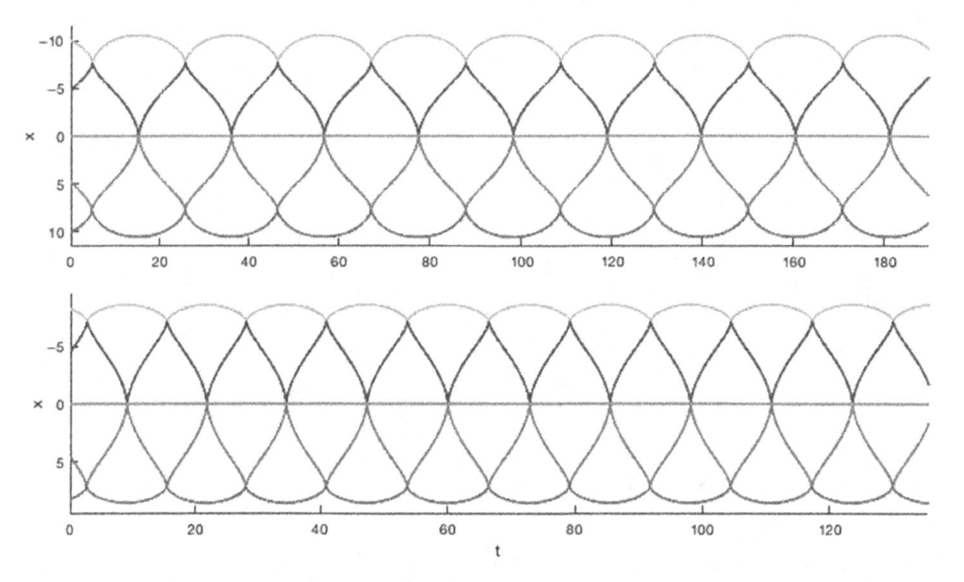

Figure 6. Five-body Schubart orbits with equal masses (top), and with larger central and outermost masses $[m_1 : m_2 : m_0] = [1 : 0.5 : 2]$ (bottom).

G approximates an absolutely periodic orbit that occurs when the angle of rotation is zero, modulo 360°.

For the case of equal masses, the family of orbits has been described by Chopovda & Sweatman (2018). This family contrasts with the unequal-mass family presented here in that the equal-mass family is symmetric with the same orbits occurring on either side of the double choreography orbit. The equal-mass family begins and ends at the same Schubart orbit: that shown in Figure 2 (bottom). The equal-mass family has an absolutely periodic orbit similar to orbit G, however, this occurs on both sides of the double choreography. The orbit presented by Sweatman (2014) is close in appearance to this orbit and was a starting point for a more exact calculation by Chopovda & Sweatman (2018). Families of orbits with near equal masses have two absolutely periodic interplay orbits, similar to orbit G. These occur on either side of the double choreography orbit as in the equal-mass case. However, for more disparate mass ratios there is only one such absolutely periodic interplay orbit, and for this the larger masses must be the inner masses of the system, as is the case for orbit G.

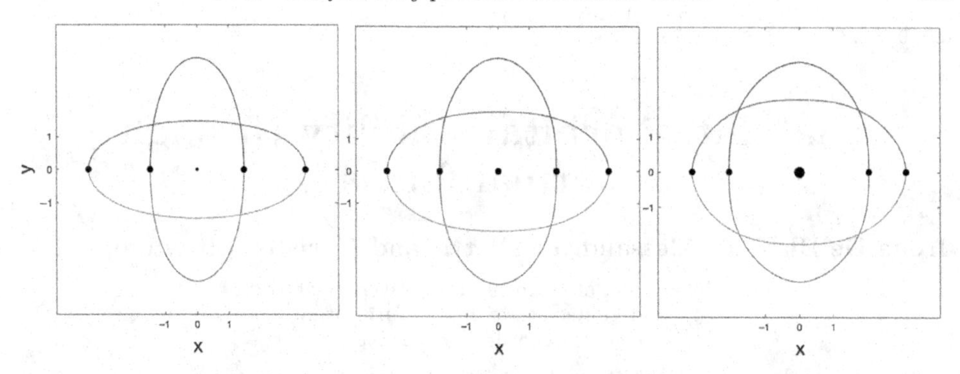

Figure 7. Five-body double choreographies. The outer masses are all equal and the outer to central $[m_1 : m_0]$ mass ratios are from left to right [35:20], [3:4] and [1:4], respectively.

Figure 5 shows three absolutely periodic interplay orbits with different mass ratios. For orbits Z1 and Z2 the outer masses are smaller than the inner ones. Orbit Z3 does have larger outer masses but they are of comparable size to the inner masses. For more extreme mass ratios, such as the 3:1 ratio above (Figure 4), there is no such absolutely periodic interplay orbit with the larger masses on the outside.

4. Caledonian five-body orbits

Similar families of orbits will occur in the Caledonian five-body problem when a central mass m_0 is added to the system. The limit $m_0 = 0$ gives the four-body families already explored. Figure 6 shows two five-body Schubart orbits. Above is the equal masses case. Below is a case with relatively larger central and outermost masses, m_0 and $m_1 = m_3$, respectively. The masses between, $m_2 = m_4$, are smaller. Figure 7 shows three double-choreography orbits including a central mass m_0 and four equal outer masses $m_1 = m_2 = m_3 = m_4$. The effect of adding the central mass is to make the periodic orbits rounder.

References

Chen, K. 2001, *Arch. Ration. Mech. Anal.* 158(4), 293

Chopovda, V. 2019, *Computational studies in the few-body problem*, PhD thesis, Massey University, Auckland, New Zealand

Chopovda, V., & Sweatman, W.L. 2018, *Cel. Mech. Dyn. Astron.*, 130, 39

Heggie, D.C. 1974, *Cel. Mech.*, 10, 217

Hénon, M. 1976, *Cel. Mech. Dyn. Astron.*, 13, 267

Roy, A.E., & Steves, B.A. 2000, *Cel. Mech. Dyn. Astron.*, 78, 299

Schubart, J. 1956, *AN*, 283, 17

Sekiguchi, M., & Tanikawa, K. 2004, *PASJ*, 56, 235

Sivasankaran, A., Steves, B.A., & Sweatman, W.L. 2010, *Cel. Mech. Dyn. Astron.*, 107, 157

Steves, B.A., Shoaib, M., & Sweatman, W.L. 2020, *Cel. Mech. Dyn. Astron.*, 132, 53

Sweatman, W.L. 2002, *Cel. Mech. Dyn. Astron.*, 82, 179

Sweatman, W.L. 2006, *Cel. Mech. Dyn. Astron.*, 94, 37

Sweatman, W.L. 2014, in: Z. Knežević & Anne Lemaître (eds.), *Complex Planetary Systems*, Proc. IAU Symposium No. 314 , p. 106

Sweatman, W.L. 2015, in: M. Cojocaru, I.S. Kotsireas, R.N. Makarov, R. Melnik & H. Shodiev (eds.), *Interdisciplinary Topics in Applied Mathematics, Modeling, and Computational Science*, Springer Proc. in Maths. and Stats., 117, p. 439

Multi-scale (time and mass) dynamics of space objects
Proceedings IAU Symposium No. 364, 2022
A. Celletti, C. Galeş, C. Beaugé, A. Lemaitre, eds.
doi:10.1017/S174392132100137X

Satellites' orbital stability through normal forms

Irene De Blasi[1], Alessandra Celletti[2] and Christos Efthymiopoulos[3]

[1]Department of Mathematics, University of Turin
and Politecnico of Turin, Turin, Italy
email: irene.deblasi@unito.it

[2]Department of Mathematics, University of Roma Tor Vergata, Rome, Italy
email: celletti@mat.uniroma2.it

[3]Department of Mathematics, University of Padova, Padua, Italy
email: cefthym@math.unipd.it

Abstract. A powerful tool to investigate the stability of the orbits of natural and artificial bodies is represented by perturbation theory, which allows one to provide normal form estimates for nearly-integrable problems in Celestial Mechanics. In particular, we consider the orbital stability of point-mass satellites moving around the Earth. On the basis of the J_2 model, we investigate the stability of the semimajor axis. Using a secular Hamiltonian model including also lunisolar perturbations, the so-called *geolunisolar* model, we study the stability of the other orbital elements, namely the eccentricity and the inclination. We finally discuss the applicability of Nekhoroshev's theorem on the exponential stability of the action variables. To this end, we investigate the non-degeneracy properties of the J_2 and geolunisolar models. We obtain that the J_2 model satisfies a "three-jet" non-degeneracy condition, while the geolunisolar model is quasi-convex non-degenerate.

Keywords. Celestial mechanics; Normal forms; Satellite dynamics.

1. Introduction

The study of the stability of celestial bodies is one of the major goals in Celestial Mechanics, especially in view of the growing importance taken, since the mid-20th century and in particular in the last decades, by the problem of the control of satellites and space debris orbiting around the Earth.

The dynamics of a small body moving around our planet is influenced by a number of factors, among which the most important are the attraction of the Earth, taking into account its non-spherical shape, and of the Sun and the Moon, which, for altitudes ranging from about 100 to 10^5 km, can be considered as third body perturbations to the dominant effect determined by the geopotential.

This work summarizes some stability results presented in De Blasi et al. 2021, where, to describe the small body's motion, two different models are considered: the J_2 *model*, which takes into account only the gravitational force of the Earth by including the associated geopotential, expanded in spherical harmonics and suitably truncated up to its dominant term, and the *geolunisolar* model, where the effects of Sun and Moon are added to the geopotential (see §2). The Hamiltonian formalism, as well as a particular set of action-angle variables (the so-called *modified Delaunay variables*), are used to describe both models.

Through a suitable sequence of canonical changes of coordinates (see §3), the Hamiltonians describing the J_2 and geolunisolar models, denoted respectively with $\mathcal{H}_{J_2}$ and $\mathcal{H}_{gls}$, are transformed to assume the form of quasi-integrable Hamiltonian functions, where one or more actions can be considered as *quasi-integrals* of motion (namely integrals for a truncated normal form Hamiltonian). Their stability can be estimated by means of suitable perturbation theory procedures as in §4 (compare with Steichen & Giorgilli 1997). We also mention that other tools can be used to obtain stability estimates for quasi-integrable systems, such as Nekhoroshev's theorem (see Nekhoroshev 1962), which, under suitable hypotheses, provides exponentially long stability times. One of the main hypotheses required is a non-degeneracy of the integrable part of the Hamiltonian. In §5 we discuss the non-degeneracy condition of the normalized Hamiltonians describing the J_2 and geolunisolar models.

2. The models

Let us consider a point-mass particle, say a debris S, moving around the Earth along an elliptic orbit with parameters $(a, e, i, M, \omega, \Omega)$, that is, semimajor axis, eccentricity, inclination, mean anomaly, argument of perigee and longitude of the ascending node. The motion of S is governed by the Earth's gravitational influence, whose associated potential can be expressed as an expansion in spherical harmonics (see Kaula 1966) which takes into account the non-spherical shape of our planet. Additionally to Earth's Keplerian attraction, the gravitational influences of Sun and Moon are taken into account.

Both models introduced in §1 are analysed by considering the associated Hamiltonian functions, denoted respectively with $\mathcal{H}_{J_2}$ and $\mathcal{H}_{gls}$, first expressed in Cartesian coordinates and then in the modified action-angle Delaunay variables

$$\begin{cases} L=\sqrt{\mu_E a} \\ P=\sqrt{\mu_E a}\left(1-\sqrt{1-e^2}\right) \\ Q=\sqrt{\mu_E a}\sqrt{1-e^2}\,(1-\cos i) \end{cases} \qquad \begin{cases} \lambda=M+\omega+\Omega \\ p=-\omega-\Omega \\ q=-\Omega, \end{cases} \tag{2.1}$$

where $\mu_E=\mathcal{G}M_E$ is the Earth's mass parameter.

The Hamiltonian $\mathcal{H}_{J_2}$ can be expressed as the sum of a zero-order Keplerian term and the J_2 term, which indicates the deformation in the geopotential due to Earth's oblateness. In Delaunay variables, having defined $\delta L=L-L_*=\sqrt{\mu_E a}-\sqrt{\mu_E a_*}$ with a_* taken as a reference value for the semimajor axis, $\mathcal{H}_{J_2}$ can be expanded in powers of $\sqrt{\delta L}$, $\sqrt{P}$ and $\sqrt{Q}$ obtaining the Hamiltonian

$$\mathcal{H}_{J_2}=n_*\delta L+\omega_1^*P+\omega_2^*+\sum_{s=1}^{2N}\sum_{\substack{k_1,k_2,k_3\in\mathbb{Z}\\ 0<|k_1|+|k_2|+|k_3|\leqslant s}}\mathcal{P}_{\substack{s,k_1\\k_2,k_3}}(\delta L,P,Q)\cos\left(k_1\lambda+k_2p+k_3q\right), \tag{2.2}$$

where $\mathcal{P}_{\substack{s,k_1\\k_2,k_3}}$ are polynomials of degree s in the actions and the sum is truncated up to order $2N$† for computational reasons.

The geolunisolar Hamiltonian $\mathcal{H}_{gls}$ is obtained by adding to $\mathcal{H}_{J_2}$ the third-body gravitational potentials due the presence of Sun and Moon, assuming the latter to be strictly on the ecliptic plane. Since in this case the attention is focused on the *secular* stability of the parameters (e, i), an average over the satellite's, Sun's and Moon's fast angles is performed, implying the constancy of the semimajor axis $a=a_*$. Moreover, the presence of two perturbing bodies on the ecliptic produces a shift in the equilibrium orbit of S (leading to the so-called *forced elements*), which modifies $(e,i)=(0,0)$ into $(e,i)=(0,i_*)$,

† In the actual computations performed in order to obtain the stability estimates provided in §4, the index N is set equal to 15.

with $i_* \neq 0$: a new set of action-angle variables $(I_1, I_2, \phi_1, \phi_2)$, centered in the corresponding forced values, is considered. Similarly to $\mathcal{H}_{J_2}$, the final geolunisolar Hamiltonian can be expressed as a trigonometric expansion in the new variables:

$$\mathcal{H}_{gls} = \nu_1 I_1 + \nu_2 I_2 + \sum_{s=3}^{N} \sum_{\substack{s_1,s_2 \in \mathbb{N} \\ s_1+s_2=s}} \sum_{\substack{k_1,k_2 \in \mathbb{Z} \\ |k_1|+|k_2| \leq s}} h_{\substack{s_1,s_2 \\ k_1,k_2}} I_1^{s_1/2} I_2^{s_2/2} \cos(k_1\phi_1 + k_2\phi_2), \qquad (2.3)$$

where we notice that $\nu_1 \simeq \nu_2$.

3. Hamiltonian normalization

Taking into account the models described in §2, we can obtain estimates on the stability times of the orbital parameters of the satellite, in particular of the semimajor axis in the J_2 model and of the quantity $\sqrt{1-e^2}\,(1-\cos i)$ in the geolunisolar case. The approach taken to achieve this goal is heavily based on normal form methods (see Efthymiopoulos 2011, to which we also refer for the definition of Lie series transformations): in particular, using the Lie series technique, a formal elimination of the fast angle in the J_2 model and of a particular quasi-resonant combination of the angles within the geolunisolar framework is performed, reducing to new quasi-integrable Hamiltonian functions for which the stability estimates are obtained.

In the case of the model described by $\mathcal{H}_{J_2}$, expressed by the expansion 2.2, a composition of near-identity canonical transformations in the form of Lie series allows to remove the dependence on the fast angle λ up to a prefixed polynomial order in the actions' square roots, leading to the normalized Hamiltonian†

$$\mathcal{H}_{J_2}^{norm}(\delta L, P, Q, \lambda, p, q) = \mathcal{Z}_{J_2}(\delta L, P, Q, p, q) + \mathcal{R}_{J_2}(\delta L, P, Q, \lambda, p, q),$$

where $\mathcal{Z}_{J_2}$, the so-called *normal part*, does not depend on λ and $\mathcal{R}_{J_2}$, the *remainder*, is of total order $M = N-3$ in $\sqrt{\delta L}$, $\sqrt{P}$, $\sqrt{Q}$.

It is straightforward from the independence of $\mathcal{Z}_{J_2}$ on λ that the variation of the first Delaunay action L (and, as a consequence, of the satellite's semimajor axis a) depends only on the remainder, as the former is a first integral for the normal part: in this sense, it can be considered a *quasi-integral* of the motion induced by the whole Hamiltonian function $\mathcal{H}_{J_2}^{norm}$.

In the case of $\mathcal{H}_{gls}$ in the form of 2.3, the presence of the $1:1$ resonance determined by the relation $\nu_1 \simeq \nu_2$ translates in the unfeasibility of a normalization procedure which simply removes the dependence on the angles of selected terms, as it would imply the uncontrolled growth in the size of the remainder due to the presence of small divisors. We opt instead for a normalized Hamiltonian in *quasi-resonant* form, in the sense that the composition of Lie series transformations removes the dependence on all combinations of the angles ϕ_1, ϕ_2 except for the resonant one $\phi_1 - \phi_2$. The resulting normalized geolunisolar Hamiltonian takes then the form

$$\mathcal{H}_{gls}^{norm}(I_1, I_2, \phi_1, \phi_2) = \mathcal{Z}_{gls}^{sec}(I_1, I_2) + \mathcal{Z}_{gls}^{res}(I_1, I_2, \phi_1 - \phi_2) + \mathcal{R}_{gls}(I_1, I_2, \phi_1, \phi_2),$$

where the sum $\mathcal{Z}_{gls}^{sec} + \mathcal{Z}_{gls}^{res}$ represents the normal part, divided into its secular and resonant terms, and the remainder $\mathcal{R}_{gls}$ is again of total order $M = N-3$ in $\sqrt{I_1}$, $\sqrt{I_2}$. The

† From a rigorous point of view, the variables of $\mathcal{H}_{J_2}^{norm}$ are not the original Delaunay variables $(\delta L, P, Q, \lambda, p, q)$. However, since we are dealing with near-identity transformations, they differ from them only by short-term small oscillations which do not affect the secular stability of the orbital elements: for this reason, with an abuse of notation, we keep the original notation for the new Delaunay variables as well.

dynamics induced by the normal part alone admits the integral

$$I_1 + I_2 = L_* \left(1 - \sqrt{1-e^2}\,(1 - \cos i)\right),$$

which, similarly to the J_2 case, can be considered a quasi-integral for the whole Hamiltonian $\mathcal{H}_{gls}^{norm}$ and whose stability will be investigated in §4.

4. Stability estimates

Taking advantage of the peculiar structure of the normalized functions $\mathcal{H}_{J_2}^{norm}$ and $\mathcal{H}_{gls}^{norm}$, stability estimates based on the size of $\mathcal{R}_{J_2}$ and $\mathcal{R}_{gls}$ can be produced.

The size of the remainders can be quantified by means of the sup norm $\|\cdot\|_{\infty,D}$ over a suitable bounded domain D in the variables, which depends on the model. In particular, if the remainders are *small* in the above sense with respect to the corresponding normal parts, the dynamics induced by $\mathcal{H}_{J_2}^{norm}$ (respectively $\mathcal{H}_{gls}^{norm}$) can be considered as a small perturbation of that defined by $\mathcal{Z}_{J_2}$ (respectively, $\mathcal{Z}_{gls}^{sec} + \mathcal{Z}_{gls}^{res}$).

In the case of the J_2 model, the derivative of the first Delaunay action L depends only on the λ–derivative of $\mathcal{R}_{J_2}$:

$$\frac{d}{dt}L = \frac{d}{dt}\delta L = -\frac{\partial \mathcal{H}_{J_2}^{norm}}{\partial \lambda} = -\frac{\partial \mathcal{R}_{J_2}}{\partial \lambda}. \tag{4.1}$$

Let us consider a bounded domain $D \subset [0,1) \times [0,\pi/2]$ in eccentricity and inclination; recalling the dependence of the Delaunay actions on the orbital parameters, define the sup-norm

$$\|f\|_{\infty,D} = \sup_{\substack{(e,i)\in D\\ (\lambda,p,q)\in \mathbb{T}^3}} \left| f\,(e,i,\lambda,p,q) \right|. \tag{4.2}$$

Suppose now that at time $t=0$ the initial value of L is L_0; defining by $L(T)$ the evolution of L at time T, from (4.1) and the mean value theorem, one obtains

$$\left|L(T) - L_0\right| \leqslant \left\|\frac{dL}{dt}\right\|_{\infty,D} T = \left\|\frac{\partial \mathcal{R}_{J_2}}{d\lambda}\right\|_{\infty,D} T.$$

Fixing a constant value ΔL, one has that a lower bound for the time T such that $|L(t) - L_0| \leq \Delta L$ is given by

$$T \geqslant T_{J_2} \equiv \frac{\Delta L}{\|\partial \mathcal{R}_{J_2}/\partial \lambda\|_{\infty,D}}.$$

With the same reasoning, recalling the definition of Poisson brackets and their relation to the time evolution of functions under the dynamics induced by a Hamiltonian (see for example Giorgilli 2002), an analogous stability estimate can be obtained for the geolunisolar model: in particular, if Γ is the maximal variation allowed for the quantity $I_1 + I_2 = L_* \left(1 - \sqrt{1-e^2}\,(1-\cos i)\right)$ over the time T, one has

$$T \geqslant T_{gls} = \frac{\Gamma}{\|\{I_1 + I_2, \mathcal{R}_{gls}\}\|_{\infty,D}},$$

where $\|\cdot\|_{\infty,D}$ is the supremum of the absolute value of the argument for $(e,i) \in D$ and $(\phi_1, \phi_2) \in \mathbb{T}^2$.

The theoretical considerations leading to the definition of the stability times T_{J_2} and T_{gls} are the basis of the numerical investigations on the stability for the J_2 and geolunisolar models for different reference values a_* of the semimajor axis. Such estimates are obtained by implementing the following procedure:

Table 1. Stability times T_{gls} for different values of altitude in the geolunisolar model.

Altitude	Stability time in D
3 000 km	$4.615\ 51 \times 10^{13}$
20 000 km	$2.201\ 44 \times 10^{12}$
35 790 km	$3.512\ 66 \times 10^{10}$
50 000 km	$1.072\ 63 \times 10^{8}$
100 000 km	$3.366\ 09 \times 10^{4}$

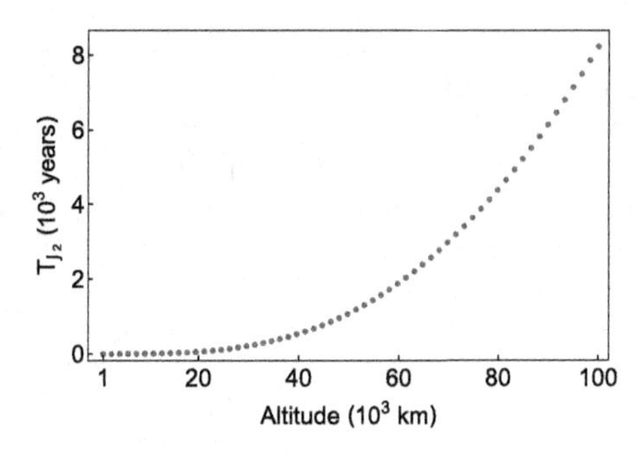

Figure 1. Stability times T_{J_2} for different values of altitude (e.g. $a_* - R_E$) in the J_2 model.

• compute the initial Hamiltonians $\mathcal{H}_{J_2}^{\leqslant N}$ and $\mathcal{H}_{gls}^{\leqslant N}$, obtained as finite trigonometric expansions in the action-angle variables;
• normalize up to order $M = N - 3$ to obtain $\mathcal{H}_{J_2}^{norm}$ and $\mathcal{H}_{gls}^{norm}$;
• compute the stability times T_{J_2} and T_{gls} on suitable domains and for chosen values of ΔL and Γ, provided that the remainders are small enough to represent small perturbations of the respective normal parts.
Given a reference value a_* for the semimajor axis, the maximal excursions for $L(T)$ and $(I_0 + I_1)(T)$ are chosen to be

$$\Delta L = 0.05 R_E \sqrt{\frac{\mu_E}{a_*}}, \quad \Gamma = 0.05 \sqrt{\frac{\mu_E}{a_*}},$$

where the first choice corresponds to a maximal variation of $\Delta a = 0.1 R_E$ for the semimajor axis. Furthermore, for computational reasons, the supremum norm defined in 4.2 is replaced by an appropriate norm $\|\cdot\|_{\infty, D_*}$ over a finite grid $D_* \subset D$, which turns out to be an upper bound of $\|\cdot\|_{\infty, D}$. To ensure the smallness of the remainders over the normal parts, D is set equal to $[0, 0.15] \times [0, \pi/2]$ for the J_2 model and to $[0, 0.1] \times [0, 0.1]$ in the geolunisolar case.

Figure 1 shows the behaviour of T_{J_2} for altitudes ranging from 10^3 to 10^5 km: as expected by the fact that the Earth's oblateness is more relevant for small altitudes, the stability time increases with a_*, when the J_2 model approaches the classical Keplerian one.

Table 1 lists the values of T_{gls} for selected values from small to high altitudes; since the lunisolar influence is stronger far from the Earth, the stability times tend to decrease with the altitude.

5. Non-degeneracy conditions

One of the most important results of the XX century on the stability of nearly-integrable Hamiltonian systems is represented by Nekhoroshev's theorem (Nekhoroshev

1962; Poshel 1993), which is a fundamental tool to provide exponential stability estimates for quasi-integrable systems of the form

$$\mathcal{H}(\mathbf{I},\boldsymbol{\phi}) = h(\mathbf{I}) + \varepsilon f(\mathbf{I},\boldsymbol{\phi}), \tag{5.1}$$

where $(\mathbf{I},\boldsymbol{\phi}) \in U \times \mathbb{T}^n$, $U \subset \mathbb{R}^n$, n being the number of degrees of freedom of the system and ε a small parameter. Typically, h is called the *unperturbed* function, and is straightforwardly integrable as it does not depend on the angles $\boldsymbol{\phi}$, while f is the *perturbing* function.

A fundamental assumption in the original version of Nekhoroshev's theorem is a non-degeneracy of the unperturbed function $h(\mathbf{I})$ called *steepness condition*, which is a geometric assumption rather complex to verify in practice. However, the steepness condition is implied by stronger non-degeneracy conditions, much simpler to check, such as the convexity, the quasi-convexity and the three-jet non-degeneracy (see, e.g., Chierchia, et al. 2018).

As a preliminary investigation of the applicability of Nekhoroshev's theorem to our cases, we establish whether these conditions are verified or not in the J_2 and geolunisolar models: to this end, a suitable splitting of $\mathcal{H}^{norm}_{J_2}$ and $\mathcal{H}^{norm}_{gls}$ into unperturbed and perturbed parts is needed. In particular, as the non-degeneracy conditions involve only $h(\mathbf{I})$, one has to define precisely the unperturbed functions, denoted respectively as h_{J_2} and h_{gls}. In the case of the J_2 model, we impose $h_{J_2}(P,Q)$ to be the sum of all the angle-independent terms of $\mathcal{H}^{norm}_{J_2}$, where, given the practical stability of L shown in §4, δL is supposed to be constant and equal to 0. In the geolunisolar model, $h_{gls}(P,Q)$ is given by the sum of all the terms of $\mathcal{H}^{norm}_{gls}$ which are independent on the angles and at most quadratic in the actions.

The verification of the non-degeneracy conditions for h_{J_2} and h_{gls} is performed numerically in De Blasi et al. 2021 for different values of the altitudes (3000 km, 20000 km, 35790 km and 50000 km) and for $(e,i) \in D = [0,0.1] \times [0,0.1]$. The different non-degeneracy assumptions are computed as follows:

- convexity: check if the Hessian matrices associated to h_{J_2} and h_{gls} are positively (or negatively) defined;
- quasi-convexity: it is equivalent to Arnold isoenergetic non-degeneracy condition, involving first and second order derivatives of the unperturbed Hamiltonian;
- three-jet non-degeneracy: it is a condition involving up to third order derivatives and it was verified numerically on a grid of 10000 points in the actions.

For the considered altitudes and in the regime of eccentricities and inclinations defined by D, we find that h_{J_2} is three-jet non-degenerate, while h_{gls} is quasi-convex; this result is of particular relevance, as it implies that the presence of the lunisolar part removes the degeneracy of the J_2 model.

References

Chierchia, L., Faraggiana, M.E. and Guzzo, M. 2018, *Annali di Matematica Pura e Applicata*, 198(6), 2151–2165

De Blasi, I., Efthymiopoulos, C. and Celletti, A. 2021, *to appear in J. Nonlinear Sci.*

Giorgilli, A. 2002, *Notes on exponential stability of Hamiltonian systems*

Kaula, W.M. 1966, *Theory of Satell. Geodesy*, 30–37

Nekhoroshev, N.N. 1962, *Uspekhi Matematicheskikh Nauk*, 32(6), 5–66

Pöshel, J. 1993, *Mathematische Zeitschrift*, 213(1), 187–216

Steichen, D. and Giorgilli, A. 1997, *Celest. Mech. and Dynamical Astron.*, 69(3), 317–330

Multi-scale (time and mass) dynamics of space objects
Proceedings IAU Symposium No. 364, 2022
A. Celletti, C. Galeş, C. Beaugé, A. Lemaitre, eds.
doi:10.1017/S1743921321001484

Noise, friction and the radial-orbit instability in anisotropic stellar systems: stochastic $N-$body simulations

Pierfrancesco Di Cintio[1,2,3] **and Lapo Casetti**[2,3,4]

[1]Enrico Fermi Research Center (CREF), Via Panisperna 89A, I-00184, Rome, Italy
email: pierfrancesco.dicintio@unifi.it

[2]INFN, Sezione di Firenze, via G. Sansone 1, I-50019, Sesto Fiorentino (FI), Italy

[3]Dipartimento di Fisica e Astronomia, Università di Firenze,
via G. Sansone 1, I-50019, Sesto Fiorentino (FI), Italy

[4]INAF-Osservatorio astrofisico di Arcetri, largo E. Fermi 5, I-50125, Firenze, Italy

Abstract. By means of numerical simulations we study the radial-orbit instability in anisotropic self-gravitating $N-$body systems under the effect of noise. We find that the presence of additive or multiplicative noise has a different effect on the onset of the instability, depending on the initial value of the orbital anisotropy.

Keywords. Stellar dynamics; Galaxies: kinematics and dynamics; Methods: n-body simulations; Diffusion.

1. Introduction

Spherically symmetric, self-gravitating collisionless equilibrium systems with a large fraction of the kinetic energy stored in low angular momentum orbits are known to be dynamically unstable. The associated instability is known as Radial Orbit Instability (hereafter ROI, see e.g. Polyachenko & Shukhman (2015) and references therein). Usually, the amount of radial anisotropy in a spherical system is quantified by introducing the Fridman-Polyachenko-Shukhman parameter (see Binney & Tremaine (2008))

$$\xi \equiv \frac{2K_r}{K_t}, \tag{1.1}$$

where the radial and tangential kinetic energies are given respectively by

$$K_r = 2\pi \int \rho(r)\sigma_r^2(r)r^2\mathrm{d}r, \quad K_t = 2\pi \int \rho(r)\sigma_t^2(r)r^2\mathrm{d}r, \tag{1.2}$$

ρ is the system density, and σ_r^2 and σ_t^2 are the radial and tangential phase-space averaged square velocity components, respectively. For isotropic systems $\xi = 1$. Numerical simulations show that the ROI typically occurs for $\xi \gtrsim 1.7$, even though it is well known that the "real" critical value of ξ above which the given system is unstable, depends on the specific phase-space structure of the initial condition under consideration.

The ROI it is frequently invoked as the mechanism responsible for the triaxiality of the elliptical galaxies and the formation of bars in disk galaxies. However, little is known on the effective nature of the underlying mechanism or its near- or far-field origin (see e.g. Polyachenko & Shukhman (2015); Di Cintio, Ciotti & Nipoti (2017) and references therein). Recently, Marechal & Perez (2010) introduced a novel interpretation of ROI

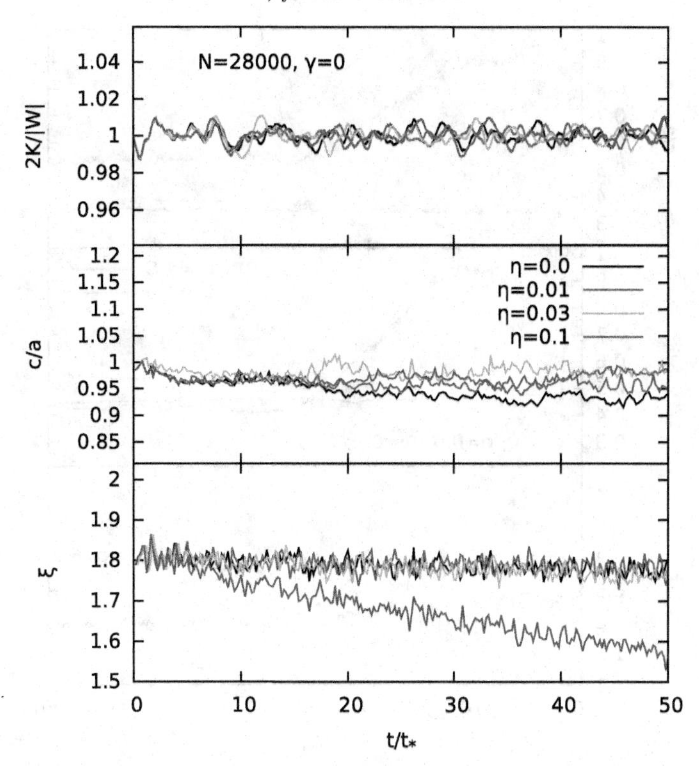

Figure 1. For an initially mildly anisotropic ($\xi_0 = 1.8$) flat-cored model ($\gamma = 0$) with $N = 28000$, evolution of the virial ratio $2K/|W|$ (top panel), axial ratio c/a (middle panel) and anisotropy parameter ξ (bottom panel).

as a (effective) dissipation-induced phenomenon. In this preliminary work we investigate their argument by means of direct $N-$body simulations with a controllable source of (external) noise and dissipation.

2. Methods

We study the stability of a family of $\gamma-$models with density profile given by

$$\rho(r) = \frac{3-\gamma}{4\pi}\frac{Mr_c}{r^{\gamma}(r+r_c)^{4-\gamma}}, \tag{2.1}$$

with total mass M, scale radius r_c and logarithmic density slope γ. In order to generate the velocities for the simulation particles we use the standard rejection technique to sample the anisotropic equilibrium phase-space distribution function $f(Q)$, obtained for a given (spherical) density-potential couple (ρ, Φ) linked by the Poisson equation $\Delta\Phi = 4\pi G\rho$, applying the usual Osipkov-Merritt reparametrization (Osipkov (1985); Merritt (1985)) of the Eddington (1916) integral inversion

$$f(Q) = \frac{1}{\sqrt{8}\pi^2}\int_Q^0 \frac{\mathrm{d}^2\rho_a}{\mathrm{d}\Phi^2}\frac{\mathrm{d}\Phi}{\sqrt{\Phi - Q}}. \tag{2.2}$$

In Equation (2.2)

$$Q = E + J^2/2r_a^2, \tag{2.3}$$

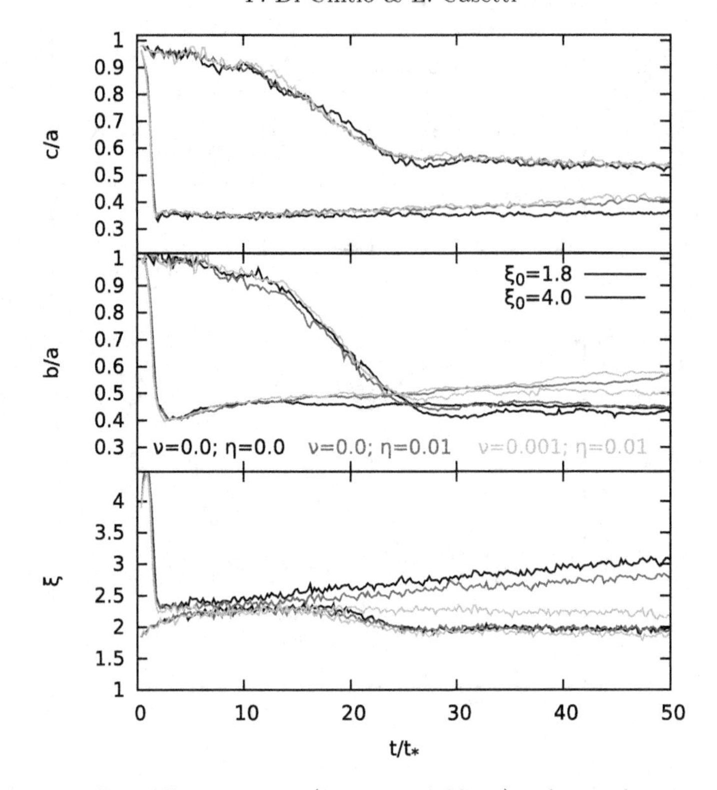

Figure 2. For initially mildly anisotropic ($\xi_0 = 1.8$, solid lines) and strongly anisotropic ($\xi_0 = 4$, dashed lines) Hernquist models ($\gamma = 1$): evolution of the axial ratios c/a (top panel), b/a (middle panel) and anisotropy parameter ξ (bottom panel).

and E and J are the particle's energy and angular momentum per unit mass, respectively. The quantity r_a is the so-called anisotropy radius, and ρ_a the augmented density, defined by

$$\rho_a(r) \equiv \left(1 + r^2/r_a^2\right) \rho(r). \tag{2.4}$$

For our specific choice of $\rho(r)$ in Eq. (2.1), the model's potential is given by

$$\begin{aligned} \Phi(r) &= -\frac{GM}{(2-\gamma)r_c}\left[1 - \left(\frac{r}{r+r_c}\right)^{2-\gamma}\right] \quad \text{for} \quad \gamma \neq 2; \\ \Phi(r) &= \frac{GM}{(2-\gamma)r_c} \ln \frac{r}{r+r_c} \quad \text{for} \quad \gamma = 2. \end{aligned} \tag{2.5}$$

The anisotropy radius r_a controls the extent of anisotropy of the model so that, the velocity-dispersion tensor is nearly isotropic for $r < r_a$, and increasingly radially anisotropic for $r > r_a$, thus small values of r_a are associated to more radially anisotropic systems, i.e. larger values of ξ.

Throughout this work we assume units such that $G = M = r_c = 1$, so that the dynamical time and the scale velocity become $t_* = \sqrt{r_c^3/GM}$ and $v_* = r_c/t_*$ and are both equal to unity. Individual particle masses are therefore $m = 1/N$.

In order to consider the effect of noise and dissipation, we express the particles' dynamics in terms of Langevin-like equations (e.g. see Kandrup (1980)) of the form

$$\ddot{\mathbf{r}}_i = -\nabla\Phi(\mathbf{r}_i) - \nu\mathbf{v}_i + \mathbf{F}(\mathbf{r}_i), \tag{2.6}$$

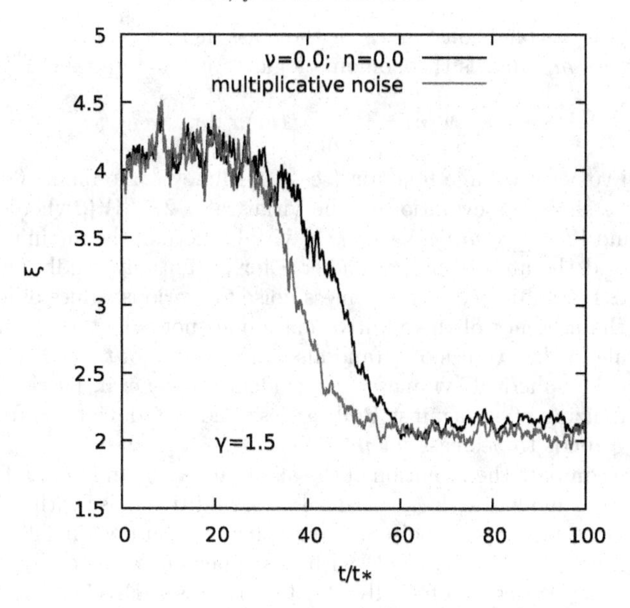

Figure 3. Evolution of the anisotropy parameter ξ for a system starting with $\gamma = 1.5$ and $\xi_0 = 4$.

where, in our case, the acceleration $-\nabla\Phi$ on each particle is evaluated self-consistently by direct sum over all other particles, ν is the dynamical friction [Chandrasekhar (1943, 1949)] coefficient, and $\mathbf{F}(\mathbf{r})$ a fluctuating force (per unit mass).

In our numerical simulations we solved Eqs.(2.6) with the so-called quasi-symplectic Mannella (2004) scheme with fixed time-step $\Delta t = 10^{-2} t_*$, in the same fashion as Pasquato & Di Cintio (2020) and Di Cintio, Ciotti & Nipoti (2020).

We note that, a similar approach could also be extended to the study of protoplanetary disks in dense environments under the effect of flyby stars, since in principle the disk hydrodynamics and the stellar dynamics have different time scales in a numerical simulation so the effect of passing stars could be simplified as a stochastic process (e.g. see Cattolico & Capuzzo-Dolcetta (2020)).

3. Numerical simulations and discussion

Following Pogorelov & Kandrup (1999), Terzic & Kandrup (2003) and Sideris & Kandrup (2004), we have implemented three different forms of noise: *i)* additive noise without friction (i.e. $\nu = 0$ in Eq. 2.6). *ii)* additive noise connected to friction via the Fluctuation-Dissipation Theorem such that

$$\eta^2 = \Theta\nu/t_c, \tag{3.1}$$

where η is the typical amplitude of the Gaussian distributed force $\mathbf{F}$, Θ is the system's temperature (proportional to the velocity dispersion σ) and t_c is the autocorrelation time of the noise. *iii)* multiplicative noise with friction where the dynamical friction coeffcient is explicitly dependent on the particle velocity as

$$\nu = 4\pi G^2 \rho_* (m + m_*) \ln \Lambda \frac{\Psi(v)}{v^3}, \tag{3.2}$$

where G is the gravitational constant, ρ_* is the mass density of a (fictitious) background of particles of mass m_*, $\ln\Lambda$ is the Coulomb logarithm, $v = ||\mathbf{v}||$, and

$$\Psi(v) = 4\pi \int_0^v f(v_*)v_*^2 dv_*, \tag{3.3}$$

is the fractional velocity volume function (see e.g. Binney & Tremaine (2008)).

In Figure 1 we show the evolution of the virial ratio $2K/|W|$, where K is the total kinetic energy and $W = \sum_N m_i\mathbf{r}_i \cdot \nabla\Phi(\mathbf{r}_i)$ the virial function; the minimum to maximum axial ratio c/a and the anisotropy parameter ξ for an initially mildly anisotropic $\gamma = 0$ system with $\xi_0 \simeq 1.8$, subjected to frictionless noise for various values of noise amplitude η. In all cases, the presence of the additive noise does not take the system out of virial equilibrium, while for low values of η (and also large values of t_c, not shown here) some deviations from the spherical symmetry are evident. In general, large values of η have a somewhat stabilizing effect against ROI, as less and less deviations from $c/a = 1$ are observable and ξ tends to decrease for $\eta > 0.05$.

In Figure 2 we compare the evolution of the axial ratios c/a and b/a and the anisotropy parameters for $\gamma = 1$ models with $\xi_0 = 4$ and 1.8 for additive noise with and without friction. In general, the presence of noise or noise plus friction does not have a significant effect on the onset of ROI for models with a steeper cusp (generally more unstable, as they admit a larger degree of wildly chaotic orbits see Di Cintio & Casetti (2019, 2020)) and low values of the initial anisotropy (i.e. $\xi_0 = 1.8$), while for larger values of ξ_0, corresponding to a more violent instability, the evolution of the triaxiality and the anisotropy are affected by the presence of noise, with systematically less anisotropic and more "triaxial" end states associated to the presence of larger amounts of noise and friction. Introducing a multiplicative noise (with velocity dependent friction coefficient) complicates the picture even further with as it apparently it does not alter significantly the evolution of the axial ratios, nor the final values attained by ξ even for extremely anisotropic models with steep cusps, while it seems to somewhat anticipate the time at which the anisotropy parameter starts moving to lower values (i.e. unstable models become more isotropic earlier), as shown in Figure 3 for a system with $\gamma = 1.5$.

From these preliminary results we speculate that the mechanisms leading to ROI might work differently in real systems subjected to different form of internal or environment-related sources of noise, as well as different central concentrations (see e.g. Trenti & Bertin (2006)). In particular, we speculate that multicomponent systems with different anisotropy profiles for each component could develop the ROI in a substantially different fashion as their single component counterparts. We will explore this matter further in a forthcoming publication (Di Cintio, Zocchi & Casetti (2021)), studying the stability of a family of Gieles & Zocchi (2015) models with more mass components with tunable degree of radial anisotropy.

References

Binney, J.; Tremaine, S. 2008 Galactic dynamics (2nd edition, Princeton University Press)
Cattolico, R.S.; Capuzzo-Dolcetta, R. 2020 *A&SS*, 365, 10
Chandrasekhar S. 1943, *ApJ*, 97, 255
Chandrasekhar S. 1949, *Reviews of Modern Physics*, 21, 383
Di Cintio, P.F.; Casetti, L. 2019 *MNRAS*, 489, 5876
Di Cintio, P.F.; Casetti, L. 2020 *MNRAS*, 494, 1027
Di Cintio, P.F.; Ciotti, L.; Nipoti, C. 2017 *MNRAS*, 468, 2222
Di Cintio, P.F.; Ciotti, L.; Nipoti, C. 2020 *IAUS Proceedings*, 351, 426
Di Cintio, P.F.; Zocchi, A.; Casetti, L. 2021 *In preparation*
Eddington, A. 1916, *MNRAS* 76, 525
Gieles, M.; Zocchi, A. 2015, *MNRAS* 454, 576

Habib, S.; Kandrup, H.E.; Mahon, M.E. 1997 *ApJ* 64, 56209
Kandrup H. E. 1980, *Phys. Rep.*, 63, 1
Kandrup, H.E.; Sideris, I.V. 2001 *Phys. Rev. E* 64, 56209
Mannella R. 2004, *Phys. Rev. E*, 69, 041107
Marechal, L; Perez, J 2010 *MNRAS* 405, 2785
Merritt D. 1985, *AJ*, 90, 1027
Osipkov L.P. 1985, *Sov. Astr. Letters*, 5, 42
Pasquato, M.; Di Cintio, P. 2020, *A&A* 640, 79
Polyachenko, E.V.; Shukhman, I.G. 2015, *MNRAS* 451, 601
Pogorelov, I.V.; Kandrup, H.E. 1999 *Phys. Rev. E* 60, 1567
Sideris, I.V.; Kandrup, H.E. 2004 *ApJ* 602, 678
Terzic, B.; Kandrup, H.E. 2003 *ArXiv Preprint: arXiv:astro-ph/0312434*
Trenti, M.; Bertin, G. 2006 *ApJ* 637, 717

Multi-scale (time and mass) dynamics of space objects
Proceedings IAU Symposium No. 364, 2022
A. Celletti, C. Galeş, C. Beaugé, A. Lemaitre, eds.
doi:10.1017/S1743921321001344

A cartographic study of spin-orbit coupling in binary asteroids

Mahdi Jafari Nadoushan

K. N. Toosi University of Technology, Tehran, Iran

Abstract. In the spin-orbit resonances, we assume that the orbit of the secondary asteroid around the primary is invariant, which is a reasonable assumption at first glance. Owing to the irregularity of asteroids' geometry and their effect on the mutual orbit, this assumption should be revised. Therefore, we focus on a binary asteroid with a spherical primary and a secondary with an irregular shape. When the shape of a secondary asteroid is not a sphere, the gravitational interaction is important, and we should consider the interaction of orbit and spin. We generate fast Lyapunov indicator (FLI) maps for both spin-orbit resonance and spin-orbit coupling problems and investigate the effect of orbit alternation on the structure of phase space.

Keywords. celestial mechanics, methods: numerical, minor planets, asteroids

1. Introduction

A spin-orbit resonance is a situation where the rotation period of a secondary asteroid and its motion around a primary asteroid are commensurate. When we consider a fixed orbit for the secondary and study its spin, we call the problem spin-orbit resonance (SOR). While, when we investigate the simultaneous alternation of orbit and spin of the secondary due to perturbations, we call it spin-orbit coupling (SOC). In the seminal work, Goldreich and Peale (1966) have formulated SOR in fixed eccentric orbits for planets and satellites. Wisdom *et al.* (1984) have studied chaotic motion in the spin-orbit problem through resonance overlapping criterion. Celletti and Chierchia (2000) have considered a nearly-integrable Hamiltonian model describing the conservative spin-orbit interaction. Nadoushan and Assadian (2016) have studied the overlap of the first- and second-order spin-orbit resonances for different values of system parameters. Misquero and Ortega (2020) have analytically investigated a dissipative spin-orbit problem and studied capturing into the synchronous resonance. In most previous works, it was assumed that the mutual orbit is invariable. Some recent works have considered a variable mutual orbit. Naidu and Margot (2015) have considered coupled spin and orbital motions of binary asteroids and showed the existence of a chaotic motion. Hou and Xin (2017) have analytically studied spin-orbit problem with a variable orbit and showed that the resonance center changes for some values of the system parameters. Wang and Hou (2020) have examined the rotation of the secondary in a binary asteroid system by considering the influence of the secondary's rotation on the mutual orbit.

Nevertheless, some questions remain. Such as how different the two approaches are, and if the SOR model can capture dynamics of the system or essentially the SOC model should be considered. In this work, we present a comparative study between SOC and SOR by surfing the phase space of both models, and try to answer these questions.

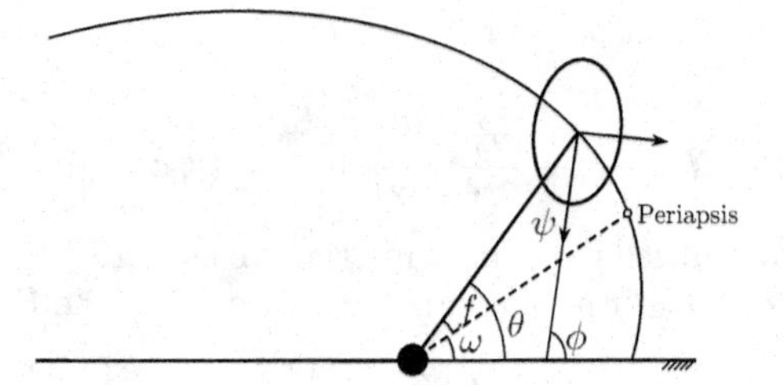

Figure 1. Geometry of spin-orbit coupling.

2. Equations of motion

Here, we provide a Hamiltonian of the SOC for the planar case. All the usual assumptions apply here as well. We utilize the gravitational potential energy function up to the fourth-order in terms of Stokes coefficients. We assume that an ellipsoidal asteroid of mass m_s is subjected to the gravitational attraction of a homogenous spherical asteroid with mass m_p, and the center of mass of the ellipsoid moves on a variable orbit, as we show in Figure 1. In the SOC, unlike the SOR, the mutual orbit varies. Therefore, the periapsis varies, and we need an appropriate reference, as depicted in Figure 1.

The gravitational potential energy for SOC model is as follows (Hou *et al.* (2017)):

$$V(r,\psi) = -Gm_pm_s\left\{\frac{1}{r}+\frac{1}{r^3}\left[-\frac{a_s^2C_{20}}{2}+\frac{a_s^2C_{22}}{4}\cos(2\psi)\right]\right.$$
$$\left.+\frac{1}{r^5}\left[\frac{3a_s^4C_{40}}{8}+\frac{a_s^4C_{42}}{24}\cos(2\psi)+\frac{a_s^4C_{44}}{192}\cos(4\psi)\right]\right\}$$

where G is the universal gravitational constant, C_{ij} are Stokes coefficients calculated as follows (Balmino, 1994):

$$C_{20} = \frac{1}{5a_s^2}\left(c_s^2 - \frac{a_s^2+b_s^2}{2}\right) \qquad C_{22} = \frac{1}{20a_s^2}\left(a_s^2 - b_s^2\right)$$
$$C_{40} = \frac{15}{7}\left(C_{20}^2 + 2C_{22}^2\right) \quad C_{42} = \frac{5}{7}C_{20}C_{22} \quad C_{44} = \frac{5}{28}C_{22}^2$$

where a_s, b_s and c_s are the semi-axes of the secondary with $a_s \geq b_s \geq c_s$. Hence, the mass normalized Hamiltonian of the SOC model, including orbital and rotational kinetic energies, and the above gravitational potential energy,is given below:

$$H\left(t,r,\dot{r},\vartheta,\dot{\vartheta},\phi,\dot{\phi}\right) = \frac{1}{2}\left(\dot{r}^2 + r^2\dot{\vartheta}^2\right) - \frac{\mu}{r} + \frac{I_3}{2m}\dot{\phi}^2 - \mu\left\{-\frac{a_s^2C_{20}}{2r^3} + \frac{3a_s^4C_{40}}{8r^5}\right.$$
$$\left.+\left[\frac{a_s^2C_{22}}{4r^3} + \frac{a_s^4C_{42}}{24r^5}\right]\cos(2\phi - 2\vartheta) + \frac{a_s^4C_{44}}{192r^5}\cos(4\phi - 4\vartheta)\right\}$$

where I_3 is the moment of inertia, m is $\frac{m_pm_s}{m_p+m_s}$ and is called reduced mass, $\mu = G(m_p + m_s)$ and ϑ, measured from the reference, is the sum of true anomaly f and argument of periapsis ω. The first two terms describe the Keplerian motion, and the third term represents the free rotational motion. The last term is the asphericity perturbation resulting in the orbital and rotational alteration of the secondary asteroid motion. The Hamiltonian is a conserved quantity.

We define the unit of mass such that $m = 1$, and take a_s as the unit of length. We also choose the unit of time such that $\mu = 1$. Now, let us introduce the generalized coordinates and momentums $(r, p_r, \vartheta, p_\vartheta, \phi, p_\phi)$ where r , ϑ and ϕ are the generalized coordinates, as shown in Figure 1 and p_r, p_ϑ and p_ϕ are the conjugated generalized momentums. Therefore, we can write

$$H = \frac{1}{2}p_r^2 + \frac{p_\vartheta^2}{2r^2} + \frac{1}{2\bar{I}_3}p_\phi^2 + V\left(r, \phi - \vartheta\right)$$

where $\bar{I}_3$ is the mass normalized moment of inertia. We use an F_2-type generating function $F_2(\vartheta, \phi, p_\psi, p_\vartheta) = (\phi - \vartheta)\, p_\psi + \phi p_\vartheta$, for canonical transfer of the Hamiltonian (2) to

$$H\left(t, r, p_r, \psi, p_\psi, \vartheta, p_\vartheta\right) = \frac{p_r^2}{2} + \left(\frac{1}{2r^2} + \frac{1}{2\bar{I}_3}\right)p_\psi^2 + \frac{p_\vartheta^2}{2\bar{I}_3} + \frac{p_\psi p_\vartheta}{\bar{I}_3} + V\left(r, \psi\right)$$

Because the new generalized coordinate ϑ does not appear in the Hamiltonian, its conjugated generalized momentum p_ϑ is conserved. Finally, we can derive the equations of motion from the above Hamiltonian using the canonical formulation:

$$\dot{\vec{q}} = \frac{\partial H}{\partial \vec{p}} \quad \text{and} \quad \dot{\vec{p}} = -\frac{\partial H}{\partial \vec{q}} \tag{6}$$

3. Cartography of Resonances

In this section, we compare the phase space of the SOC and the SOR through cartography of resonances. To this end, we utilize the Fast Lyapunov Indicator (FLI) maps (Froeschlé *et al.* (1997)). For a regular motion, the value of FLI grows linearly, whereas, for a chaotic motion, the value of FLI increases exponentially. Hence, by computing FLI, we could distinguish between chaotic and regular motions. We report the FLI via a color scale, such that the highest FLI values are shown in brown, and the lowest values are shown in blue. We consider a set of 300x300 initial conditions placed on a regular grid in the plane $\left(\dot{\phi}/n, \phi\right)$, where normalized angular velocity ranges from 0 to 2.5, while the angle ϕ ranges from 0 to π. To produce the FLI maps, we find that the proper integration time is 100 times the non-dimensional orbital period. We choose different values of non-dimensional orbital semi-major axis, eccentricity, argument of latitude, mass ratio of bodies, and non-dimensional semi-axes of the secondary to generate a cartographic image of the phase space, and compare the resulting phase space for both models.

In the first case, we choose $a = 5$, $e = 0$, $f + \omega = 0$, $m_s/m_p = 0.75$, $\left(\bar{a}_s : \bar{b}_s : \bar{c}_s\right) = (1 : 0.95 : 0.90)$, and numerically integrate the equations, including equations of spin-orbit problems and their corresponding variational equations. As evident in Figure 2, for a circular orbit, the 1:1 resonance is the only resonance. The synchronous resonance plays a significant role in the dynamics of both models. However, there are some differences between the models. First, the maximum value of FLIs near separatrix of synchronous resonance is 20 for the SOC and 7 for the SOR. That means, near separatrix, spin of the secondary in the SOC model could be more chaotic than the SOR model. Second, an invariant torus with high FLI values presents near 2:3 resonance in the SOC (Figure 2).

We consider an eccentric orbit with $a = 10$, $e = 0.01$, $f + \omega = \pi/4$, $m_s/m_p = 0.25$, $\left(\bar{a}_s : \bar{b}_s : \bar{c}_s\right) = (1 : 0.90 : 0.45)$ as the second case. As can be seen from Figure 3, the values of FLIs are in the same order. According to the value of the argument of latitude, we expect that stable and unstable configurations of both systems are at $\phi = \pi/4$, $5\pi/4$ respectively, as demonstrated in Figure 3. In both models, the homoclinic intersection of stable and unstable manifolds results in a chaotic layer around the synchronous resonance, where the onset of chaos is. Nevertheless, in the SOR model, there is a chain of three islands around the synchronous resonance. Also, the separatrix of 2:3 resonance is visible in this model. It seems that in the SOR model, more initial grids have high FLIs values. In the SOC model, there are no secondary islands and the chaotic layer around the synchronous resonance is narrower than the other model. Instead of the 2:3 resonance island, we see a chaotic layer.

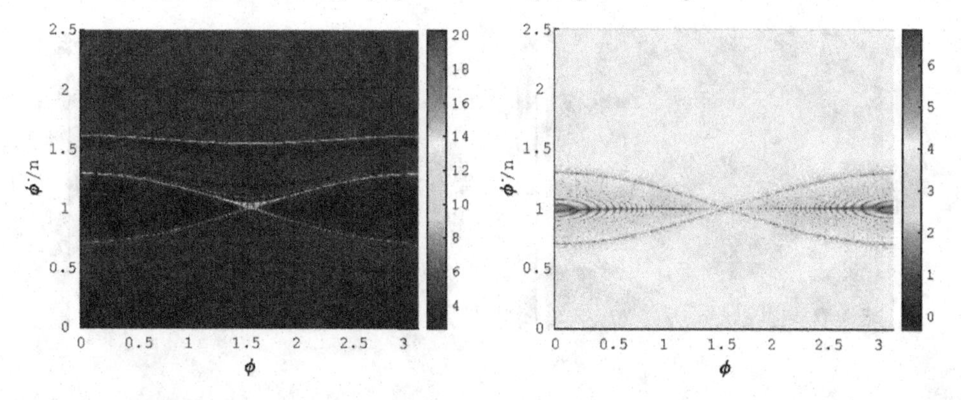

Figure 2. FLI maps for spin-orbit coupling (left) and spin-orbit resonance (right). The Initial values are $a=5$, $e=0$, $f+\omega=0$, $m_s/m_p=0.75$, $(\bar{a}_s:\bar{b}_s:\bar{c}_s)=(1:0.95:0.90)$.

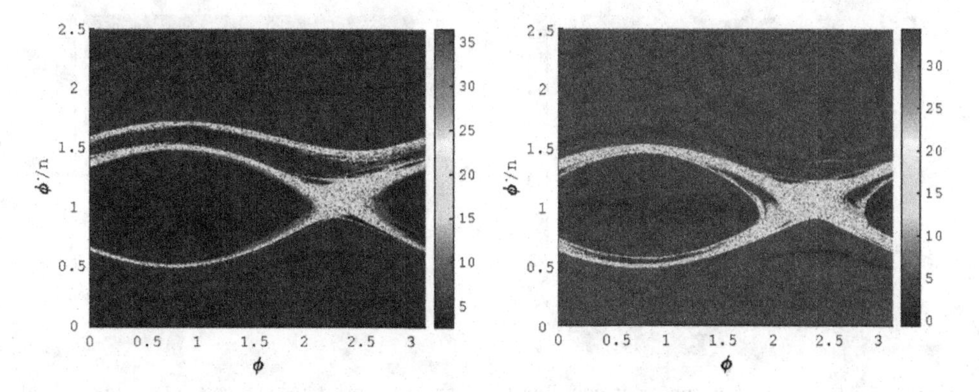

Figure 3. FLI maps for spin-orbit coupling (left) and spin-orbit resonance (right). The Initial values are $a=10$, $e=0.01$, $f+\omega=\pi/4$, $m_s/m_p=0.25$, $(\bar{a}_s:\bar{b}_s:\bar{c}_s)=(1:0.90:0.45)$.

We calculate the FLI map for $a=15$, $e=0.05$, $f+\omega=\pi/2$, $m_s/m_p=0.10$, $(\bar{a}_s:\bar{b}_s:\bar{c}_s)=(1:0.80:0.60)$. Figure 4 shows the resulting FLI maps. The eccentric orbit of the secondary causes the appearance of other resonances, of which some are overlapping. However, the synchronous resonance has sizeable resonant islands and still prevails in the phase space. The overlapping of nearby resonances with the synchronous resonance leads to a thick chaotic layer around it. The 3:1 secondary resonance exists in both models, although its separatrix begins to be destroyed. In Figure 4, it is quite obvious that in the SOR model compared with the SOC model, more initial grids have high FLIs values. Three persistence islands in the thick chaotic layer in the SOC model versus four islands in the SOR model is another difference. Also, in the left panel of Figure 4, the 7:4 resonance is evident.

If we consider $a=20$, $e=0.15$, $f+\omega=\pi$, $m_s/m_p=0.05$, $(\bar{a}_s:\bar{b}_s:\bar{c}_s)=(1:0.70:0.35)$ we can see the resulting maps are dramatically different, as indicated in Figure 5. While the islands of 2:1 and 2:3 resonances are persevering for the SOC model, the rotation of the secondary is chaotic for the normalized rotation rate of 2 in the SOR model.

Now we take $a=50$, $e=0.25$, $f+\omega=5\pi/4$, $m_s/m_p=0.01$, $(\bar{a}_s:\bar{b}_s:\bar{c}_s)=(1:0.50:0.50)$ and generate FLI maps (see Figure 6). In this case, although the phase space portraits are dissimilar, we face widespread chaos in both models. We can see a considerable regular region for the rotation of the secondary with a low normalized

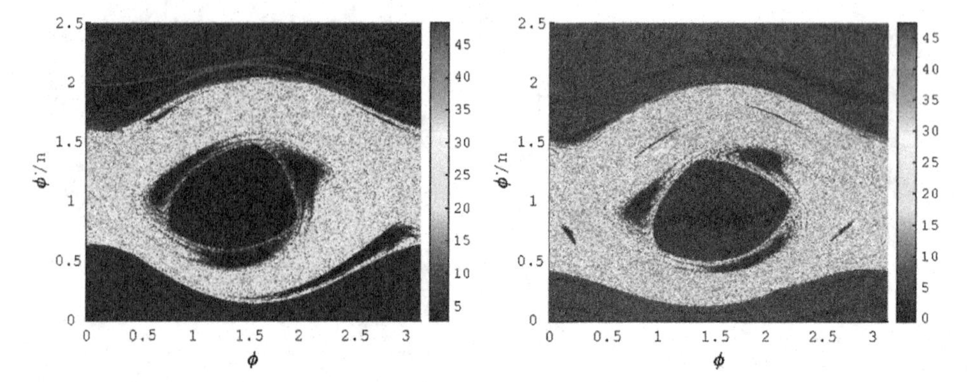

Figure 4. FLI maps for spin-orbit coupling (left) and spin-orbit resonance (right). The Initial values are $a = 15$, $e = 0.05$, $f + \omega = \pi/2$, $m_s/m_p = 0.10$, $(\bar{a}_s : \bar{b}_s : \bar{c}_s) = (1 : 0.80 : 0.60)$.

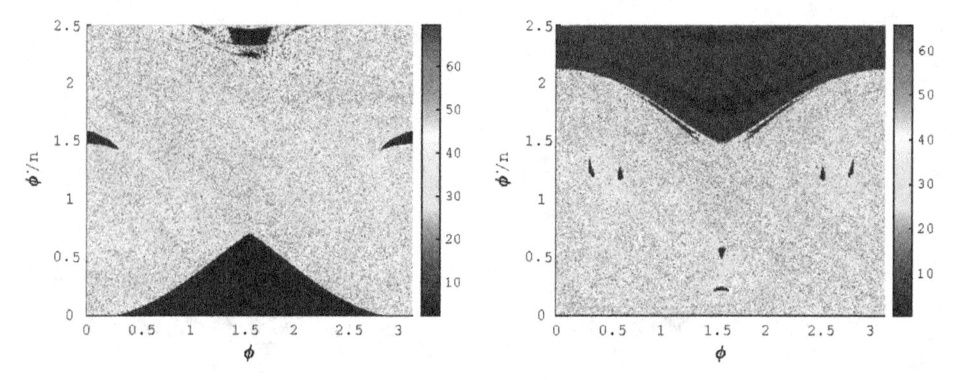

Figure 5. FLI maps for spin-orbit coupling (left) and spin-orbit resonance (right). The Initial values are $a = 20$, $e = 0.15$, $f + \omega = \pi$, $m_s/m_p = 0.05$, $(\bar{a}_s : \bar{b}_s : \bar{c}_s) = (1 : 0.70 : 0.35)$.

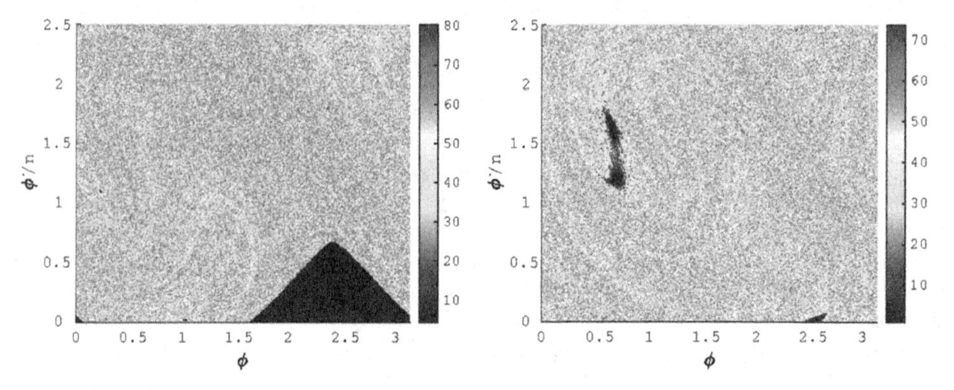

Figure 6. FLI maps for spin-orbit coupling (left) and spin-orbit resonance (right). The Initial values are $a = 50$, $e = 0.25$, $f + \omega = 5\pi/4$, $m_s/m_p = 0.01$, $(\bar{a}_s : \bar{b}_s : \bar{c}_s) = (1 : 0.50 : 0.50)$.

rotation rate in the SOC model. Comparingly, a tiny island, in the vicinity of 2:3 resonance, surrounded by the chaotic sea, is noticeable in the SOR model.

Since it may be inferred that the large chaotic region in Figure 6 is because of the high asphericity of the secondary, we consider the previous case with a different semi-axis, i.e., $(\bar{a}_s : \bar{b}_s : \bar{c}_s) = (1 : 0.90 : 0.90)$. It is evident that the phase space is occupied with a large

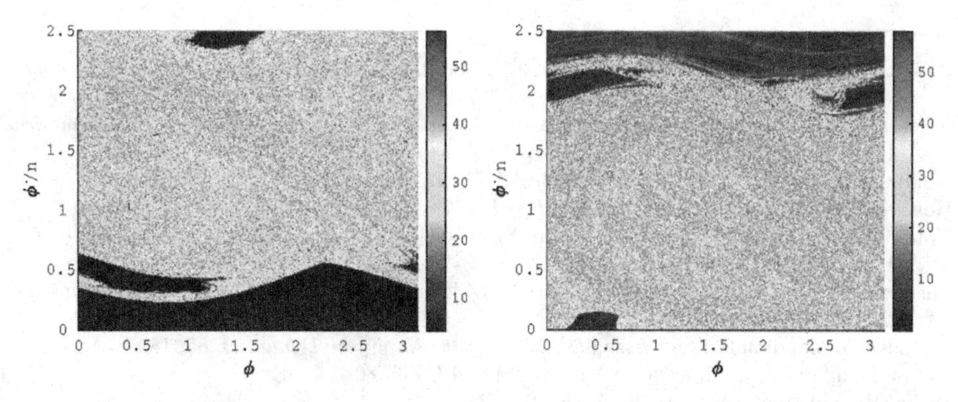

Figure 7. FLI maps for spin-orbit coupling (left) and spin-orbit resonance (right). The Initial values are $a = 50$, $e = 0.25$, $f + \omega = 5\pi/4$, $m_s/m_p = 0.01$, $(\bar{a}_s : \bar{b}_s : \bar{c}_s) = (1 : 0.90 : 0.90)$.

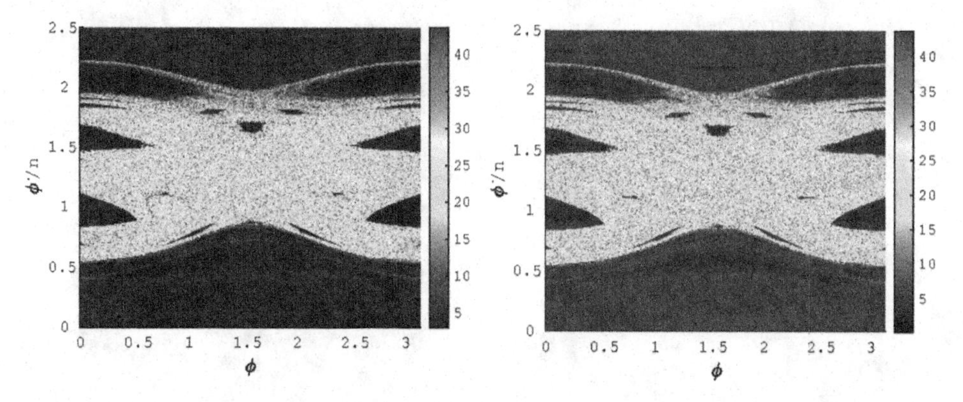

Figure 8. FLI maps for spin-orbit coupling (left) and spin-orbit resonance (right). The Initial values are $a = 20$, $e = 0.15$, $f + \omega = 0$, $m_s/m_p = 0.01$, $(\bar{a}_s : \bar{b}_s : \bar{c}_s) = (1 : 0.95 : 0.90)$.

chaotic region (Figure 7). However, there is a major difference between the two models. While the 2:1 resonance island exists in the phase space of the SOC model, and rotation of the secondary for a low normalized rotation rate is regular, the 1:2 resonances region is preserved in the phase space of the SOR model. In addition, the high normalized rotation rate of the secondary is regular in the SOR model.

There may be some systems where using both models have the same phase space portrait. For instance, we consider a system with $a = 20$, $e = 0.15$, $f + \omega = 0$, $m_s/m_p = 0.01$, $(\bar{a}_s : \bar{b}_s : \bar{c}_s) = (1 : 0.95 : 0.90)$. The FLI maps are depicted in Figure 8. There is no difference between them, though more initial grids still have high FLIs values in the SOR model.

4. Conclusion

We explore the phase space of spin-orbit problem in both models of SOC and SOR. The comparison of phase space structures shows noticeable differences in the models. That is to say, the SOR model cannot correctly capture the dynamics of the system, and the SOC model is more appropriate. However, this depends on the orbital and physical properties of the system. Because there may be some systems that show same results in both models, the question of which model is suitable should be examined in each case.

References

Celletti, Alessandra and Chierchia, Luigi 2000, *Celestial Mechanics and Dynamical Astronomy*, 76, 229

Froeschlé, Claude and Lega, Elena and Gonczi, Robert 1997, *Celestial Mechanics and Dynamical Astronomy*, 67, 41

Goldreich, Peter and Peale, Stanton 1966, *AJ*, 71, 425

Hou, Xiyun and Xin, Xiaosheng 2017, *AJ*, 154, 257

Hou, Xiyun and Scheeres, Daniel J and Xin, Xiaosheng 2017, *Celestial Mechanics and Dynamical Astronomy*, 127, 369

Misquero, Mauricio and Ortega, Rafael 2020, *SIAM Journal on Applied Dynamical Systems*, 19, 2233

Nadoushan, Mahdi Jafari and Assadian, Nima 2016, *Nonlinear Dynamics*, 85, 1837

Naidu, Shantanu P and Margot, Jean-Luc 2015, *AJ*, 149, 80

Wang, HS and Hou, XY 2020, *MNRAS*, 493, 171

Wisdom, Jack and Peale, Stanton J and Mignard, François 1984, *Icarus*, 58, 137

Multi-scale (time and mass) dynamics of space objects
Proceedings IAU Symposium No. 364, 2022
A. Celletti, C. Galeş, C. Beaugé, A. Lemaitre, eds.
doi:10.1017/S1743921321001290

Probabilistic evolution of pairs of trans-Neptunian objects in close orbits

Eduard Kuznetsov, Omar Al-Shiblawi and Vladislav Gusev

Department of Astronomy, Geodesy, Ecology and Environmental Monitoring,
Ural Federal University, Lenina Avenue, 51, Yekaterinburg, 620000, Russia
emails: eduard.kuznetsov@urfu.ru, themyth_24@yahoo.com, vlad06gusev@gmail.com

Abstract. We have studied the probabilistic evolution of four candidates for young pairs of trans-Neptunian objects: 2003 QL_{91} – 2015 VA_{173}, 1999 HV_{11} – 2015 VF_{172}, 2002 CY_{154} – 2005 EW_{318} and 2013 SD_{101} – 2015 VY_{170} over 10 Myr in the past. All pairs belong to cold Classical Kuiper Belt objects. We concluded that the age of the considered pairs exceeds 10 Myr.

Keywords. Celestial mechanics, methods: numerical, Kuiper Belt.

---

1. Introduction

Apart from Pluto and Charon, the first trans-Neptunian object (TNOs) was found in 1992. Although many TNOs were found on quite elliptic orbits, some of them had roughly circular orbits on a plane near the ecliptic (or the invariant solar system plane), today about 3 500 objects have been recognized and indexed. The distribution of the orbits of asteroids in the Solar system is the result of various processes that affect for a long time.

A candidate collisional family in the outer Solar system was proposed by Chiang (2002). The first asteroid family identified in the outer Solar system was the one associated with dwarf planet Haumea (Brown et al. 2007). The subject of finding collisional families of trans-Neptunian objects has been studied by Chiang et al. (2003) and Marcus et al. (2011). de la Fuente Marcos & de la Fuente Marcos (2018) perform a systematic search for statistically significant pairs and groups of dynamically correlated objects through those with a semi-major axis greater than 25 au, applying a technique that uses the angular separations of orbital poles and perihelia together with the differences in time of perihelion passage to single out pairs of relevant objects from which groupings can eventually be uncovered. They confirm the reality of the candidate collisional family of TNOs associated with the pair 2000 FC_8 – 2000 GX_{146} and initially proposed by Chiang (2002). They find four new possible collisional families of TNOs associated with the pairs (134860) 2000 OJ_{67} – 2001 UP_{18}, 2003 UT_{291} – 2004 VB_{131}, 2002 CU_{154} – 2005 CE_{81} and 2003 HF_{57} – 2013 GG_{137}. They find several unbound TNOs that may have a common origin, the most significant ones are (135571) 2002 GG_{32} – (160148) 2001 KV_{76} and 2005 GX_{206} – 2015 BD_{519}.

Kuznetsov et al. (2021) performed a search for statistically significant pairs and groups of dynamically correlated objects through those with a semi-major axis greater than 30 au, applying a novel technique that uses Kholshevnikov metrics (Kholshevnikov et al. 2016, 2020) in the space of Keplerian orbits. Found 27 pairs of TNOs in close orbits, 22 pairs in which one of the TNO is binary, and 11 pairs of binary trans-Neptunian objects. All pairs belong to cold classical Kuiper belt objects. Among the dynamically

Table 1. Candidates for young TNO pairs.

TNO pair	ϱ_2 [au$^{1/2}$]	ϱ_5 [au$^{1/2}$]	$\varrho_2 - \varrho_5$ [au$^{1/2}$]
2003 QL_{91} – 2015 VA_{173}	0.0412	0.0368	0.0044
1999 HV_{11} – 2015 VF_{172}	0.0432	0.0409	0.0023
2002 CY_{154} – 2005 EW_{318}	0.0484	0.0391	0.0093
2013 SD_{101} – 2015 VY_{170}	0.0496	0.0434	0.0062

cold population of the classical Kuiper belt, during the evolution of the protoplanetary disk and the migration of planets, conditions are implemented for the preservation of close binary or contact TNOs with components of approximately equal masses (Nesvorný & Vokrouhlický 2019). On the other hand, the evolution of wide binary trans-Neptunian objects turns out to be unstable due to frequent encounters with other TNOs, which lead to the decay of binary systems (Campbell 2021) and the formation of TNO pairs in close orbits.

We perform a study of the dynamical evolution of pairs of TNOs in which one of the component is binary. This paper is organized as follows. Section 2 reviews the methods which we used to studied the probabilistic evolution of the TNO pairs. The results of the probabilistic evolution study are presented in Section 3. In Section 4, we discuss the results and summarize our conclusions.

2. Method

To search for candidates for young TNO pairs, we used the metrics $\varrho(\mathcal{E}_1, \mathcal{E}_2)$ in the space of Keplerian orbits (Kholshevnikov et al. 2016, 2020). The metric ϱ_2 defines the distance between two orbits in the five-dimensional space of Keplerian orbits $\mathcal{E} = (a,\ e,\ i,\ \omega,\ \Omega)$ (where $a,\ e,\ i,\ \omega,\ \Omega$ are the semi-major axis, eccentricity, inclination, argument of the pericentre and longitude of the ascending node of the orbit, respectively) and shows the current distance between the Keplerian orbits. The metric ϱ_5 defines the distance in the three-dimensional factor-space of the positional elements $\mathcal{E}' = (a,\ e,\ i)$ and gives the minimum metric ϱ_2 among all possible positions of the nodes and pericenter of the orbits and therefore $\varrho_5 \leq \varrho_2$. Analyzing the metrics will help identify candidates for young pairs. The positions of the lines of nodes and apses of the TNO orbits in young pairs should be close because the orientation of the orbits has changed slightly since the formation of the pair due to the secular drift of nodes and pericenter. If the metrics ϱ_2 and ϱ_5 are small (for TNO pair, one can limit ourselves to 0.05 au$^{1/2}$) and have close values (e.g. $\varrho_2 - \varrho_5 < 0.01$ au$^{1/2}$), then such a pair of TNOs can be considered a candidate for young pair. The criteria for the metrics ϱ_2 and ϱ_5 correspond to two or three values of the Hill sphere radius for TNO. However, this is only a necessary condition for the youth of the pairs because the precession of the nodes and pericenter of the orbits has a conditionally periodic type.

We have used both numbered and multiopposition objects from the Asteroids Dynamic Site (AstDyS, https://newton.spacedys.com/astdys/). We calculated the Kholshevnikov metrics ϱ_2 and ϱ_5 using the osculating orbital elements for the epoch MJD 59000 (00^h 00^m 00.000^s BDT 31.05.2020). We selected four pairs of TNOs satisfying the conditions: $\varrho_2 < 0.05$ au$^{1/2}$, $\varrho_5 < 0.05$ au$^{1/2}$ ($\varrho^2 < 0.0025$ au $= 3.7 \cdot 10^5$ km) and $\varrho_2 - \varrho_5 < 0.01$ au$^{1/2}$ ($(\varrho_2 - \varrho_5)^2 < 0.0001$ au $= 1.5 \cdot 10^4$ km) (see Tab. 1). Tab. 2 gives orbital elements and absolute magnitude H for TNOs in pairs.

To model the dynamical evolution of TNOs, we have performed numerical integrations of the orbits of TNOs in pairs backward in time (a period of 10 Myr) with the code known as Orbit9 (the OrbFit Software Package, http://adams.dm.unipi.it/orbfit/). The four giant planets were integrated consistently. The mean ecliptic of J2000.0 was taken as reference plane for the output. We used heliocentric coordinates.

Table 2. Orbital elements of TNOs in pairs.

TNO	*a* [au]	*e*	*i* [deg]	*Ω* [deg]	*ω* [deg]	*H* [mag]
2003 QL_{91}	43.246	0.01397	1.540	164.595	186.171	6.87
2015 VA_{173}	42.923	0.01091	1.689	169.363	184.498	8.41
1999 HV_{11}	43.114	0.02100	3.158	160.952	275.115	7.61
2015 VF_{172}	43.319	0.01866	2.924	162.939	276.482	8.87
2002 CY_{154}	44.229	0.07940	0.978	120.902	235.871	6.68
2005 EW_{318}	44.407	0.07423	1.060	128.503	225.712	6.35
2013 SD_{101}	43.430	0.02450	1.585	44.632	302.693	7.43
2015 VY_{170}	43.064	0.02097	1.765	43.830	311.497	7.73

We used two methods to estimate the age of TNO pairs in close orbits: 1) search for low relative-velocity close encounters of TNOs (e.g., Pravec et al. (2019)), 2) search for the minimum distances between the orbits of TNOs (e.g., Kuznetsov et al. (2020)).

The condition of convergence of orbits does not yet guarantee the convergence of objects moving in these orbits. Therefore, to estimate the age of pairs, it is also necessary to analyze the possibility of the onset of low relative-velocity close encounters, at which the distance between objects r_{rel} is comparable to the radius of the Hill sphere R_H of a more massive body, and the relative velocity v_{rel} is of the order of the escape velocity V_{esc} relative to a more massive body. Pravec et al. (2019) used follow the criteria for low-speed encounters for asteroids in the main belt are: $r_{rel} < (5 \,\text{or}\, 10) R_H$, $v_{rel} < (2 \,\text{or}\, 4) V_{esc}$, where V_{esc} is the escape velocity on the surface of a more massive body.

For each close approach of TNOs in pair we determined the relative distance r_{rel} between TNOs and relative velocity v_{rel}, as well as the Hill sphere radius R_H and escape velocity V_{esc} of the primary body. The radius of the Hill sphere was estimated as:

$$R_H = \frac{1}{2} r_1 D_1 \left(\frac{4\pi}{9} \frac{G\rho_1}{\mu} \right)^{1/3}, \tag{2.1}$$

where r_1 is the heliocentric distance of the primary's TNO, D_1 is its diameter, ρ_1 is its bulk density, G is the gravitational constant and μ is the gravitational parameter of the Sun. The escape velocity of primary body for relative distance r_{rel} was estimated as:

$$V_{esc} = \sqrt{\frac{\pi}{3} \frac{G D_1^3 \rho_1}{r_{rel}}}. \tag{2.2}$$

The diameter D of the TNO can be estimated from the absolute magnitude H and the geometric albedo p_v (Bowell et al. 1989):

$$D = 1329\,\text{km}\, 10^{-H/5} \frac{1}{\sqrt{p_v}}. \tag{2.3}$$

We need to know the physical parameters of the TNO to estimate the radius of the Hill sphere R_H (2.1) and the escape velocity V_{esc} (2.2). Since the objects included in the studied TNO pairs belong to the dynamically cold population of the classical Kuiper belt and have dimensions not exceeding several hundred km, we used the same density values $\rho = 0.5$ g cm^{-3} and geometric albedo $p_v = 0.13$ for all TNOs (Müller et al. 2020).

Estimates of the single TNO density range from 0.5 to 2 g cm^{-3} (Lacerda & Jewitt 2007; Grundy et al. 2008; Fernández 2020) and grow with an increase in the TNO diameter. For TNOs several hundred km in size, the density estimates are 0.5 – 0.6 g cm^{-3} (Lacerda & Jewitt 2007; Grundy et al. 2008; Fernández 2020). We used the minimum density value $\rho = 0.5$ g cm^{-3}, which will give the minimum estimates for the radius of the Hill sphere R_H (2.1) and the escape velocity V_{esc} (2.2). If the density value is 2 g cm^{-3}, the value of the radius of the Hill sphere R_H will be underestimated by $4^{1/3} \approx 1.6$ times, and the value of the escape velocity V_{esc} by $4^{1/2} = 2$ times. This can be taken into account when establishing the criteria for close encounters of the TNO pair.

To study the probabilistic evolution and estimate the ages of the TNO pairs, we consider 1000 clones for each TNO in pair. Using the Monte Carlo method, it is possible to generate distributions of clones' equivalent to those of observational results. Consequently, the simulated distribution represents the actual propagation of errors. Covariance matrix values and element errors were taken from AstDyS database. Based on this data, 1000 clones with a $\pm 3\sigma$ dispersion were generated for each nominal orbit. Such a strategy allows relatively good coverage of the whole probability space. Clones covering a 6-dimensional error ellipsoid were generated using a random number generator, with the following assumptions: the dispersion of each element has a normal distribution, the distribution coverage limit is $\pm 3\sigma$, the errors of each element are the same for clones as for real observational ones, and the distribution of all clones reproduces the original covariance matrix.

3. Results

2003 QL_{91} – 2015 VA_{173} Analysis of the results of probabilistic evolution shows that in the considered interval of 10 Myr, there is no noticeable concentration of close approaches to any selected time interval. The distribution of minimum distances Δr_{min} for close encounters up to a distance of less than 4 R_H is also uniform and does not allow identifying time intervals with prevailing close encounters. The relative velocity at close encounters exceeds 15.5 V_{esc}. All this allows us to conclude that the age of the pair 2003 QL_{91} – 2015 VA_{173} exceeds 10 Myr.

1999 HV_{11} – 2015 VF_{172} There is no noticeable concentration of close approaches to any selected time in the considered interval of 10 Myr. The distribution of minimum distances Δr_{min} for close encounters up to a distance of less than 4 R_H is also uniform and does not allow identifying time intervals with prevailing close encounters. The relative velocity at close encounters exceeds 59 V_{esc}. We conclude that the age of the pair 1999 HV_{11} – 2015 VF_{172} exceeds 10 Myr.

2002 CY_{154} – 2005 EW_{318} There is no noticeable concentration of close approaches to any selected time in the considered interval of 10 Myr. The distribution of minimum distances Δr_{min} for close encounters up to a distance of less than 4 R_H is also uniform and does not allow identifying time intervals with prevailing close encounters. The relative velocity at close encounters exceeds 16.1 V_{esc}. The minima of the metric $\varrho_{2\mathrm{min}}$ are concentrated in the intervals from 0 to 0.25 and from 0.75 to 1.5 Myr in the past (see Fig. 1). The minimum metric $\varrho_{2\mathrm{min}}$ values are 0.01 $\mathrm{au}^{1/2}$, which exceeds the expected metric value near the moment of pair formation 0.001 $\mathrm{au}^{1/2}$. We conclude that the age of the pair 2002 CY_{154} – 2005 EW_{318} exceeds 10 Myr.

2013 SD_{101} – 2015 VY_{170} There is no noticeable concentration of close approaches to any selected time in the considered interval of 10 Myr. The distribution of minimum distances Δr_{min} for close encounters up to a distance of less than 4 R_H is also uniform and does not allow identifying time intervals with prevailing close encounters. The relative velocity at close encounters exceeds 22 V_{esc}. The minima of the metric $\varrho_{2\mathrm{min}}$ are concentrated to the present (see Fig. 2). The minimum metric $\varrho_{2\mathrm{min}}$ values are 0.019 $\mathrm{au}^{1/2}$. We conclude that the age of the pair 2013 SD_{101} – 2015 VY_{170} exceeds 10 Myr.

4. Discussion and Conclusions

The pairs studied belong to the dynamically cold population of the classical Kuiper belt. This region has favorable conditions for the preservation of close binary TNO systems (Nesvorný & Vokrouhlický 2019). However, at the same time, wide TNO binary

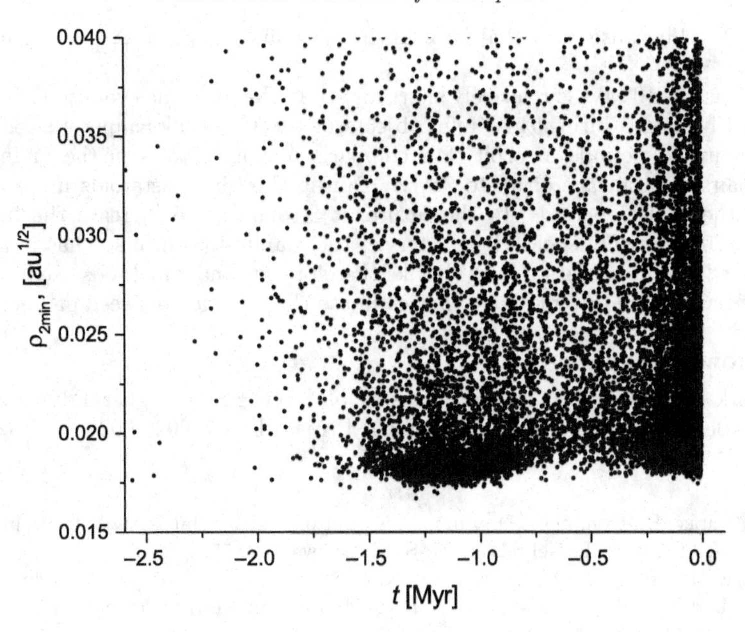

Figure 1. Minimum metric $\varrho_{2\min}$ vs time t for pair 2002 CY_{154} – 2005 EW_{318}.

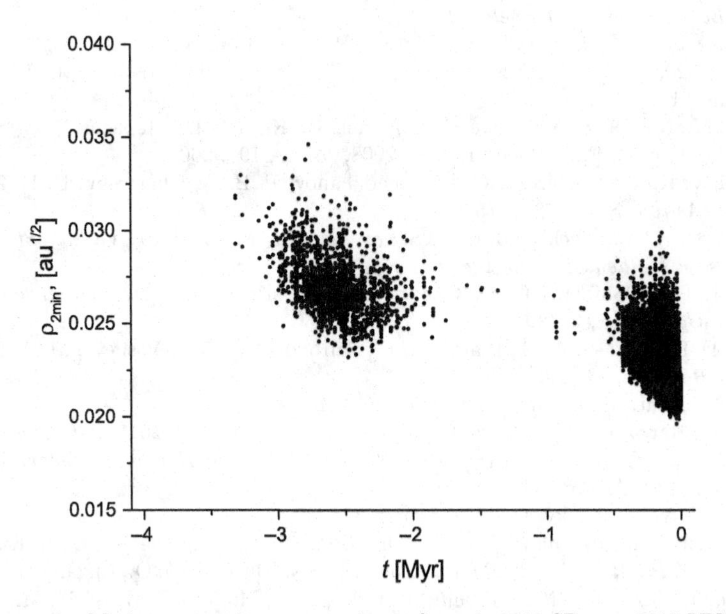

Figure 2. Minimum metric $\varrho_{2\min}$ vs time t for pair 2013 SD_{101} – 2015 VY_{170}.

systems disintegrate due to encounters with other objects (Campbell 2021). The most probable source of TNO pairs in the cold classical Kuiper belt is the decay of binary TNO systems.

Analysis of the results of probabilistic evolution shows the absence of low relative-velocity close encounters of TNO in pairs. Approaches to distances less than 4 R_H occur, but the relative velocities exceed 15.5 V_{esc}. We cannot estimate the moments of formation of pairs of TNO because the distribution of the minimum distances Δr_{min} in time is close to uniform.

The interval of 10 Myr is relatively short for study the dynamic evolution of the young TNO pairs because, during this time, objects of the Classical Kuiper Belt make only $33-36$ thousand periods in orbit. For comparison, young pairs in the main asteroid belt are pairs with an age of up to 2 Myr. During this time, asteroids make $400-600$ thousand the orbital periods. In the future, it is planned to increase the integration interval to 200 Myr. On such a long interval, the manifestation of stochastic properties of the TNO dynamic evolution is inevitable; therefore, the main methods used to estimate the age of pairs should be methods that estimate the distance between orbits.

5. Acknowledgments

The work was supported by the Ministry of Science and Higher Education of the Russian Federation via the State Assignment Project FEUZ-2020-0038.

References

Bowell, E., Hapke, B., Domingue, D., Lumme, K., Peltoniemi, J., Harris, A.W. 1989, in *Asteroids II*, ed. R. P. Binzel, T. Gehrels, & M. S. Matthews, 524–556

Brown, M. E., Barkume, K. M., Ragozzine, D., & Schaller, E. L. 2007, *Nature*, 446, 294

Campbell, H. 2021, *AAS/Division of Dynamical Astronomy Meeting*, 53, 501.04

Chiang, E. I. 2002, *Astrophys. J. Lett.*, 573, L65

Chiang, E. I., Lovering, J. R., Millis, R. L., Buie, M. W., Wasserman, L. H., & Meech, K. J. 2003, *Earth Moon and Planets*, 92, 49

de la Fuente Marcos, C. & de la Fuente Marcos, R. 2018, *Mon. Not. R. Astron. Soc.*, 474, 838

Fernández, J. 2020, *The Trans-Neptunian Solar System*, ed. D. Prialnik, M. A. Barucci, & L. Young, 1

Grundy, W. M., Noll, K. S., Virtanen, J., Muinonen, K., Kern, S. D., Stephens, D. C., Stansberry, J. A., Levison, H. F., & Spencer, J. R. 2008, *Icarus*, 197, 260

Kholshevnikov, K. V., Kokhirova, G. I., Babadzhanov, P. B., & Khamroev, U. H. 2016, *Mon. Not. R. Astron. Soc.*, 462, 2275

Kholshevnikov, K. V., Shchepalova, A. S., & Jazmati, M. S. 2020, *Vestnik St. Petersburg University: Mathematics*, 53, 108

Kuznetsov, E. D., Al-Shiblawi, O. M., Gusev, V. D., & Ustinov, D. S. 2021, *Lunar and Planetary Science Conference*, 2548, 1859

Kuznetsov, E. D., Rosaev, A. E., Plavalova, E., Safronova, V. S., & Vasileva, M. A. 2020, *Solar System Research*, 54, 236

Lacerda, P. & Jewitt, D. C. 2007, *Astronomical Journal*, 133, 1393

Marcus, R. A., Ragozzine, D., Murray-Clay, R. A., & Holman, M. J. 2011, *Astrophys. J.*, 733, 40

Müller, T., Lellouch, E., Fornasier, S. 2020, in *The Trans-Neptunian Solar System*, ed. D. Prialnik, M. A. Barucci, & L. Young, 153

Nesvorný, D. & Vokrouhlický, D. 2019, *Icarus*, 331, 49

Pravec, P., Fatka, P., Vokrouhlický, D., Scheirich, P., Ďurech, J., Scheeres, D. J., Kušnirák, P., Hornoch, K., Galád, A., Pray, D. P., Krugly, Yu. N., Burkhonov, O., Ehgamberdiev, Sh. A., Pollock, J., Moskovitz, N., Thirouin, A., Ortiz, J. L., Morales, N., Husárik, M., Inasaridze, R. Ya., Oey, J., Polishook, D., Hanuš, J., Kučáková, H., Vraštil, J., Világi, J., Gajdoš, Š., Kornoš, L., Vereš, P., Gaftonyuk, N. M., Hromakina, T., Sergeyev, A. V., Slyusarev, I. G., Ayvazian, V. R., Cooney, W. R., Gross, J., Terrell, D., Colas, F., Vachier, F., Slivan, S., Skiff, B., Marchis, F., Ergashev, K. E., Kim, D. -H., Aznar, A., Serra-Ricart, M., Behrend, R., Roy, R., Manzini, F., & Molotov, I. E. 2019, *Icarus*, 333, 429

Multi-scale (time and mass) dynamics of space objects
Proceedings IAU Symposium No. 364, 2022
A. Celletti, C. Galeş, C. Beaugé, A. Lemaitre, eds.
doi:10.1017/S1743921322000734

Multiple bifurcations around 433 Eros with Harmonic Balance Method

Leclère Nicolas[1], Kerschen Gaëtan[1] and Dell'Elce Lamberto[2]

[1]Space Structures and Systems Laboratory, Department of Aerospace and Mechanical Engineering, Université de Liège, Belgique

[2]Inria & Université Côte Azur, McTAO team, Sophia Antipolis, France.

Abstract. The objective of this paper is to carry out periodic orbital propagation and bifurcations detection around asteroid 433 Eros. Specifically, we propose to exploit a frequency-domain method, the harmonic balance method, as an efficient alternative to the usual time integration. The stability and bifurcations of the periodic orbits are also assessed thanks to the Floquet exponents. Numerous periodic orbits are found with various periods and shapes. Different bifurcations, including period doubling, tangent, real saddle and Neimark-Sacker bifurcations, are encountered during the continuation process. Resonance phenomena are highlighted as well.

Keywords. Harmonic balance, asteroid, Eros, bifurcations, resonance

---

1. Introduction

Interest in asteroids and small celestial bodies significantly increased in the last two decades, and, consequently, the number of space exploration missions increased as well. Concerning the orbital propagation around asteroids and the study of the bifurcations of their periodic orbits, one of the methods considered is the grid searching method proposed by Yu & Baoyin (2012). It was applied to compute families of orbits around 216 Kleopatra Yu & Baoyin (2012); a summary of all the classifications of bifurcations was proposed by Jiang (2015). This paper focuses on orbital dynamics about 433 Eros, which was subject of recent studies Ni (2016), Scheeres (2000). The rotation period of 433 Eros is of 5.270 hours and the density is equal to 2.673 g/cm^3. The focus of this study is made on the method applied for the computation of periodic orbits and the detection of the bifurcation. The harmonic balance method is applied to the equation of motion offering a new and unique approach to compute periodic solutions. The paper is organized as follows. First, details on the polyhedron method for the gravitational modeling of the asteroid are briefly recalled. Then, the harmonic balance method, used for orbital propagation, is introduced. Finally, the obtained results are exposed.

2. Polyhedron method

The polyhedron method introduced by Werner (1994) is extensively used to model the gravitational field of irregular celestial bodies like asteroids. The method is based on the construction of a surface mesh with the assumption that the density of the body is constant. Some meshes were generated from radar observations, others are the direct result of on-site observations during specific missions, e.g., NASA mission *Osiris-REX* on 101955 Bennu, ESA mission *Rosetta* on 67P Churyumov-Gerasimenko and JAXA

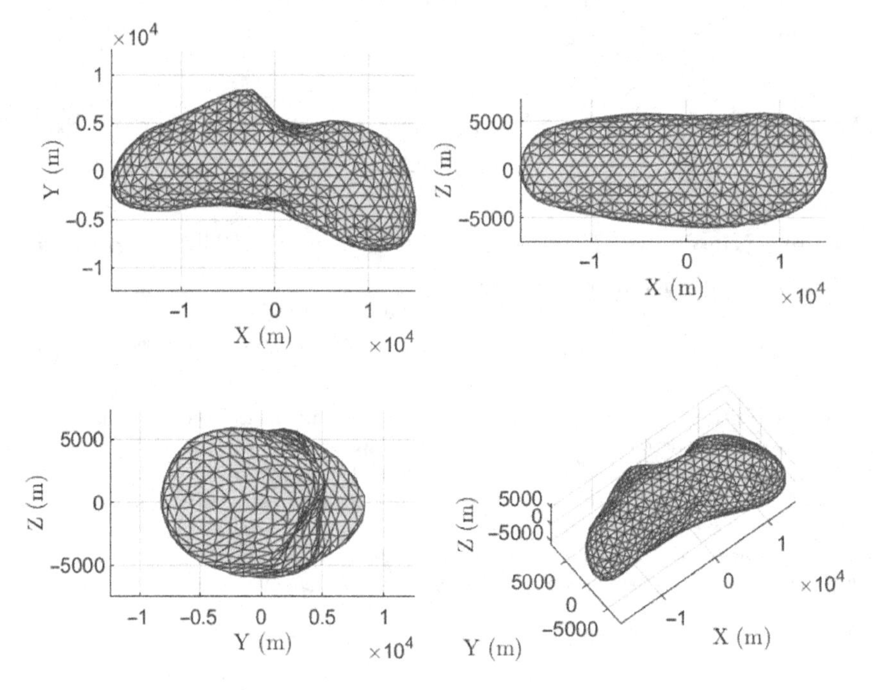

Figure 1. Polyhedron model of 433 Eros.

mission *Hayabusa* on 25143 Itokawa Tsuchiyama (2011). The polyhedron model of 433 Eros is the result of NASA mission *NEAR Shoemaker* Veverka (2001). Figure 1 shows the mesh of 433 Eros, consisting of 856 vertices and 1708 faces.

The main advantage of the polyhedron method is the simple computation of the gravitational potential of the celestial body, U, namely

$$U = \frac{1}{2} G\rho \sum_{edges} (\mathbf{r}_e \cdot \mathbf{E}_e \cdot \mathbf{r}_e) \cdot L_e - \frac{1}{2} G\rho \sum_{faces} (\mathbf{r}_f \cdot \mathbf{F}_f \cdot \mathbf{r}_f) \cdot \omega_f \tag{1}$$

where ρ is the bulk density of the body, G is the gravitational constant, the two vectors $\mathbf{r}_e$ and $\mathbf{r}_f$ are body-fixed vectors from the particle to the edge e and the face f, respectively. Matrices $\mathbf{E}_e$ and $\mathbf{F}_f$ gather the geometric parameters of the edges e and faces f. $\mathbf{L}_e$ denotes the integration factor of the particle position and the edge e whereas ω_f corresponds to the solid angle of the face f relative to the particle.

The gradient of the gravitational potential is then easily obtained

$$\nabla U = -G\rho \sum_{edges} (\mathbf{E}_e \cdot \mathbf{r}_e) \cdot L_e + G\rho \sum_{faces} (\mathbf{F}_f \cdot \mathbf{r}_f) \cdot \omega_f \tag{2}$$

This expression of the gravitational potential yields the equation of motion in the co-rotating frame

$$\ddot{\mathbf{x}} + 2\omega_{\mathbf{a}} \times \dot{\mathbf{x}} + \omega_{\mathbf{a}} \times (\omega_{\mathbf{a}} \times \mathbf{x}) + \nabla U(\mathbf{x}) = 0 \tag{3}$$

The body-fixed vector that links the asteroid body's center of mass to the particle is denoted $\mathbf{x}$, ω_a represents the angular velocity of the asteroid. The harmonic balance method is used to compute solutions to this equation.

3. Harmonic Balance

The harmonic balance method aims at approximating periodic solutions of the equations of motion, $x(t)$, by means of a Fourier series truncated to the N-th harmonic, namely

$$x(t) = \frac{c_0^x}{\sqrt{2}} + \sum_{k=1}^{N_H} (s_k \sin(k\omega t) + c_k \cos(k\omega t)) \tag{4}$$

A similar decomposition is carried out for the nonlinear force ∇U. Vectors s_k and c_k are the Fourier coefficients associated to the sine and cosine, respectively. ω corresponds to the frequency of the periodic orbit (which is not correlated to the angular velocity of the asteroid, ω_a). The Fourier coefficients are gathered in a new vector $\mathbf{z}$ for the displacement and $\mathbf{b}$ for the nonlinear force of dimension $(2N_H + 1)\,n \times 1$, with n the degrees of freedom of the studied system. Equation 3 can eventually be rewritten as

$$\mathbf{h}(\mathbf{z}, \omega) = \mathbf{A}(\omega)\mathbf{z} - \mathbf{b}(\mathbf{z}) = \mathbf{0} \tag{5}$$

where matrix $\mathbf{A}$ describes the linear dynamics. A predictor-corrector algorithm is used to solve Equation 5, as proposed in Detroux (2014). This approach presents numerous advantages over the classical time integration method. Working in the frequency domain provides a fast and efficient alternative from time domain methods to solve the equation of motion. The predictor-corrector algorithm is a great tool to compute orbit families.

The stability of orbits, as well as the detection of bifurcations, can be determined through the Floquet multipliers which are the eigenvalues of the monodromy matrix Peletan (2013). In the frequency domain, an alternative method known as Hill's method exists. It consists in introducing the periodic solution $\mathbf{x}^*(t)$ perturbed with another periodic solution $\mathbf{s}(t)$ modulated by an exponential decay into the equation of motion, Eq. 3.

$$\mathbf{p}(t) = \mathbf{x}^*(t) + e^{\lambda t}\mathbf{s}(t) \tag{6}$$

which eventually leads to the simple quadratic eigenvalue problem

$$\left(\Delta_2\lambda^2 + \Delta_1\lambda + \mathbf{h_z}\right)\mathbf{u} = \mathbf{0} \tag{7}$$

that provides the Floquet mutlipliers, λ. If two multipliers cross at $+1$ on the unit circle the bifurcation is a tangent bifurcation. If the crossing happens at -1, it is a period doubling bifurcation. We refer to a real saddle if two multipliers leave the real axis as complex conjugates; if they leave the unit circle as complex conjugates, there is the presence of a Neimark Sacker bifurcation. A graphical depiction summarizes the different cases in Figure 2. Δ_1 and Δ_2 also describe the linear dynamic. $\mathbf{h_z}$ is the derivative of equation 5 with respect to $\mathbf{z}$. The vector $\mathbf{u}$ is the equivalent of $\mathbf{z}$ but for the Fourier coefficients of $\mathbf{s}$.

4. Results

We consider the continuation process between the periodic ratios, $\frac{\omega}{\omega_a}$ starting from approximately 2 up to 3. The results of the continuation are displayed in Figure 3, each point in this plot correspond to a unique periodic orbit. We refer to each group of orbits separated by bifurcations as families. They are gathered in Table 1.

Most bifurcations are located around resonant periods, associated to the nominal period ratio 2:1 and 3:1, where Jacobi's constant changes abruptly, whereas it remains

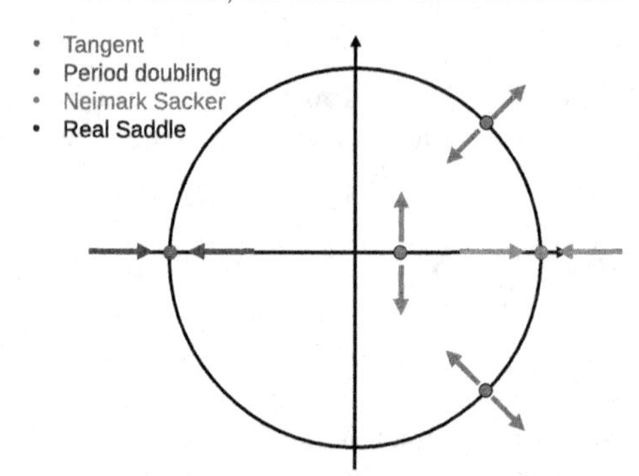

Figure 2. Bifurcations classified thanks to Floquet multipliers

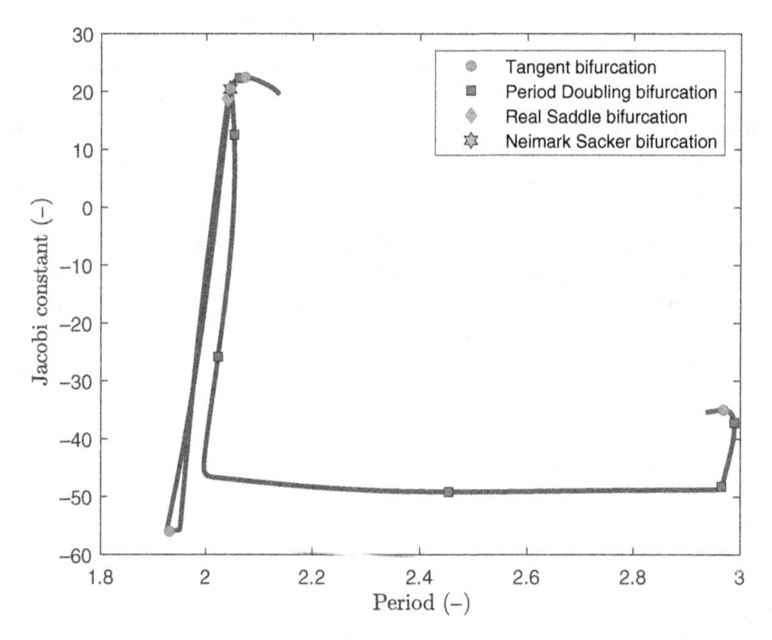

Figure 3. Continuation process between the periodic ratio of 2 and 3 with regard to the Jacobian constant

roughly constant in between. There is just a period doubling bifurcation between the two resonances around the ratio 2.5:1. It is in fact a quasi-period doubling bifurcation, meaning that the Floquet multipliers simply cross the value -1 on the unit circle and remain on it afterwards Kang (2018).

The stable region identified during the continuation process, corresponding to family 6, is displayed in black in Figure 4. The stability appears between a Neimark Sacker bifurcation and a tangent bifurcation.

Figure 5 displays representative orbits for each family. The unfolding of the few first families can clearly be observed. After family 4, the shape of the orbits is rotated by 90 degrees compared to those in family 3. Starting from family 8, the orbits begin to

Table 1. Bifurcations, stability and period ratio of the orbit families

Family n°	Period ratio	Jacobian constant	Bifurcation	Stability
1	]2.138;2.074]	]19.496;22.451]	Tangent	U
2	]2.074;2.063]	]22.451;22.324]	Period Doubling	U
3	]2.063;1.931]	]22.324;-55.984]	Tangent	U
4	]1.931;2.041]	]-55.984;18.689]	Real Saddle	U
5	]2.041;2.045]	]18.689;20.395]	Neimark Sacker	U
6	]2.045;2.0455]	]20.395;20.411]	Tangent	S
7	]2.0455;2.053]	]20.411;12.504]	Period Doubling	U
8	]2.053;2.023]	]12.504;-25.824]	Period Doubling	U
9	]2.023;2.454]	]-25.824;-49.136]	Period Doubling	U
10	]2.454;2.965]	]-49.136;-48.199]	Period Doubling	U
11	]2.965;2.989]	]-48.199;-37.213]	Period Doubling	U
12	]2.989;2.969]	]-37.213;-34.969]	Period Doubling	U
13	]2.969;2.935]	]-34.969;-35.409]	Tangent	U

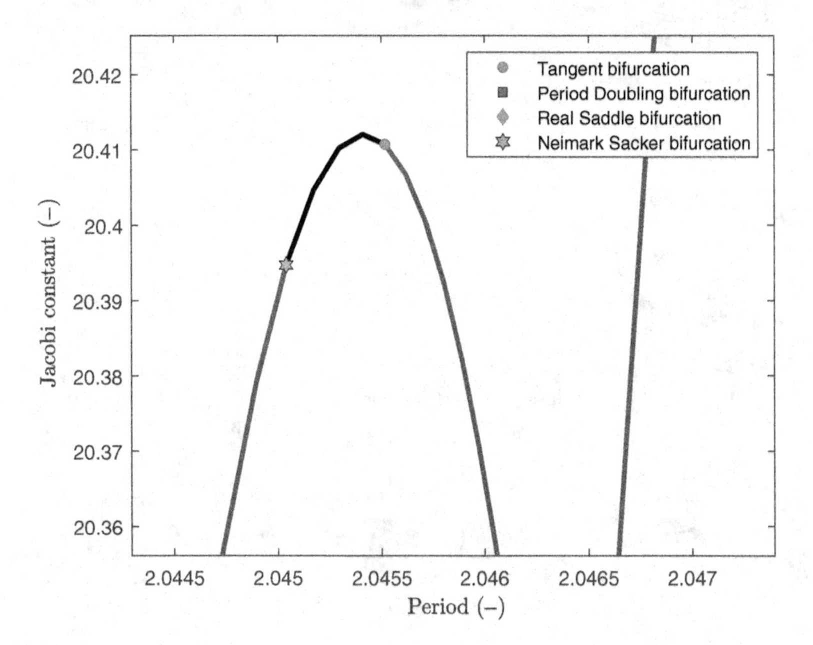

Figure 4. Focus on the stable orbits during the continuation.

flatten to the point that family 10 has orbits with almost zero inclination. Families 11, 12 and 13 evolve into more complex orbits.

5. Conclusion

In this paper, a new approach for the computation of periodic orbits around asteroids is proposed. The harmonic balance method is introduced, and its application to the detection of bifurcation and search for periodic orbits around the asteroid 433 Eros is presented. Twelve bifurcations of different types are encountered mainly around the resonances 2:1 and 3:1. The stability of the orbits is also studied and only one family of orbits is found to be stable.

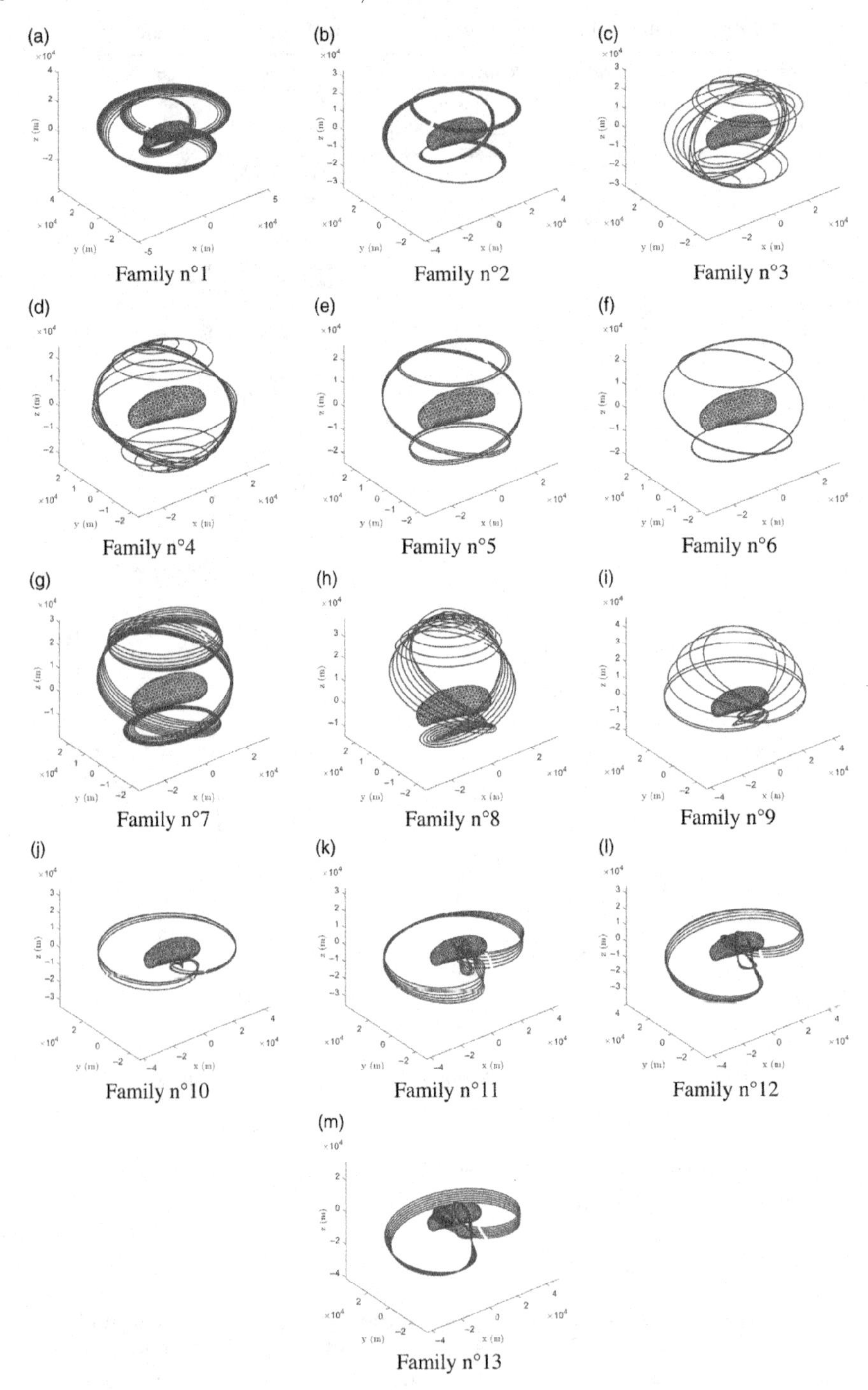

Figure 5. Families of periodic orbits around 433 Eros.

References

Detroux, T., Renson, L., Kerschen, G. 2014 The harmonic balance method for advanced analysis and design of nonlinear mechanical systems. Conf. Proc. Soc. Exp. Mech. S. 2, 19–34

Jiang, Y., Yu, Y., Baoyin, H. 2015 Topological classifications and bifurcations of periodic orbits in the potential field of highly irregular-shaped celestial bodies. Nonlinear Dyn. 81(1–2), 119–140

Kang, H., Jiang,Y., Li, H. 2018 Pseudo Bifurcations and Variety of Periodic Ratio for Periodic Orbit Families Close to Asteroid (22) Kalliope. Planetary and Space Sci.

Ni, Y., Jiang, Y., Baoyin, H. 2016 Multiple bifurcations in the periodic orbit around Eros. Astriohys Space Sci. 361:170

Peletan, L., Baguet, S., Torkhani, M., Jacquet-Richardet G. 2013 A comparison of stability computational methods for periodic solution of nonlinear problems with application to rotordynamics. Nonlinear Dynamics. 72(3), 671:682

Scheeres, D.J., Williams,B.G., Miller, J.K. 2000 Evaluation of the Dynamic Environment of an Asteroid: Applications to 433 Eros. Journal of Guidance, Control and Dynamics Vol. 23 No. 3

Tsuchiyama, A., Uesugi,M., Matsushima, T., et al. 2011 Three-dimensional structure of Hayabusa samples: origin and evolution of Itokawa regolith. Science 333(6046), 1125–1128

Veverka, J., Farquhar, B., Robinson, M., et al. 2001 The landing of the NEAR-Shoemaker spacecraft on asteroid 433 Eros. Nature 413(6854), 390–393

Werner, G.A. 1994 The gravitational potential of a homogeneous polyhedron or don't cut corners. Celest. Mech. Dyn. Astron. 59(3), 253–278

Yu, Y., Baoyin, H., Jiang, Y. 2015 Constructing the natural families of periodic orbits near irregular bodies. Mon. Not. R. Astron. Soc. 453(1), 3269–3277

Yu, Y., Baoyin, H. 2012 Generating families of 3D periodic orbits about asteroids. Mon. Not. R. Astron. Soc. 427(1), 872–881

Multi-scale (time and mass) dynamics of space objects
Proceedings IAU Symposium No. 364, 2022
A. Celletti, C. Galeş, C. Beaugé, A. Lemaitre, eds.
doi:10.1017/S1743921321001381

The effect of the passage of Gliese 710 on Oort cloud comets

Birgit Loibnegger, Elke Pilat-Lohinger, Max Zimmermann and Sharleena Clees

Department of Astrophysics, University of Vienna,
Türkenschanzstraße 17, 1180 Vienna, Austria
email: birgit.loibnegger@univie.ac.at

Abstract. Based on observations by Bailer-Jones *et al.* (2018) who propose a close fly-by of the K-type star Gliese 710 in approximately 1.36 Myr we investigate the immediate influence of the stellar passage on trajectories of Oort cloud objects. Using a newly developed GPU-based N-body code (Zimmermann (2021)) we study the motion of 3.6 million testparticles in the outer Solar system where the comets are distributed in three different "layers" around the Sun and the 4 giant planets. We study the immediate influence of Gliese 710 at three passage distances of 12000, 4300, and 1200 au. Additionally, different inclinations of the approaching star are considered. Depending on the passage distance a small number of comets (mainly from the disk and flared disk) is scattered into the observable region (< 5 au) around the Sun. In addition, a huge number of comets (mainly the ones directly in the path of the passing star) shows significant changes of their perihelia. But, they will enter the inner Solar system a long time after the stellar fly-by depending on their dynamical evolution.

Keywords. Comets: general; Oort cloud; Solar system: general

1. Introduction

Most stars are born in clusters and thus, gravitational interactions between the cluster members can have significant influence on the planet formation process and small body distributions (Bancelin *et al.* (2019)). Even after having left the cluster fly-bys with other stars can happen.

Recent observations show that the Sun will experience a close fly-by of another star in about 1.36 Myr when Gliese 710 will probably pass inside the Oort cloud (Bailer-Jones *et al.* (2018)). Gliese 710 is a 0.6 $M_{\odot}$ star which today is approximately 64 ly away from the Sun. Berski & Dybczyński (2016) state that in 1.35±0.05 Myr, Gliese 710 will be 13 366±6250 au from the Sun. De la Fuente Marcos (2018) find a slightly closer mean distance of 10 721±2114 au in 1.28±0.04 Myr in the future. Using older input data they show that the closest approach might even reach 0.021 pc or 4303 au, in 1.29 Myr. As in either case Gliese 710 will pass right through the Oort cloud it will possibly influence the orbits of the small bodies orbiting there.

As we know from e.g., Sizova et al. (2020), Torres *et al.* (2019), Vokrouhlický *et al.* (2019), Rickman et al. (2008), the influence of passing stars on Oort cloud objects is very high and in combination with galactic tides is responsible for the long period comes we are able to observe today.

All of this motivated us to investigate the influence of passing stars on the orbits of cometary objects residing in the Oort cloud.

Table 1. Initial values for the cometary objects used in the simulations. Each region contains 600 000 objects.

inner edge (au)	outer edge (au)	eccentricity	inclination (deg)	description
50	5 000	< 0.1	< 1 °	disk – scattered disk
5 000	10 000	< 0.1	< 45°	flared disk
10 000	25 000	< 0.1	< 180°	Outer Oort cloud – 1st shell
25 000	50 000	< 0.1	< 180°	Outer Oort cloud – 2nd shell
50 000	75 000	< 0.1	< 180°	Outer Oort cloud – 3rd shell
75 000	100 000	< 0.1	< 180°	Outer Oort cloud – 4th shell

2. Method and Setup

Our setup consists of the outer Solar system including the Sun and the 4 giant planets (Jupiter, Saturn, Uranus, and Neptune) with their known masses and properties, an Oort cloud (consisting of an inner disk like structure and an outer spherical structure), and a passing star. The latter initially moves on a straight line passing the Sun at a certain distance (impact parameter, d_i). The properties used for the passing star are summed up as follows:

- Mass of passing star: 0.6 $M_\odot$ (i.e., a K-type star as Gliese 710)
- Impact parameter d: d_1=12 000 au, d_2=4300 au, d_3=1200 au†
- Impact angle α: 0° = planar case (in the ecliptic), 30° = inclined case

The impact angle is defined as angle between the stellar orbit and the ecliptic. As there are no observational constraints we investigated the influence at the chosen angles given above. The velocity of the star is taken from Rickman et al. (2008), who give the mean relative velocity for different stellar types in the Solar neighborhood.

Cometary objects are distributed around the planetary system in the extended scattered disk (between 50 and 5000 au), the flared disk (between 5000 and 10 000 au), and the spherical Oort cloud itself (10 000 to 100 000 au). The initial positions of the objects were calculated using a Rayleigh distribution (where $F(x) = 1 - e^{-x/2\sigma}$ for $x \geq 0$ or $F(x) = 0$ for $(x < 0)$. The value of σ defines the location of the maximum of $F(x)$). For computational reasons the spherical Oort cloud was divided in 4 shells. Each shell contained 600 000 testparticles (see table 1). All simulations were conducted with the newly developed parallelized GPU-based N-body code (Zimmermann (2021)) actually designed to handle huge numbers of massive particles in interaction. Nevertheless, as the mass of the objects residing in the Oort cloud is far less than the mass of the giant planets these particles were treated as massless in our N-body computations while the Sun, the 4 giant planets, and the passing star were treated as massive. The system and the passing star are integrated for the "passing time" of approximately 20 000 yr (i.e., the time the passing star needs to cross the Oort cloud).

3. Results and Discussion

Our integrations show a significant influence of the passing star mainly on the orbits of objects close to the stellar trajectory. Fig. 2 shows a comparison between the outcome of two simulations. The left panel shows the perturbations of a planar fly-by of Gliese 710 while the right panel shows the outcome of a simulation of the inclined passage. For better visibility only the objects distributed in the disk are shown. The results indicate that a passing star in the planar case has stronger influence on the objects than a star passing at an inclined trajectory (i.e., more objects are scattered to higher eccentricities after a planar passage – see table 2). From the objects scattered due to the planar fly-by of Gliese 710 a whole of 12 objects (out of 600 000) of the inner disk are moved onto

† This distance was choosen to see if the influence is significantly stronger for such a close passage.

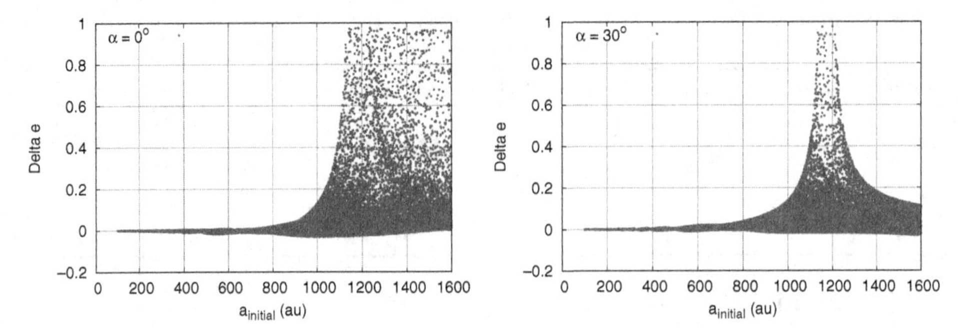

Figure 1. Comparison of different impact angles of Gliese 710 which passes at d_3=1200 au (left panel – planar case vs right panel – inclined case). The x-axis shows the initial semi-major axis of the disk objects. The y-axis shows the difference in eccentricity between the initial and final orbit.

Table 2. Comparison between the results of two integrations of Gliese 710 (planar (blue) and inclined (black) case) passing at d_3=1200 au. The table shows the absolute number of comets ending up on orbits with the eccentricity in the given ranges for the disk and the flared disk. For inclined fly-bys we observe almost no difference in the scattering process in the spherical Oort cloud.

ecc_{end}	objects D		objects FD	
$0.00 < e \leq 0.10$	573834	596177	170913	422014
$0.10 < e \leq 0.20$	12225	3092	342369	177985
$0.20 < e \leq 0.30$	5470	498	53824	1
$0.30 < e \leq 0.40$	2823	104	14457	0
$0.40 < e \leq 0.50$	1840	56	7359	0
$0.50 < e \leq 0.60$	1077	40	4243	0
$0.60 < e \leq 0.70$	864	12	2707	0
$0.70 < e \leq 0.80$	731	8	1807	0
$0.80 < e \leq 0.90$	542	10	1368	0
$0.90 < e \leq 1.00$	588	2	948	0

orbits with perihelia smaller than 5 au (observable region in the Solar system). For an inclined passage no objects were scattered into the observable area. Note that the star with d_3=1200 au passes outside of the denser part of the disk (due to the Rayleigh distribution used for the initial conditions) and inside of most of the objects residing in the flared disk. This means the dense part of the disk (i.e., most of the disk objects) is not influenced by the passage of the star which explains the huge number of objects that does not experience a big change in eccentricity (see table 2). Nevertheless, in case of an inclined passage there are some disk objects scattered to high eccentricity orbits while for the flared disk no such scattering is observed.

The long-term influence of the planets on the orbits of the scattered comets can be seen in Fig. 2 where simulations with all 4 planets show a broader distribution in the change of eccentricity (blue) than simulations where only Neptune was included (orange).

If one compares Figs. 2 (left panel) and 3 (left panel) one can see that a passing star with d_1 = 12 000 au (Fig. 3 – left panel) has almost no influence on the disk objects (note the difference in the units on the y-axis).

The influence of a passing star on the spherical Oort cloud can be seen in Fig. 3 (right panel) where the different colors depict the different shells used in the computations (see table 1). A comparison of the effects of two stellar passages at 1200 au and 4300 au can be seen in table 3 where one can observe that a closer fly-by causes stronger scattering in the disk than a fly-by with bigger impact parameter, d_i (i.e., more disk objects – even if

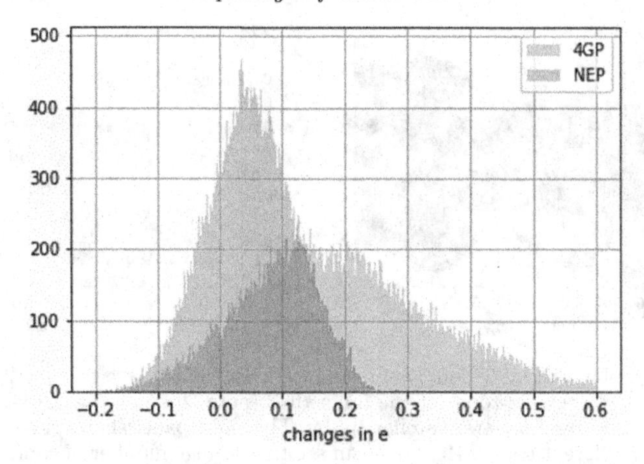

Figure 2. Influence of the giant planets on the changes in eccentricity of the comets. Blue denotes the numbers gathered from simulations with four giant planets, orange from simulations with only Neptune. Simulations in this case were run for 1 Myr showing long-term effects in the orbits of the comets. A total of 10 000 bins is used (1 bin = 100 objects) for the number of objects shown on the y-axis. Note that the negative changes in eccentricities occur due to the fact that for these long-term computations initial eccentricities for the comets were chosen to have values up to 0.2 and Δ e = $e_{end} - e_{start}$ (figure from Clees (2021)).

Table 3. Influence of the impact parameter, d_i, on the scattering of objects. The simulations represent the planar case for two different impact parameters: d_3 = 1200 au (blue) and d_2 = 4300 au (black).

eccentricity	objects D		objects FD		objects OC	
$0.00 < e \leq 0.10$	573834	598430	170913	429363	126014	145860
$0.10 < e \leq 0.20$	12225	1183	342369	114297	312020	322817
$0.20 < e \leq 0.30$	5470	266	53824	27471	144010	123060
$0.30 < e \leq 0.40$	2823	121	14457	12474	6675	6147
$0.40 < e \leq 0.50$	1840	0	7359	6696	6041	647
$0.50 < e \leq 0.60$	1077	0	4243	3882	3942	600
$0.60 < e \leq 0.70$	864	0	2707	2417	609	434
$0.70 < e \leq 0.80$	731	0	1807	1539	363	397
$0.80 < e \leq 0.90$	542	0	1368	1021	306	24
$0.90 < e \leq 1.00$	588	0	948	836	15	9

the numbers are really small – are scattered to higher eccentricity orbits for d_3=1200 au (blue numbers) than for d_2=4300 au (black numbers)).

4. Summary and Conclusions

In this investigation we numerically studied the immediate influence of a passing star on comets in the Oort cloud. The region feeling the strongest influence of the passing star changes with its impact parameter, d_i. If Gliese 710 passes very close to the Sun (i.e., d_3=1200 au) it influences all defined regions (see table 3) and a significant number of disk objects will experience higher changes in eccentricity. Nevertheless, the number thrown into the observable area is rather low (below 1 % – see table 3).

A passage of the star with closest approach at d_2=4300 au will strongly influence objects in the flared disk. However, for these objects it is less likely to be scattered on orbits which bring them closer than 50 au (planetary region) to the Sun.

A passage further away from the Sun (d_1=12 000 au in our simulations) shows almost no influence on the disk and the flared disk. The objects influenced in the Oort cloud in the path of the passing star are scattered to orbits with higher eccentricity but it is very

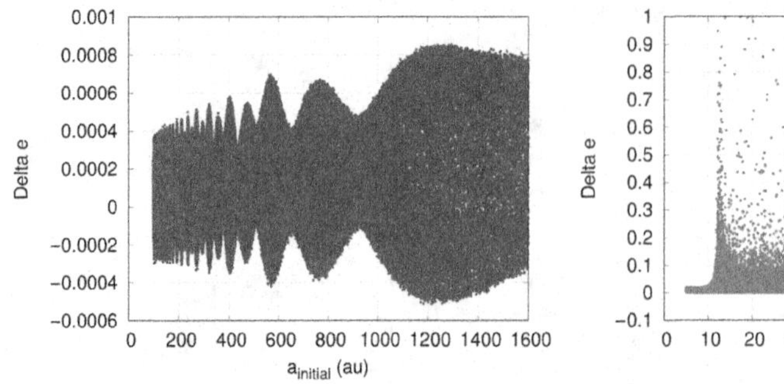

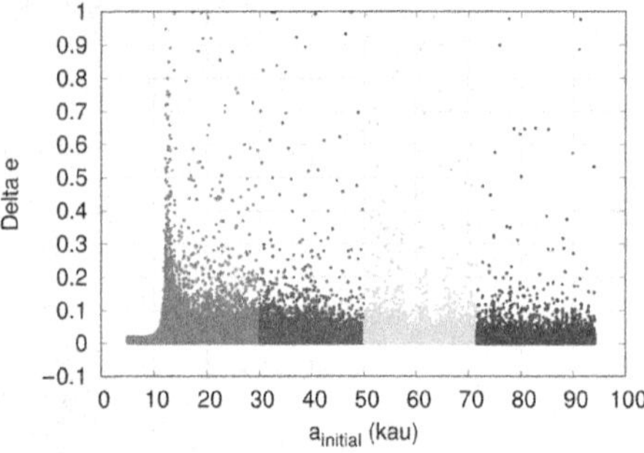

Figure 3. Same plot as Fig. 2 (left panel) but for a stellar passage at d_1=12 000 au (planar case). The left panel shows a zoom of the inner disk (note the units at the y-axis). The inner disk is not influenced by a far-away stellar fly-by. The right panel shows the different shells of the Oort cloud in different colors. Here one can see that a huge number of comets outside of the closest approach distance is scattered while the inner parts feel almost no influence.

unlikely for them to come close to the Sun. Longer computations will be needed in order to investigate the dynamical long-term influence on their orbits.

An investigation of the influence of the impact angle shows significantly different results only for the inner disk and the flared disk. A passing star penetrating the disk (flat and flared) in the ecliptic influences objects over a longer time span (the comets that are all distributed more or less in the plane where the star passes through). Thus, a high number of objects in the path of the star is scattered to orbits with high eccentricities. An inclined passing star crosses the disk at a certain distance, d_i. Thus, only objects in the vicinity of it will be influenced (see Fig. 2, right panel). Which means that only a small number of objects (with initially high inclination) is influenced by its passage. This can be seen also in table 2, where the planar and inclined passages are compared.

The influence of the passing star on the Oort cloud does not vary significantly between impact angles. This probably is due to our choice of distribution of Oort cloud objects.

Nevertheless, different masses of passing stars and longer computation times will probably yield more insight into this interesting topic.

Acknowledgements

This research was funded in whole by the Austrian Science Fund (FWF) [P33351-N]. The authors wand to thank Prof. Dr. Cristian Beaugé for useful comments to improve the paper.

References

Bailer-Jones, C.A.L., Rybizki, J., Andrae, R. & Fouesneau, M. 2018, *A&A*, 616, A37

Bancelin, D., Nordlander, T., Pilat-Lohinger, E., & Loibnegger, B. 2019, *MNRAS*, 486, 4, 4773-4780

Berski, F. & Dybczyński, P. A. 2016, *A&A*, 595, L10

Clees, S. 2021, *Master thesis*, University of Vienna

de la Fuente Marcos, R. & de la Fuente Marcos, C. 2018, *Research Notes of the American Astronomical Society*, 2, 2, 30

de la Fuente Marcos, R. & de la Fuente Marcos, C. 2020, *Research Notes of the American Astronomical Society*, 4, 12, 222

Rickman, H., Fouchard, M., Froesch'e, C., Valsecchi, G. B. 2008, *Celestial Mechanics and Dynamical Astronomy* 1405, 8077

Sizova, M.D., Vereshchagin, S.V., Shustov, B.M. & Chupina, N.V. 2020, *Astronomy Reports*, 64, 8, 711-721

Tanikawa, K. & Ito, T., 2007, *PASJ*, 59, 989
Torres, S., Cai, M.X., Brown, A.G.A. & Portegies Zwart, S. 2019, *A&A*, 629, A139
Vokrouhlický, D., Nesvorný, D., & Dones, L. 2019, *AJ*, 157, 5, 181
Zimmermann, M. 2021, *Master thesis*, University of Vienna

Multi-scale (time and mass) dynamics of space objects
Proceedings IAU Symposium No. 364, 2022
A. Celletti, C. Galeş, C. Beaugé, A. Lemaitre, eds.
doi:10.1017/S1743921322000011

Weak stability transition region near the orbit of the Moon

Zoltán Makó and Júlia Salamon

Department of Economic Sciences, Sapientia Hungarian University of Transylvania, Miercurea Ciuc, Romania

Abstract. This paper provides a study on the weak stability transition region in the framework of the planar elliptic restricted three-body problem. We define the lower boundary curve of the weak stability transition region and as a particular case, we determine this curve in the Sun-Earth system. The orbit of the Moon is near the lower boundary of the weak stability transition region.

Keywords. Weak stability boundary; Weak stability transition region; Elliptic restricted three-body problem.

1. Introduction

The capture of small bodies by major planets is an important phenomenon in planetary systems. The phenomenon has applications to the study of comets, asteroids, irregular satellites of the giant planets and different types of low energy planetary transfers, as well.

Ballistic capture (or weak capture) is analytically defined for the n-body problem, and it monitors the sign of the Kepler energy with respect to a massive primary. Weak capture occurs in special regions of the phase space, namely around one of the two primaries (e.g. the Earth in the case of the Sun–Earth system), which are referred to as the Weak Stability Boundaries (WSB). These regions are the boundary of the stable regions, delimiting stable and unstable orbits.

WSB was first introduced by Belbruno (1987) to design low energy transfers to the Moon and it is rigorously defined in Belbruno (2004). García & Gómez (2007) proposed a more general definition, which generalizes the concept of WSB given by Belbruno, and expanded the range of the WSB. The concept of WSB was extended in more accurate models by Belbruno, Topputo & Gidea (2008); Belbruno, Gidea & Topputo (2008); Belbruno, Gidea & Topputo (2013), Topputo & Belbruno (2009), Romagnoli & Circi (2009). Ceccaroni, Biggs & Biasco (2012) gave an analytical definition of the WSB. The effect of primaries' true anomaly on the structure of WSB was treated by Hyeraci & Topputo (2013) and Makó et al. (2010) in the model of elliptic restricted three-body problem (ER3BP).

In these articles, dedicated to the study of WSB properties, certain parameters (for example, the true anomaly of Earth, the initial eccentricity of the test particle, the direction of velocity vector at the initial time of the test particle) are considered constant and the initial position $\mathbf{r}_2$ of the test particle relative to the primary P_2 is considered a variable parameter. The modulus of initial velocity $\mathbf{v}_2(\mathbf{r}_2)$ of the massless particle relative to the primary P_2 is determined, and then the weak stability of the orbit with the initial values $(\mathbf{r}_2, \mathbf{v}_2(\mathbf{r}_2))$ is examined. In these cases, the variable parameter is $\mathbf{r}_2$.

The aim of this article is to study the properties of the transition zone where the weakly stable region switches over to weakly unstable region if $\mathbf{v}_2$ is considered as a variable parameter also. In Section 2 we recall the definition of the WSB. Section 3 contains the definition of the lower boundary curve of weak stability transition region. In Section 4, as a case study, we determine the lower boundary curve of the weak stability transition region around the Earth in the Sun-Earth system in the framework of ER3BP. We compare the position of the boundary curve to the Earth-Moon mean distance and to the radius of the Earth's Hill sphere. The concluding remarks are described in Section 5.

We choose the planar elliptic restricted three-body problem (PER3BP) model to study the dynamics, since even a small change of the eccentricity of the secondary influences the weak stability of the orbits. This model also takes into consideration the weak stability dependence on the true anomaly of the Earth.

In the PER3BP two massive primaries P_1 and P_2, with masses m_1 and m_2 revolve on elliptical orbits under their mutual gravitational attraction. Apart from these two bodies, the motion of a third, massless particle P_3 is investigated. The variation of the distance $R = \|P_1P_2\|$ with respect to the true anomaly f of the primary P_2 is given by $R = a\left(1-e^2\right)/(1+e\cos f)$, where a is the semi-major axis, e is the eccentricity and P is the orbital period of the elliptical orbit of P_2 around P_1. The motion of P_3 is restricted to the orbital plane of the primaries. The mass ratio is $\mu = m_2/(m_1+m_2)$, if $m_1 > m_2$.

The origin O of this system is considered to be the center of mass of the two massive primaries, where the $\tilde{\xi}$ axis is directed towards P_2, and the $\tilde{\xi}\tilde{\eta}$ coordinate-plane rotates with a variable angular velocity, in such a way that the two massive primaries are always on the $\tilde{\xi}$ axis and the period of the rotation is 2π. Beside the rotation, the system also pulsates in order to keep the primaries in fixed positions ($\tilde{\xi}_1 = -\mu$, $\tilde{\eta}_1 = 0$, $\tilde{\xi}_2 = 1-\mu$, $\tilde{\eta}_2 = 0$). To obtain a relatively simple set of equations, we use a nonuniform rotating and pulsating coordinate system (Szebehely (1967)).

To investigate the weak stability, we need a new coordinate system P_2xy, where the center P_2 is continuously moving and the axis P_2x is always parallel with the initial direction of the axis $O\tilde{\xi}$. The connections between these two reference frames are given in (Makó et al. (2010)).

If the normalized units are $a\,(1-e) = 1$ and $2\pi/P = 1$, then the Keplerian energy of the massless particle related to P_2 (Makó et al. (2010)) is

$$H_2 = \frac{v_2^2}{2} - \frac{Gm_2}{r_2} = \frac{v_2^2}{2} - \frac{1}{(1-e)^3}\frac{\mu}{r_2} = \frac{(1+e\cos f)^2}{2\,(1+e)\,(1-e)^3}\cdot$$
$$\left[\left(\tilde{\xi}' + D(f)\left(\tilde{\xi}+\mu-1\right) - \tilde{\eta}\right)^2 + \left(\tilde{\eta}' + D(f)\tilde{\eta} + \tilde{\xi} + \mu - 1\right)^2 - \frac{1}{1+e\cos f}\cdot\frac{2\mu}{\tilde{r}_2}\right],$$

where the derivatives are taken with respect to the true anomaly f of the primary P_2, $D(f) = e\sin f/(1+e\cos f)$, and $\tilde{r}_2 = \sqrt{\left(\tilde{\xi}+\mu-1\right)^2 + \tilde{\eta}^2}$.

2. The algorithmic definition of WSB in PER3BP

We recall the construction procedure of WSB introduced by Belbruno (2004) and improved by García & Gómez (2007) and Ceccaroni, Biggs & Biasco (2012).

For a fixed initial value of the true anomaly $f = f_0$ of P_2, we consider the half-line $l(\alpha, f_0)$ starting from P_2 and making an angle $\alpha \in [0, 360^o)$ with the axis P_1P_2 (see Fig. 1). The initial position of P_3 is on $l(\alpha, f_0)$. The initial eccentricity e_3 of the test particle P_3 is fixed. The direction of velocity vector of P_3 at the initial time is perpendicular to the line $l(\alpha, f_0)$. The initial distance between P_2 and P_3 in the coordinate system P_2xy is r_2 and the semi-major axis is $a_3 = r_2/(1-e_3)$.

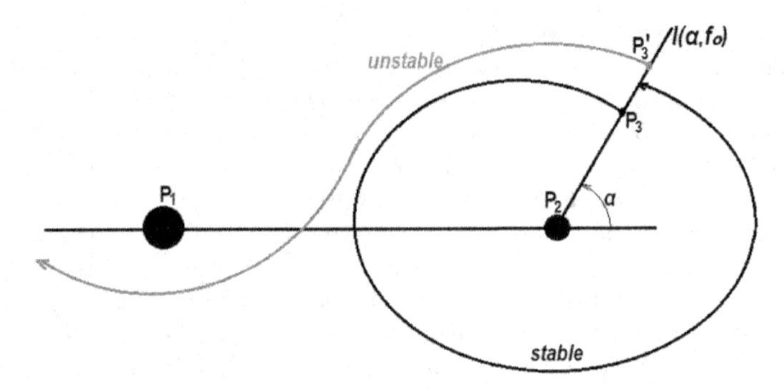

Figure 1. Construction procedure of WSB.

The modulus of velocity vector of P_3 at the initial time with respect to the reference frame P_2xy is

$$v_2^2(r_2, e_3) = \frac{Gm_2(1+e_3)}{r_2} = \frac{\mu(1+e_3)}{(1-e)^3 \cdot r_2} \in \left[v_c^2(r_2), v_e^2(r_2)\right],$$

where $v_c(r_2) = v_2(r_2, 0)$ is the circular initial velocity and $v_e(r_2) = v_2(r_2, 1)$ is the escape initial velocity with respect to primary P_2 in the context of the two-body problem.

We transform the initial position $(r_2 \cos\alpha, r_2 \sin\alpha)$ and initial velocity $\pm(-v_2 \sin\alpha, v_2 \cos\alpha)$ of P_3 (+ for direct, − for retrograde direction) from reference frame P_2xy into the reference frame $O\tilde{\xi}\tilde{\eta}$, where $(\tilde{\xi}_0, \tilde{\eta}_0)$ is the initial position and $(\tilde{\xi}_0', \tilde{\eta}_0')$ is the initial velocity of P_3 given by formulas (7) and (9) in Makó et al. (2010).

The motion of a particle is said to be weakly stable relative to P_2, under the PER3BP dynamics, if after leaving $l(\alpha, f_0)$ the particle P_3 makes a full cycle around P_2 without going near to P_1 or crashing into P_2 and returns to a point on $l(\alpha, f_0)$ with Kepler energy $H_2(\gamma(T)) \leq 0$ at the first return time T to half-line $l(\alpha, f_0)$. Otherwise, the motion will be weakly unstable.

García & Gómez (2007) show, that for each fixed $(\alpha, e_3) \in [0^o, 360^o) \times [0, 1)$ there are many changes from weak stability to weak instability and that the set of weakly stable points is a Cantor set.

The García & Gómez (2007) definition of WSB in the PER3BP model, taking into consideration Ceccaroni, Biggs & Biasco (2012) is the following.

Definition 1. For all fixed $(f_0, \alpha, e_3) \in [0^o, 360^o) \times [0^o, 360^o) \times [0, 1)$ we set

$$\begin{aligned} S^*(f_0, \alpha, e_3) = \{r_2 > 0: &\text{ the orbit of test particle with initial condition} \\ &(\mathbf{r}_2, \mathbf{v}_2(\mathbf{r}_2, e_3)) \text{ is weakly stable and the angular velocity } \dot{\vartheta}(T) > 0 \\ &\text{at the first return time } T \text{ to half-line } l(\alpha, f_0)\}. \end{aligned}$$

Then the WSB for fixed value of $f_0 \in [0^o, 360^o)$ and $e_3 \in [0, 1)$ is defined as:

$$W(f_0, e_3) = \left\{(r_2 \cos\alpha, r_2 \sin\alpha) \in \mathbb{R}^2 : \alpha \in [0, 360^o) \text{ and } r_2 \in \partial S^*(f_0, \alpha, e_3)\right\}.$$

The WSB lies in the transition zone from the connected part of the weakly stable region to the connected part of the weakly unstable region.

Definition 2. The weak stability transition region (WSTR) is the transition zone from the connected part of the weakly stable region to the connected part of the weakly unstable region in the fixed reference frame P_2xy.

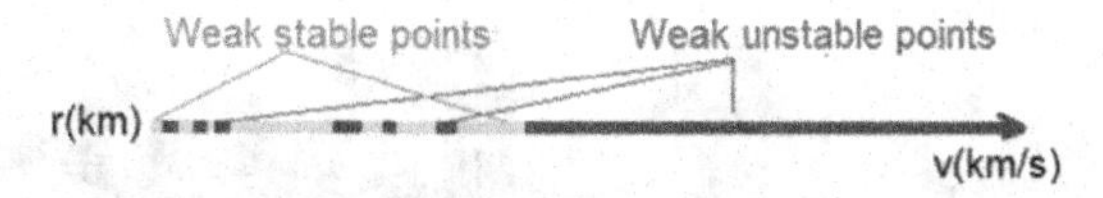

Figure 2. Stable and unstable points for a given true anomaly f_0 and distance r_2 with respect to the value of initial velocities in the system P_2xy.

3. The lower boundary curve in the weak stability transition region

Let f_0 and e_3 be given, and α, r_2 be the variable parameters. In the procedure of constructing the WSB, first, we calculate $v_2\,(r_2, e_3)$, the magnitude of the velocity and then we examine the weak stability of trajectories for initial values $(\mathbf{r}_2, \mathbf{v}_2\,(\mathbf{r}_2, e_3))$. In this case, the variable parameter is r_2.

The question is: what are the properties of WSTR if v_2 is a variable parameter and belongs to an interval $\left[v_2^{\min}(r_2), v_2^{\max}(r_2)\right]$?

For a given r_2, the smallest value of v_2 is $v_2^{\min}\,(r_2)$. In Fig. 2 it can be observed that for a given true anomaly f_0 and distance r_2, as the velocity increases, the orbits are weakly stable for a while (green dots), and after reaching a certain value, the orbits become weakly unstable (red dots). We can also observe that neither the weakly stable domain nor the weakly unstable domain are connected. By increasing the velocity, in the weakly stable region, they will appear weakly unstable parts.

We observe the following property of the WSTR (see Fig. 2).

Property A: There exist distances r_2 from P_2, such that the maximum velocity where the orbit remains weakly stable is strictly larger than the smallest velocity where the orbit becomes weakly unstable.

Next, for a fixed real anomaly f_0, we define the lower boundary curve of WSTR in the P_2xy coordinate system.

Definition 3. The lower boundary curve $LB\,(f_0)$ of the WSTR is

$$LB\,(f_0) = \left\{(r_2^* \cos\alpha, r_2^* \sin\alpha) \in \mathbb{R}^2 : \alpha \in [0^o, 360^o) \text{ and } r_2^* \text{ is the smallest distance from } P_2 \text{ on half-line } l(\alpha, f_0), \text{ where property A appears}\right\}.$$

For a fixed value of f_0, the defined set $LB\,(f_0)$ is two dimensional in WSTR, and it has two components corresponding to the direct and retrograde motions.

The Hill sphere (or Roche sphere) of primary P_2 is the region where the gravitational attraction of P_2 dominates. In the model of PER3BP, with normalized unit $a\,(1-e) = 1$, the radius of the Hill sphere of primary P_2 can be approximated (Hamilton & Burns (1992)) by $R_{P_2} = \sqrt[3]{\mu/\left[3\,(1-\mu)\right]}$. The Hill circle HB around P_2 is

$$HB = \left\{(R_{P_2}\cos\alpha, R_{P_2}\sin\alpha) \in \mathbb{R}^2 : \alpha \in [0^o, 360^o)\right\}.$$

4. Case study for Sun-Earth system

In this section, we investigated the weak stability in the PER3BP model of the Sun-Earth system. For the Sun-Earth system, the mass ratio is $\mu = 3.003158242 \cdot 10^{-6}$ and the eccentricity of the elliptical orbit of Earth is $e = 0.0167$.

First, we classify the weak stability of the orbits based on the initial distance r_2 and initial velocity v_2, where $f_0 = 0^o$ and $\alpha = 45^o$ (Fig. 3). The initial velocities are directly guided. Initial conditions for weakly unstable orbits are marked with red, the initial values for weakly stable orbits with green. The yellow points give the initial values of the collision orbits with the Earth (the trajectories where the distance between the massless particle and Earth will be equal to the radius of the Earth at some time: $\|P_3P_2\| = R_E$ $= 6378$ km). The second panel of Fig. 3 is an enlarged part of the first panel.

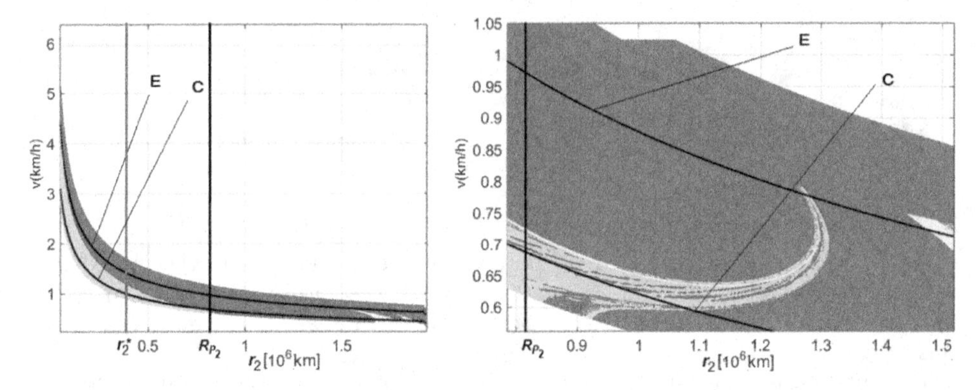

Figure 3. Weak stability in system (r_2, v_2). Initial values for weakly unstable orbits are marked with the color red, the initial values for weakly stable orbits with the color green. The yellow points give the initial values of the collision orbits with the Earth. The second panel is the enlarged part of the first panel.

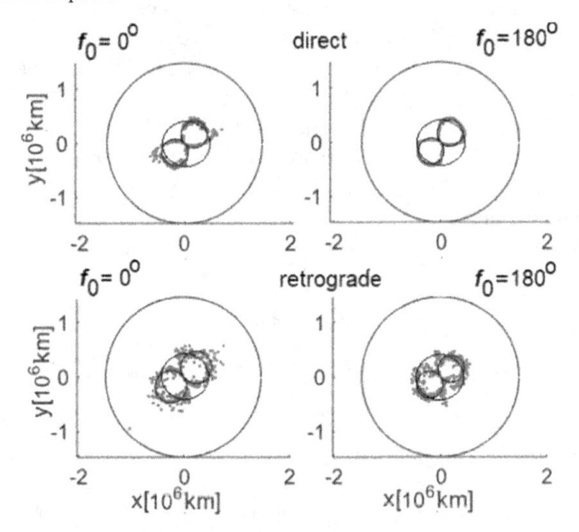

Figure 4. The lower boundary curve of WSTR in the Sun-Earth system for direct and retrograde initial velocities in P_2xy coordinate system, where $f_0 \in \{0^o, 180^o\}$. The points of $LB(f_0)$ are marked by color red. The inner circle is the orbit of the Moon and the outer circle plots the Hill circle. The curve near the red dots shows the best fitting lemniscate.

The two black curves represent the circular (C) and escape (E) velocity for a given distance in two body problem regarding to the planet Earth. The first vertical line (red line) indicates the position of boundary point corresponding to the angle $\alpha = 45^o$ of $LB(0^o)$. At this distance r_2^*, the property A appears for the first time. The second vertical line indicates the position corresponding to the radius of the Hill sphere.

Fig. 4 illustrates the lower boundary curve $LB\,(f_0)$ in coordinate system P_2xy, when $f_0 \in \{0^o, 180^o\}$ and the initial conditions give direct or retrograde motions. The points of $LB(f_0)$ are marked by color red. The inner circle is the orbit of the Moon and the outer circle plots the Hill circle.

Remark 4. In Fig. 4 it can be observed, that the approximation of the lower boundary curve in analytical form can be obtained by fitting a quadratic plane curve to the determined points (red dots). Moreover, in the case of direct motion, the red dots are more

Table 1. The optimal fitting parameters

	f_0	L	L_{inf}	L_{sup}	L_{width}	δ	δ_{inf}	δ_{sup}	δ_{width}
D.	0°	495318	483949	506538	22589	42.78	41.48	44,08	2.6
D.	180°	476020	468690	483500	14810	43.11	42.22	44	1.78
R.	0°	590462	574605	606320	31715	36.44	34.91	37,97	3.06
R.	180°	490681	476020	505491	29471	35.78	34.06	37.5	3.44

dispersed for $f_0 = 0^o$ than for $f_0 = 180^o$. In the case of retrograde motion, the boundary curves become blurred (with greater dispersion). The points of the $LB\,(f_0)$ are very important in the design of ballistic escape trajectories in the Sun-Earth system. In the graph of direct motion on Fig. 4, we observe weakly unstable points (red dots) around the Moon's orbit in the vicinity of $\alpha = 45^o$.

The $LB(f_0)$ is approximated by a Bernoulli lemniscate (Fig. 4). Thus, an analytical formula is obtained to estimate the lower bound of WSTR. If we fit the lemniscate of Bernoulli : $r^2 = L^2 \cos^2\,(\alpha - \delta)$ to the points of $LB(f_0)$ then we obtain the optimal fitting parameters reported in Tab. 1. When d is the half-distance between the focal points F_1 and F_2 of the lemniscate then $L = d\sqrt{2}$. The δ is the angle between the line F_1F_2 and axis Ox.

The first column of Table 1 indicates that the motion is direct (D.) or retrograde (R.). In the second column the value of the true anomaly and in the third column the length (L[km]) of the best-fitting lemniscate are given. The fourth and fifth columns give the upper and lower limits of the confidence interval of L at a 95% confidence level. The sixth column gives the width of the confidence interval in km.

The seventh column gives the angle $\delta[^o]$ between the semi major-axis of the lemniscate and the axis Ox. The eighth and ninth columns represent the upper and lower limits of the confidence interval of δ at 95% confidence level. The last column gives the width of the confidence interval of δ in degrees.

5. Concluding remarks

In this paper, the velocity is also considered as a variable parameter in the study of weak stability. The lower boundary curve is defined in the weak stability transition-region (WSTR). As an application, the lower boundary curve of WSTR is numerically determined in the PER3BP model of the Sun-Earth system. The location of the lower boundary curve is compared to the Earth-Moon mean distance. We show that the lower boundary curve can be approximated by a Bernoulli lemniscate.

The analysis shows that, if we ignore the mean inclination of Moon's orbit from the ecliptic ($i = 5.15^o$), the orbit of the Moon is near to the lower boundary of the WSTR in the system P_2xy (the first two panel of Fig. 4). The orbit of the Moon intersects $LB\,(f_0)$ at four points, approximately: $\alpha_1 = 41^0$, $\alpha_2 = 45^0$, $\alpha_3 = 221^0$, $\alpha_4 = 225^0$. If the true anomaly of the Moon relative to Earth is approximately in $[41^0, 45^0] \cup [221^0, 225^0]$, then the Moon is below the lower boundary of the WSTR (i.e. the motion is weakly stable), otherwise, it is in the weak stability transition region.

Another interesting situation is when the angle between the directions of Sun-Earth and Earth-test particle is 135°. In this case, the weakly unstable points are very close to Earth. From these points escape trajectories can be designed even if the initial velocity is smaller than the escape velocity.

The simplicity and enough precise estimation of the lower boundary curve raises the question on whether the equation of the lower boundary curve can be analytically derived in the ER3BP or CR3BP model.

References

Belbruno, E. 1987, in *Proc. of AIAA/DGLR/JSASS Inter. Propl. Conf.* AIAA paper No. 87-1054
Belbruno, E., 2004, *Capture Dynamics and Chaotic Motions in Celestial Mechanics*, Princeton Univ. Press
Belbruno, E., Topputo, F., & Gidea, M. 2008, *Adv Space Res*, 42, 1330
Belbruno, E., Gidea, M., & Topputo, F. 2010, *SIAM J Appl Dyn Syst*, 3, 1061
Belbruno, E., Gidea, M., & Topputo, F. 2013, *Qual Theor Dyn Syst*, 12, 53
Ceccaroni, M., Biggs, J., & Biasco, L. 2012, *Celest Mech Dyn Astr*, 114, 1
García, F., & Gómez, G. 2007, *Celest Mech Dyn Astr*, 97, 87
Hamilton, D.P., & Burns, J.A. 1992, *Icarus*, 96, 43
Hyeraci, N., & Topputo, F. 2013, *Celest Mech Dyn Astr*, 116, 175
Makó, Z., Szenkovits, F., Salamon, J., & Oláh-Gál, R. 2010, *Celest Mech Dyn Astr*, 108, 357
Romagnoli, D., & Circi, C. 2009, *Celest Mech Dyn Astr*, 103, 79
Szebehely, V. 1967, *Theory of orbits*, Academic Press, New-York
Topputo, F., Belbruno, E. 2009, *Celest Mech Dyn Astr*, 3, 17

Multi-scale (time and mass) dynamics of space objects
Proceedings IAU Symposium No. 364, 2022
A. Celletti, C. Galeş, C. Beaugé, A. Lemaitre, eds.
doi:10.1017/S1743921321001368

Secular dynamics in extrasolar systems with two planets in mutually inclined orbits

Rita Mastroianni[1] **and Christos Efthymiopoulos**[2]

[1]Dep. of Mathematics Tullio Levi-Civita, University of Padua
via Trieste 63, 35121 Padova, Italy
email: rita.mastroianni@math.unipd.it

[2]Dep. of Mathematics Tullio Levi-Civita, University of Padua
via Trieste 63, 35121 Padova, Italy
email: cefthym@math.unipd.it

Abstract. We revisit the problem of the secular dynamics in two-planet systems in which the planetary orbits exhibit a high value of the mutual inclination. We propose a 'basic hamiltonian model' for secular dynamics, parameterized in terms of the system's Angular Momentum Deficit (AMD). The secular Hamiltonian can be obtained in closed form, using multipole expansions in powers of the distance ratio between the planets, or in the usual Laplace-Lagrange form. The main features of the phase space (number and stability of periodic orbits, bifurcations from the main apsidal corotation resonances, Kozai resonance etc.) can all be recovered by choosing the corresponding terms in the 'basic Hamiltonian'. Applications include the semi-analytical determination of the actual orbital state of the system using Hamiltonian normalization techniques. An example is discussed referring to the system of two outermost planets of the ν-Andromedae system.

Keywords. Celestial mechanics; Planetary systems.

---

1. Introduction

The dynamics of exoplanetary systems with two planets in orbits with non-zero mutual inclination is a very interesting topic in view of the discovery, in the last 20 years, of several such systems (see Naoz (2016) and references therein, as well as Libert & Tsiganis (2009)). In the present short review we do a preliminary discussion of the phase space structures observed in such systems; a more complete study will be presented in a separate work. In particular we discuss below two distinct characteristic regimes. The first, called the *nearly-planar regime*, is characterized by small values of the mutual inclination, high values of the planetary eccentricities and by the predominance of orbits (periodic or quasi-periodic) linked to apsidal corotations. The second regime, called *Kozai-Lidov regime*, is characterized, instead, by a high value of the mutual inclination, small values of the eccentricities and by the instability of the circular orbit for the inner planet. We will finally discuss the sequence of bifurcations that connect the two regimes.

2. Hamiltonian model

The Hamiltonian of the three-body problem in Poincaré heliocentric canonical variables takes the form:

$$\mathcal{H}(\mathbf{r_2},\mathbf{r_3},\mathbf{p_2},\mathbf{p_3}) = \underbrace{\frac{\mathbf{p_2}^2}{2\,m_2} - \frac{\mathcal{G}\,m_0\,m_2}{r_2} + \frac{\mathbf{p_3}^2}{2\,m_3} - \frac{\mathcal{G}\,m_0\,m_3}{r_3}}_{\textit{Keplerian part}} + \underbrace{\frac{(\mathbf{p_2}+\mathbf{p_3})^2}{2\,m_0}}_{\textit{"Indirect" part}} - \underbrace{\frac{\mathcal{G}\,m_2\,m_3}{|\mathbf{r_2}-\mathbf{r_3}|}}_{\textit{"Direct" part}}, \tag{2.1}$$

where m_0 = mass of the star, and m_i, $\mathbf{p_i}$, $\mathbf{r_i}$, $i=2,3$ are the masses, barycentric momenta and heliocentric position vectors of the planets. Starting from (2.1), a secular Hamiltonian is arrived at by averaging the above Hamiltonian with respect to all short period terms. We consider two distinct types of expansion, i.e. i) series expansions in powers of the planets' eccentricities and inclinations and ii) closed-form averaging. In both cases, we end up with a secular Hamiltonian model of the form

$$\mathcal{H}_{sec} = -\frac{\mathcal{G}m_0m_2}{2a_2} - \frac{\mathcal{G}m_0m_3}{2a_3} + \mathcal{R}_{sec}(a_2,a_3,e_2,e_3,i_2,i_3,\varpi_2,\varpi_3,\Omega_2,\Omega_3). \tag{2.2}$$

In the Hamiltonian (2.2) the 'fast' angles $\lambda_i = M_i + \varpi_i$, $i=2,3$, are ignorable, a fact implying the constancy of the semi-major axes under the secular model. The total angular momentum $\mathbf{L} = \mathbf{r}_2 \times \mathbf{p}_2 + \mathbf{r}_3 \times \mathbf{p}_3$ is an exact first integral of the system, a fact implying that the Hamiltonians (2.1) and (2.2) depend on the angles Ω_2, Ω_3 only through the difference $\Omega_2 - \Omega_3$. Using modified Delaunay canonical variables

$$\begin{aligned}
\Lambda_j &= L_j = m_j\sqrt{\mathcal{G}\,m_0\,a_j}\,, & \lambda_j &= l_j + g_j + h_j = M_j + \varpi_j\,,\\
\Gamma_j &= L_j - G_j = \Lambda_j\left(1-\sqrt{1-e_j^2}\right), & \gamma_j &= -g_j - h_j = -\varpi_j\,,\\
\Theta_j &= G_j - H_j = \Lambda_j\sqrt{1-e_j^2}\,\left(1-\cos(i_j)\right), & \vartheta_j &= -h_j = -\Omega_j\,,
\end{aligned} \tag{2.3}$$

the secular Hamiltonian $\mathcal{H}_{sec}(\Gamma_2,\Gamma_3,\Theta_2,\Theta_3,\gamma_2,\gamma_3,\vartheta_2,\vartheta_3)$ has 4 degrees of freedom. However, the existence of two independent integrals in involution (i.e. the components L_z and L_{plane} of the total angular momentum $\mathbf{L}$) allows to reduce the number of degrees of freedom by two, a process known as 'Jacobi's reduction of the nodes'. We propose a novel method to perform Jacobi's reduction with respect to the one followed by Libert & Henrard (2007). Our method leads to an explicit analytic control of all small parameters appearing in the problem, by taking advantage of the symmetries of the Hamiltonian expressed in Keplerian heliocetric elements with respect to the Laplace reference frame. In particular setting, as usual, $\Omega_3 - \Omega_2 = \pi$ and observing that the Hamiltonian depends only on the mutual inclination, $\mathcal{H}_{sec} = \mathcal{H}_{sec}(a_2,a_3,e_2,e_3,\cos(i_2+i_3),\omega_2,\omega_3,\Omega_3-\Omega_2=\pi)$, we introduce two book-keeping parameters, ε and η (with numerical values $\varepsilon=\eta=1$), via the relations

$$\cos(i_2)\cos(i_3) = \varepsilon^2\eta\,(\cos(i_2)\cos(i_3)-1)+1\,, \quad \sin(i_2)\sin(i_3) = \varepsilon^2\eta\sin(i_2)\sin(i_3)\,. \tag{2.4}$$

In view of these book-keeping parameters, the Hamiltonian takes the form

$$\begin{aligned}
\mathcal{H}_{sec} &= \mathcal{H}_{plane} + \eta\,\mathcal{H}_{space}\\
&= \sum_{s_2\geqslant 0} \varepsilon^{s_2} h_{s_2}(a_2,a_3,e_2,e_3,\omega_2-\omega_3)\\
&+ \eta\sum_{s_2\geqslant 2} \varepsilon^{s_2}\widehat{h}_{s_2}(a_2,a_3,e_2,e_3,\cos(i_2)\cos(i_3),\sin(i_2)\sin(i_3),\omega_2,\omega_3)\,,
\end{aligned}$$

where, in the last passage, all powers η^{s_1} with $s_1 \geqslant 1$ are replaced simply by η (this is a possible simbolic operation since $\eta=1$). Finally, setting

$$\sin(i_2)\sin(i_3) = \cos(i_2)\cos(i_3) + \frac{G_2}{2\,G_3} + \frac{G_3}{2\,G_2} - \frac{C^2}{2\,G_2\,G_3}\,, \tag{2.5}$$

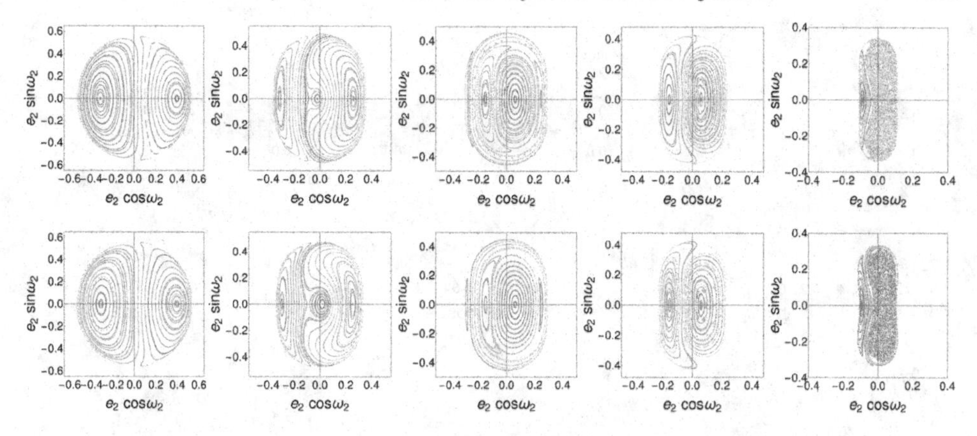

Figure 1. Poincaré surfaces of section in the plane $(e_2\cos(\omega_2), e_2\sin(\omega_2))$ with L_z fixed and different values of energy (from left to right), top $\mathcal{E} = -6.62\cdot 10^{-5}, -2.94\cdot 10^{-5}, -1.92\cdot 10^{-5}, -1.18\cdot 10^{-5}, -2.73\cdot 10^{-6}$, bottom $\mathcal{E} = -6.67\cdot 10^{-5}, -2.53\cdot 10^{-5}, -1.9\cdot 10^{-5}, -1.17\cdot 10^{-5}, -2.61\cdot 10^{-6}$. The top row shows the surfaces of section as computed by a numerical integration of trajectories in the Hamiltonian expanded with Laplace coefficients up to order 10 in the eccentricities. The bottom rows shows the corresponding sections with a Hamiltonian averaged in closed form with a multipolar expansion truncated at degree 5 and expanded up to order 10 in the eccentricities.

where $C = L_z$ and $G_j = m_j\sqrt{\mathcal{G}\, m_0\, a_j}\sqrt{1-e_j^2}\ j=2,3$, the Hamiltonian takes the form $\mathcal{H}_{sec}(a_2,a_3,e_2,e_3,\omega_2,\omega_3;L_z) = \mathcal{H}_{plane}(\Gamma_2,\Gamma_3,\omega_2-\omega_3) + \eta\,\mathcal{H}_{space}(\Gamma_2,\Gamma_3,\omega_2,\omega_3;L_z)$. We note that Eq. (2.4) explicitly keeps track of all the small quantities of the problem: these are the planetary eccentricities e_2, e_3 and the inclinations i_2, i_3. However, in our approach the dependence of the Hamiltonian on the inclinations is accounted for through powers of the quantities $\sin(i_2)\sin(i_3)$ and $\cos(i_2)\cos(i_3)-1$. Finally it is possible to express the secular Hamiltonian in the form of a polynomial $\mathcal{H}_{sec}(X_2,X_3,Y_2,Y_3;\mathrm{AMD})$, where the canonical pairs $(X_2,\,Y_2)$, $(X_3,\,Y_3)$ are Poincaré variables $X_i = -\sqrt{2\Gamma_i}\cos\omega_i$, $Y_i = -\sqrt{2\Gamma_i}\sin\omega_i$, while the 'Angular Momentum Deficit' $\mathrm{AMD} = \Lambda_2+\Lambda_3-L_z$ is constant. We checked the precision of each type of expansion against numerical experiments; in particular we compared the Poincaré section $Y_3=0$, $\dot{Y}_3 \geqslant 0$ obtained from the Hamiltonian expanded as in both methods above. This allows to specify which is a sufficient order of truncation of the multipolar expansion and of the small parameters; Figure 1 shows an example of such comparison, allowing to see that we need to reach degree 5 of the multipolar expansion to obtain the same phase portraits as in the Laplace expansion up to order 10 in the eccentricities.

3. Dynamics

3.1. *Phase portraits*

In order to analyze the most important phenomena in the phase portrait of the Hamiltonian $\mathcal{H}_{sec}(X_2,X_3,Y_2,Y_3;\mathrm{AMD})$, we consider, as a representative example, the value $\mathrm{AMD} = 0.016$ in a fictitious system with two planets same as the planets c and d of the system ν-Andromedae. In particular we chose $m_0 = 1.31\,M_\odot$, $m_2 = 13.98\,M_J$, $m_3 = 10.25\,M_J$, $a_2 = 0.829\,AU$, $a_3 = 2.53\,AU$ (according to Table 13 of McArthur et al. (2010)) and $e_2 = 0.2445$, $e_3 = 0.316$, $i_2 = 11.347°$ and $i_3 = 25.609°$ (according to Table 1 of Deitrick et al. (2015)). Figure 2 shows the Poincaré section $Y_3 = 0$, $\dot{Y}_3 \geq 0$ for different values of the energy $\mathcal{E} = \mathcal{H}_{sec}$. Since the AMD is fixed, altering the values of the planets' eccentricities implies that the inclinations also change to keep the AMD constant to its

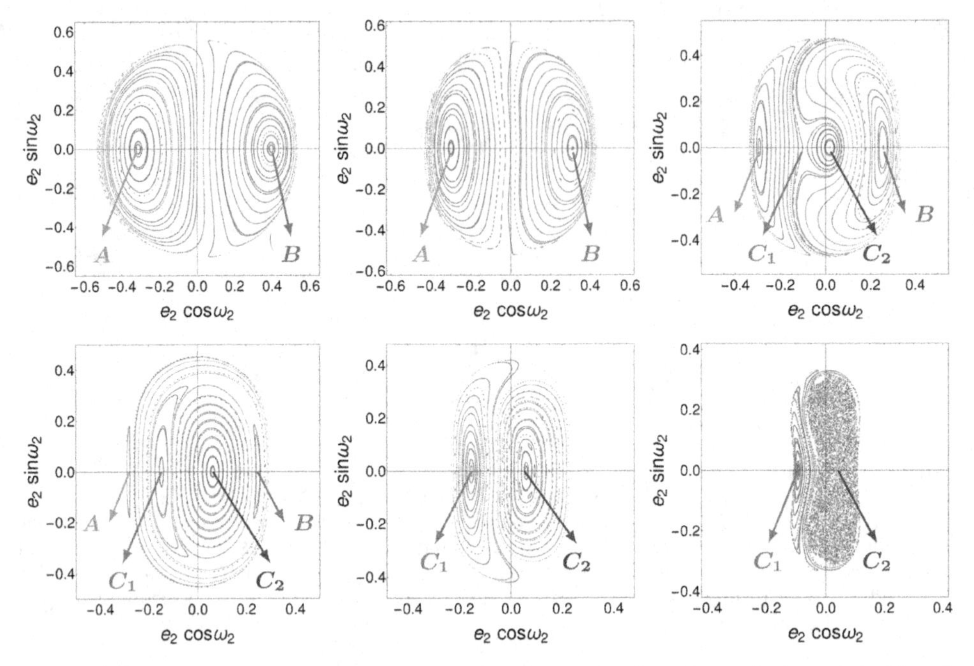

Figure 2. Poincaré surfaces of section in the representative plane $(e_2\cos(\omega_2), e_2\sin(\omega_2))$ with L_z fixed and for different values of the energy (from left to right) $\mathcal{E} = -6.67\cdot10^{-5}, -4.26\cdot10^{-5}, -2.53\cdot10^{-5}, -1.9\cdot10^{-5}, -1.17\cdot10^{-5}, -2.61\cdot10^{-6}$.

pre-selected value. The maximum values of e_2 and e_3 allowed for a specific value of the AMD can be computed using the Lagrange multiplier method. In the same way we find the allowed region $(X_{2_{min}}, X_{2_{max}})$ for the phase portrait corresponding to each section.

3.2. *Nearly-planar regime*

The first two panels in the Figure 2 are representative of the phase portrait obtained in the *nearly-planar regime*. This is similar to the phase portrait found in the exact planar case, in which the averaged Hamiltonian turns to be integrable (see Figure 3c). In fact the 3D Hamiltonian can be decomposed as

$$\mathcal{H}_{sec} = \underbrace{\mathcal{H}_{plane}(\Gamma_2,\Gamma_3,\omega_2-\omega_3) + \eta\,\mathcal{H}_{0_{space}}(\Gamma_2,\Gamma_3,\omega_2-\omega_3)}_{\textit{Integrable part}:=\mathcal{H}_{int}} + \eta\,\mathcal{H}_{1_{space}}(\Gamma_2,\Gamma_3,\omega_2,\omega_3)\,.$$

Of particular importance in the integrable part $\mathcal{H}_{int}$ are the periodic orbits called anti-aligned (mode A) and aligned (mode B) apsidal corotation (see Laughlin et al. (2002), Lee & Peale (2003), Beaugé et al. (2003)). In order to compute all possible (symmetric or asymmetric) apsidal corotations for given energy and angular momentum it is particularly convenient to express the integrable part of the Hamiltonian $\mathcal{H}_{int}$ in *Weyl variables*, defined as

$$\sigma_0 = \frac{1}{2}\left(X_2^2+Y_2^2+X_3^2+Y_3^2\right), \qquad \sigma_1 = X_2X_3+Y_2Y_3\,,$$

$$\sigma_2 = Y_2X_3 - Y_3X_2\,, \qquad \sigma_3 = \frac{1}{2}\left(X_2^2+Y_2^2-X_3^2-Y_3^2\right),$$

where X_i, Y_i $i=2,3$ are the canonical Poincaré variables defined above. These variables satisfy the Poisson algebra relations $\{\sigma_i,\sigma_j\} = -2\varepsilon_{ijk}\sigma_k$, where ε_{ijk} is the Levi-Civita

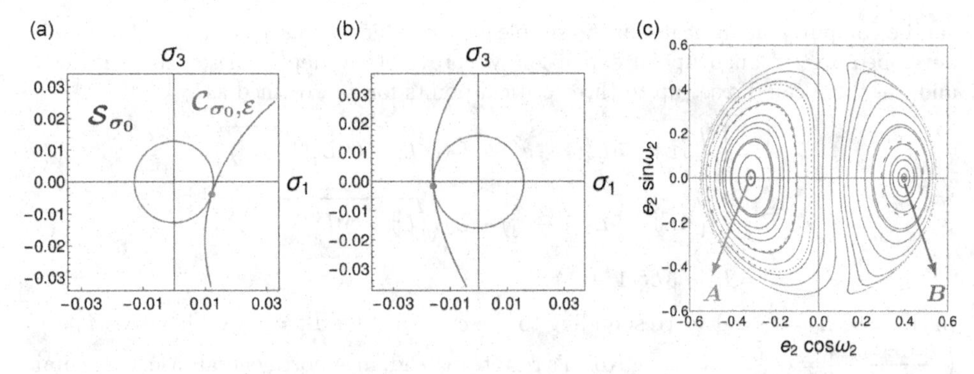

Figure 3. Examples of apsidal corotations calculated by imposing the tangency condition between the sphere $\mathcal{S}_{\sigma_0}$ and the cylindrical energy surface $\mathcal{C}_{\sigma_0,\mathcal{E}}$. The right panel shows the positions of the apsidal corotations in the surface of section $Y_3 = 0$, $\dot{Y}_3 \geqslant 0$ in the representative plane $(e_2 \cos(\omega_2), e_2 \sin(\omega_2))$, obtained considering only the $\mathcal{H}_{int}$ term in the Hamiltonian. Note the absence of a separatrix between the two stable modes. This is due to the fact that the system's reduced phase-space has the topology of the 3D-sphere.

symbol, when $i, j, k = 1, 2, 3$ and $\varepsilon_{ijk} = 0$ if one of the i, j, k is equal to 0. We easily verify that the Hamiltonian $\mathcal{H}_{int}$ does not depend on σ_2. Consider the surfaces

$$\mathcal{S}_{\sigma_0} = \{(\sigma_1, \sigma_2, \sigma_3) \in \mathbb{R}^3 \, : \, \sigma_1^2 + \sigma_2^2 + \sigma_3^2 = \sigma_0^2\}$$
$$\mathcal{C}_{\sigma_0,\mathcal{E}} = \{(\sigma_1, \sigma_2, \sigma_3) \in \mathbb{R}^3 \, : \, \mathcal{H}_{int}(\sigma_0, \sigma_1, \sigma_3) = \mathcal{E}\} \, . \tag{3.1}$$

It can be proved that the periodic orbits of $\mathcal{H}_{int}$ are given by the points of tangency of the surfaces $\mathcal{S}_{\sigma_0}$ and $\mathcal{C}_{\sigma_0,\mathcal{E}}$. Using this property we can compute analytically all the apsidal corotations corresponding to a fixed energy $\mathcal{E}$ (varying the angular momenta of the planets, i.e. σ_0), or fixed σ_0 (varying $\mathcal{E}$). Figure 3 shows a particular example.

Having specified, now, the coordinates of the apsidal corotations, it is possible to apply a Birkhoff normal form of the Hamiltonian in order to compute the quasi-periodic orbits around one of the two modes. Details of this construction are given in a separate study. We emphasize the method's utility in order to provide a semi-analitycal representation of the long term time series of the orbital elements for the planetary trajectories.

3.3. *Sequences of bifurcations*

As shown in Figure 2, the nearly-planar regime is followed by a sequence of bifurcations occurring for a fixed value of the AMD as the energy increases (from left to right), implying that the maximum allowed mutual inclination between the planets also increases. Starting with the basic apsidal corotations (A, B), a saddle-mode bifurcation generates the orbits C_1, C_2 which correspond to an orbital configuration with non-zero mutual inclination. Furthermore, as the mutual inclination increases, the orbit C_2 becomes unstable by the "Kozai mechanism"(see Kozai (1962)), as shown in the last picture of Figure 2.

3.4. *Kozai-Lidov regime*

In the last two pictures of Figure 2 we can see the transition of C_2 from stable to unstable, accompanied by a large volume of the trajectories around C_2 becoming chaotic. The transition occurs at critical values of L_z, or, equivalently, of the mutual inclinations; for example the limiting inclination in the case of an inner test particle and an outer planet in circular orbits is $\cos^{-1}\sqrt{\frac{3}{5}} \sim 39^\circ.2$ (see Naoz (2016) for a review). These limits can be computed analytically in the simple case in which we have a secular Hamiltonian developed up to a quadrupolar expansion; we proved that, depending on the value of L_2 and e_3, there could exist up to three critical points for L_z^2, defined as

$$A = \frac{1}{5}\left(4L_2^2 + 5L_3^2(1-e_3^2) - L_2\sqrt{L_2^2 + 60L_3^2(1-e_3^2)}\right),$$

$$B = \frac{1}{5}\left(4L_2^2 + 5L_3^2(1-e_3^2) + L_2\sqrt{L_2^2 + 60L_3^2(1-e_3^2)}\right),$$

$$C = L_2^2 - 3L_3^2(1-e_3^2).$$

In the case $L_2 = e_3 = 0$, corresponding to the critical value $L_z^2 = B$, we have $\cos(i_{mut}) = \left(\frac{B - L_2^2 - L_3^2}{2L_2L_3}\right) = \sqrt{\frac{3}{5}}$, recovering Kozai's result. Instead, in a more general case, these limits can be computed numerically through the Jacobian matrix of the quadratic Hamiltonian vector field.

Ackowledgements

Useful discussions with U. Locatelli are greatfully acknowledged. C.E. was partially supported by the MIUR-PRIN 20178CJA2B 'New Frontiers of Celestial Mechanics: theory and Applications'.

References

Beaugé, C., Ferraz-Mello, S., Michtchenko, T.A. 2003, *ApJ*, 593, 1124

Deitrick, R., Barnes, R., McArthur, B., Quinn, T. R. Luger, R., Antonsen, A., Benedict, G. F. 2015, *ApJ*, 798, 46

Henrard, J., Libert, A.S. 2005, *Dynamics of Populations of Planetary Systems* IAU Colloq. 197, 49–54

Jacobi, M. 1842, *Astr. Nachrichten*, 20, 81

Kozai, Y. 1962, *AJ*, 67, 591–598

Laskar, J., Robutel, P. 1995, *Cel. Mec. and Dyn. Astr.*, 62, 193–217

Laughlin, G., Chambers, J., Fischer, D. 2002, *ApJ*, 579, 455–467

Lee, M.H., Peale, S.J. 2003, *ApJ*, 592, 1201–1216

Libert, A.S., Henrard, J. 2007, *Icarus*, 191, 469–485

Libert, A.S., Tsiganis, K. 2009, *A&A*, 493, 677–686

Marchesiello, A., Pucacco, G. 2016, *Inter. Jour. of Bif. and Chaos*, 26, 1630011

McArthur, B.E., Benedict, G.F., Barnes, R., Martioli, E., Korzennik, S., Nelan, Ed.,Butler, R.P. 2010, *ApJ*, 715, 1203

Murray, C.D. and Dermott, S.F. 1999, *Solar syst. dyn.*

Naoz, S. 2016, *ARAA*, 54, 441–489

Robutel, P. 1995, *Cel. Mec. and Dyn. Astr.*, 62, 219–261

Multi-scale (time and mass) dynamics of space objects
Proceedings IAU Symposium No. 364, 2022
A. Celletti, C. Galeş, C. Beaugé, A. Lemaitre, eds.
doi:10.1017/S1743921321001241

Dynamics around the binary system (65803) Didymos

R. Machado Oliveira[1], O. C. Winter[1], R. Sfair[1], G. Valvano[1], T. S. Moura[1] and G. Borderes-Motta[2]

[1]Grupo de Dinâmica Oribtal e Planetologia, São Paulo State University, UNESP, Guaratinguetá, CEP 12516-410, São Paulo, Brazil

[2]Bioengineering and Aerospace Engineering Department, Universidad Carlos III de Madrid, Leganés, 28911, Madrid, Spain

Abstract. Didymos and Dimorphos are primary and secondary, respectively, asteroids who compose a binary system that make up the set of Near Earth Asteroids (NEAs). They are targets of the Double Asteroid Redirection Test (DART), the first test mission dedicated to study of planetary defense, for which the main goal is to measure the changes caused after the secondary body is hit by a kinect impactor. The present work intends to conduct a study, through numerical integrations, on the dynamics of massless particles distributed in the vicinity of the two bodies. An approximate shape for the primary body was considered as a model of mass concentrations (mascons) and the secondary was considered as a massive point. Our results show the location and size of stable regions, and also their lifetime.

Keywords. Asteroids; Binary system; Computational simulations; Mascons.

---

1. Introduction

The sub-kilometer asteroid Didymos and its moon Dimorphos form a binary system classified as a Near-Earth Asteroid and member of both the Apollo and Amor group†. The system was chosen as the target for the Double Asteroid Redirection Test (DART) mission, the first one dedicated to study of planetary defense (Cheng et al. 2018). The main DART objective is to test the asteroid deflection technique by intentionally impacting Dimorphos and measuring the changes in its orbit. Ground-based observatories will also monitor these changes, and more precise information about the orbital evolution of Dimorphos after the impact will be assessed by the Hera mission‡.

Here we investigate the dynamics around the binary system, taking into account the irregular shape of the primary and also the gravitational disturbance of Dimorphos. Our goal is to analyze the orbital evolution of test particles in the vicinity of the system, and verify the existence of stable regions where particles can remain for an extended period and eventually pose a threat to the mission.

Damme et al. (2017) made a similar study, but considering spacecraft (CubeSats) instead of natural particles. They adopted a gravitational model of Didymos, including spherical harmonics up to order and degree two, and integrating the system for short timescales (a few months). Here we consider a more complex gravitational model based on a polyhedron shape model and look for orbital stability on much longer timescales (several years).

† https://ssd.jpl.nasa.gov/sbdb.cgi?sstr=2065803
‡ https://www.esa.int/Safety_Security/Hera/Hera

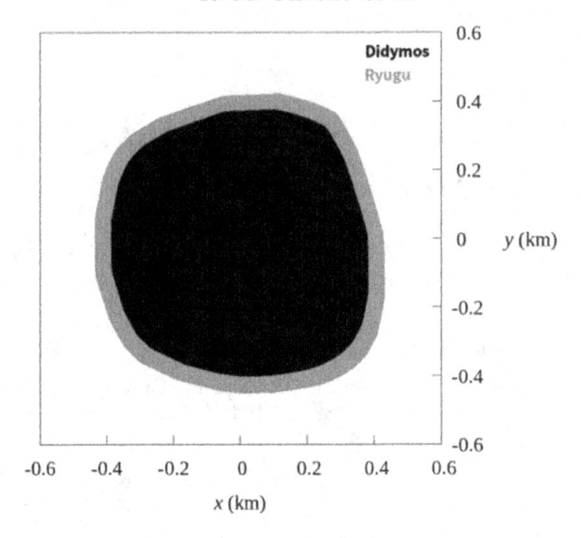

Figure 1. Comparative size between Didymos and Ryugu model.

This paper is organized as follows: in Section 2 we present the shape model adopted for Didymos, while the numerical setup and the main results are described in Section 3. The last section presents our final remarks.

2. Shape model

Given the irregular shape of Didymos, particles at the vicinity of the body are subject to a gravitational potential that cannot be reasonably modeled as originated from a point of mass or even an ellipsoid. However, the proper shape model for Didymos is not publicly available, and given that the object resembles the asteroid Ryugu, we assumed the shape as a scaled version of the latter.

From ground-based radar observations, Scheirich and Pravec (2009) determined an equivalent diameter of 780 m and Naidu et al. (2020) reported a bulk density as 2.17 gcm^{-3} for the primary. Starting from a polyhedra representation of Ryugu composed of 574 vertices and 1144 triangular faces (Müller et al. 2011), we determined the scaling factor that must be applied to Ryugu's shape to match the expected mass and volume for Didymos. A comparison between the two models is shown in Fig. 1.

Instead of computing the gravitational potential directly from the polyhedron (Werner 1994), our simulations were carried out with the more efficient, and yet precise, approach of mass concentrations – mascons (Geissler et al. 1996), implemented in the N-BOM package (Winter et al. 2020). From a regularly spaced tridimensional grid that encompasses the object, the mascons are those grid points that lie within the shape model. Assuming that Didymos is homogeneous, the total mass of the asteroid is evenly distributed among the 26070 mascons.

In our model, Didymos rotates with a period of 2.26 hours (Pravec et al. 2006). Dimorphos is represented as a point of mass with 3.45×10^9 kg orbiting the primary with a semimajor axis of 1178 m, and the eccentricity of 0.05 (Scheirich and Pravec 2009).

3. Stable regions

In order to explore the stability in the neighbourhood of the two bodies, we randomly distributed 20 thousand massless particles in a radial range between 0.45 and 1.45 km.

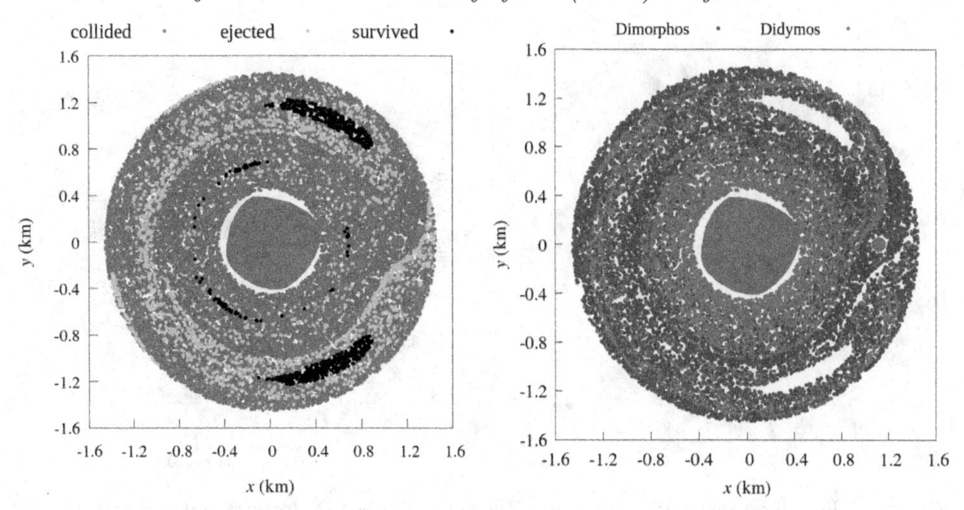

Figure 2. Fate of the particle after 5 years of simulation. The particles are shown at their initial positions. Left: The blue dots represent the particles that collided with one of the bodies, the pink dots indicate the ejected particles and the black dots the survived particles. Right: This plot shows only the collisions. The red dots represent the particles that collided with Didymos, while the green dots indicate the particles that collided with Dimorphos.

The initial conditions of the particles were determined by keplerian orbits with eccentricity and inclination equal to zero. The system was integrated for five years, which is the approximated interval between the DART† and HERA‡ missions. The criterion for the remotion from the system was a collision or an ejection. For the collision we took into account the approximated dimension of the primary's mascons model and the estimated mean radius of the secondary given by Scheirich and Pravec (2009). In the case of ejection, it was adopted 6 km as the maximum radial distance from the primary, the same as considered in Damme et al. (2017). This distance is about five times the value of the semi-major axis of Dimorphos.

The results are summarized in Fig. 2. It shows the particles at their initial positions and the colors indicate the fate of the particles at the end of the simulations. After five years of simulations, 8.74% were ejected, 30.3% collided with Didymos, 56.94% collided with Dimorphos and just 4.02% survived. From the features of this plot can be made some considerations. The particles that collided are initially spread over several regions (left plot). However, there is a pattern separating the the two groups of collision. As can be seen in the right plot, the particles that collided with Dimorphos are mainly located in two spiral arms that depart from the satellite. In the case of particles that collided with Didymos, they can be separated into two groups: one in a disk inside the orbit of Dimorphos and other involving the stable regions around the triangular lagrangian points. On the other hand, the particles that were ejected and those that survived are found initially in specific regions. The ejected particles were preferentially in places close to the orbit of Dimorphos. Their fate was determined by close encounters with the secondary.

In the case of the survivors, Fig. 2 (left plot) shows that they are initially located in the coorbital region, around the lagrangian points L_4 and L_5, and also in arcs of a ring formed between the two bodies. The particles around one of the two triangular lagrangian points present behaviours expected in binary systems. Since these are stable equilibrium

† https://dart.jhuapl.edu/Mission/index.php
‡ https://www.heramission.space/

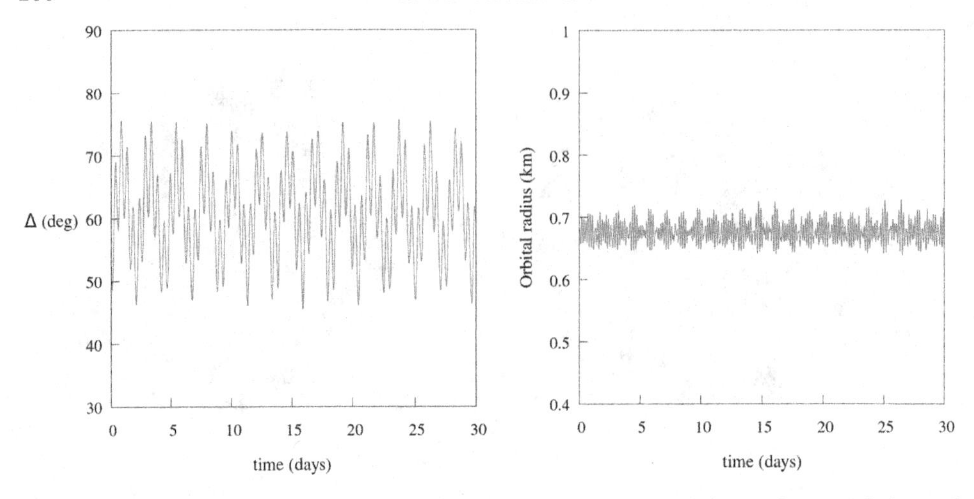

Figure 3. Left: Example of the evolution of the angle between Dimorphos and a particle located around L_4 lagrangian point. Right: Example of the radial evolution of the trajectory of a particle located in one of the stable arcs close to Didymos.

points for the given mass ratio of the binary, it is coherent to have survivors around them. These particles show a tadpole like trajectory, as in the example given in Fig. 3 (left plot). The trajectories of the particles in the stable regions close to Didymos show a very narrow amplitude of radial variation. An example is given in Fig. 3 (right plot).

An idea of the time evolution of the system as a whole can be seen in Fig. 4. This plot shows the lifetime of the particles according to their initial location. The survivors are indicated in green. The system as seen in Fig. 2 is quickly defined, since 79% of the particles are removed within 10 days. Those that were removed in just a few hours are the ones that were close to the surface of Didymos and those that were in spiral arms associated to Dimorphos. The satellite is the responsible for the large majority of this fast removal in the first orbital periods of the particles, by collision with Dimorphos or ejection of the region.

Note in the Fig. 4 that the particles which live longer (very dark color) are in the neighbourhood of the survivors (in green). Some of these particles can stay alive for even more than one year of integration. The orbital period of the particles in the dark ring shown in Fig. 4 is a value in the range between five and six hours.

A comparison of these results with those of Damme et al. (2017) shows that the general structure is similar. However, in their work was found a large stable ring of initial conditions close to Didymos which is not the case in our results. That is certainly due to the large difference in timescales considered (a few months for them, while some years in ours), but the effects due to different gravitational potential models also contributed in the evolution of the trajectories with initial conditions closer to Didymos.

4. Final comments

In this work we explored the long term stability of massless particles in the Didymos-Dimorphos system. In order to take into account the gravitational contribution of Didymos, we developed an approximated shape model in terms of a polyhedron composed of triangular faces and then, transformed it in a model of Mascons.

The results clearly show the location and size of the stable regions. An analysis of the lifetime of the particles indicated that the huge majority of the particles with unstable

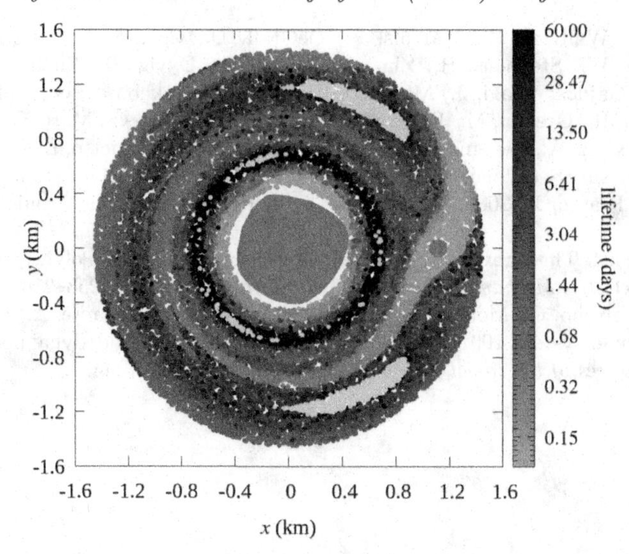

Figure 4. Lifetime of the particles according to their initial location. The particles that survived the whole integration time (5 years) are indicated in green.

trajectories are removed in just ten days. The largest amount of particles in the stable regions have tadpole like trajectories around the triangular lagrangian equilibrium points.

5. Acknowledgements

This study was financed in part by the Coordenação de Aperfeiçoamento de Pessoal de Nível Superior - Brasil (CAPES) - Finance Code 001, Fundação de Amparo à Pesquisa do Estado de São Paulo (FAPESP) - Proc. 2016/24561-0 and Proc. 2020/14307-4, Conselho Nacional de Desenvolvimento Cientifico e Tecnológico (CNPq) - Proc. 120338/2020-3 and Proc. 305210/2018-1

References

Cheng, A. F., Rivkin, A. S., Michel, P., Atchison, J., Barnouin, O., Benner, L., Chabot, N. L., Ernst, C., Fahnestock, E. G., Kueppers, M., Pravec, P., Rainey, E., Richardson, D. C., Stickle, A. M., & Thomas, C. 2018, Aida dart asteroid deflection test: Planetary defense and science objectives. *Planetary and Space Science*, 157, 104–115.

Damme, F., Hussmann, H., & Oberst, J. 2017, Spacecraft orbit lifetime within two binary near-earth asteroid systems. *Planetary and Space Science*, 146, 1–9.

Geissler, P., Petit, J. M., Durda, D. D., Greenberg, R., Bottke, W., Nolan, M., & Moore, J. 1996, Erosion and Ejecta Reaccretion on 243 Ida and Its Moon. *Icarus*, 120, 140–157.

Müller, T. G., Ďurech, J., Hasegawa, S., Abe, M., Kawakami, K., Kasuga, T., Kinoshita, D., Kuroda, D., Urakawa, S., Okumura, S., Sarugaku, Y., Miyasaka, S., Takagi, Y., Weissman, P. R., Choi, Y. J., Larson, S., Yanagisawa, K., & Nagayama, S. 2011, Thermo-physical properties of 162173 (1999 JU3), a potential flyby and rendezvous target for interplanetary missions. *Astronomy & Astrophysics*, 525, A145.

Naidu, S., Benner, L., Brozovic, M., Nolan, M., Ostro, S., Margot, J., Giorgini, J., Hirabayashi, T., Scheeres, D., Pravec, P., Scheirich, P., Magri, C., & Jao, J. 2020, Radar observations and a physical model of binary near-earth asteroid 65803 didymos, target of the dart mission. *Icarus*, 348, 113777.

Pravec, P., Scheirich, P., Kušnirák, P., Šarounová, L., Mottola, S., Hahn, G., Brown, P., Esquerdo, G., Kaiser, N., Krzeminski, Z., Pray, D., Warner, B., Harris, A., Nolan, M., Howell, E., Benner, L., Margot, J.-L., Galád, A., Holliday, W., Hicks, M., Krugly, Y.,

Tholen, D., Whiteley, R., Marchis, F., DeGraff, D., Grauer, A., Larson, S., Velichko, F., Cooney, W., Stephens, R., Zhu, J., Kirsch, K., Dyvig, R., Snyder, L., Reddy, V., Moore, S., Gajdoš, Világi, J., Masi, G., Higgins, D., Funkhouser, G., Knight, B., Slivan, S., Behrend, R., Grenon, M., Burki, G., Roy, R., Demeautis, C., Matter, D., Waelchli, N., Revaz, Y., Klotz, A., Rieugné, M., Thierry, P., Cotrez, V., Brunetto, L., & Kober, G. 2006, Photometric survey of binary near-earth asteroids. *Icarus*, 181(1), 63–93.

Scheirich, P. & Pravec, P. 2009, Modeling of lightcurves of binary asteroids. *Icarus*, 200(2), 531–547.

Werner, R. A. 1994, The Gravitational Potential of a Homogeneous Polyhedron or Don't Cut Corners. *Celestial Mechanics and Dynamical Astronomy*, 59(3), 253–278.

Winter, O. C., Valvano, G., Moura, T. S., Borderes-Motta, G., Amarante, A., & Sfair, R. 2020, Asteroid triple system 2001 SN_{263}: surfaces characteristics and dynamical environment. *Monthly Notices of the Royal Astronomical Society*, 492, 4437–4455.

Multi-scale (time and mass) dynamics of space objects
Proceedings IAU Symposium No. 364, 2022
A. Celletti, C. Galeş, C. Beaugé, A. Lemaitre, eds.
doi:10.1017/S1743921321001496

Orbit propagation around small bodies using spherical harmonic coefficients obtained from polyhedron shape models

P. Peñarroya[1] and R. Paoli[2]

[1]Deimos Space S.L.U., Ronda de Poniente, 19, 28760 Tres Cantos (Madrid), Spain
email: pelayo.penarroya@deimos-space.com

[2]Department of Mathematics, Universitatea Alexandru Ioan Cuza, Bd. Carol I, nr. 11, 700506, Iasi Romania
email: roberto.paoli@uaic.ro

Abstract. Missions to asteroids have been the trend in space exploration for the last years. They provide information about the formation and evolution of the Solar System, contribute to direct planetary defense tasks, and could be potentially exploited for resource mining. Be their purpose as it may, the factor that all these mission types have in common is the challenging dynamical environment they have to deal with. The gravitational environment of a certain asteroid is most of the times not accurately known until very late mission phases when the spacecraft has already orbited the body for some time.

Shape models help to estimate the gravitational potential with a density distribution assumption (usually constant value) and some optical measurements of the body. These measurements, unlike the ones needed for harmonic coefficient estimation, can be taken from well before arriving at the asteroid's sphere of influence, which allows to obtain a better approximation of the gravitational dynamics much sooner. The disadvantage they pose is that obtaining acceleration values from these models implies a heavy computational burden on the on-board processing unit, which is very often too time-consuming for the mission profile.

In this paper, the technique developed on [1] is used to create a validated Python-based tool that obtains spherical harmonic coefficients from the shape model of an asteroid or comet, given a certain density for the body. This validated software suite, called *AstroHarm*, is used to analyse the accuracy of the models obtained and the improvements in computational efficiency in a simulated spacecraft orbiting a small body.

The results obtained are shown offering a qualitative comparison between different order spherical harmonic models and the original shape model. Finally, the creation of a catalogue for harmonics is proposed together with some thoughts on complex modelling using this tool.

Keywords. Spherical harmonics; Shape model; Propagation; Small bodies; Polyhedron.

---

1. Introduction

The number of missions targeting small bodies, such as asteroids or comets, has risen in the past decades. The Near-Earth Asteroid Rendezvous (NEAR) mission [2] flew to Eros and rendezvoused the asteroid to an approach distance of around 500 km, to then lower that distance to 35 km. This mission included a large set of instruments to perform different measurements of the environment around Eros, including a Multispectral Imager (MSI), a Near-Infrared Spectrograph (NIS), an X ray/Gamma Ray Spectrometer (XRS/GRS), a Magnetometer (MAG), and a NEAR Laser Rangefinder (NLR). After that, the objective of the missions shifted towards sampling the small body they were focused on, by means of small lander probes or even by performing short sampling manoeuvres (often called Touch-and-Go (TaG) manoeuvres).

The Rosetta mission [3; 4], for instance, targeted comet 67P/Churyumov-Gerasimenko close to its aphelion and studied the physical and chemical properties of the nucleus and its coma, and the development of the interaction region of the solar wind and the comet. Additionally, Rosetta deployed Philae, a small lander, onto the surface of the comet. Although the landing was not successful, Philae helped retrieving information about the gravitational field of the comet and its surface properties.

With a different approach, Hayabusa [5], Origins, Spectral Interpretation, Resource Identification, and Security-Regolith Explorer (OSIRIS-REx) [6], and Hayabusa2 [7] missions collected samples by making touch-down -or TaG- sequences. Both Hayabusa missions visited asteroid Itokawa, while OSIRIS-REx visited Bennu. The retrieval of surface samples from the small body they were orbiting enriches the scientific outcome of the mission by allowing the mission to bring back to Earth, for further analysis, materials directly extracted from the asteroid. To do that, however, an extremely high level of knowledge about the asteroid's mass distribution is needed. Without a proper gravitational potential model, attempting such manoeuvres would risk the end of the mission. Thus, and due to the non-sphericity that these bodies often present, very detailed shape models are constructed from observations to feed the trajectory design process.

Future missions to asteroids include Double Asteroid Redirection Test (DART) [8] and Hera [9] missions, which are part of the Asteroid Impact and Deflection Assessment (AIDA) program [10]. This program will attempt to deflect Dimorphos, the smaller body of the Near-Earth Asteroid (NEA) binary system Didymos. Another example would be the Psyche mission [11], which will rendezvous the namesake asteroid (16 Psyche) and will investigate its metallic composition.

In this work, a method to aid future missions accomplish their objectives is presented that involves the use of a precise gravitational model using spherical harmonics, obtained from early observations of a certain celestial body. This method trades-off accuracy of the solution and computational effort, to meet the mission requirements (on-board modelling, navigation uncertainty...).

Throughout this document, Section 2 will introduce the fundamentals of the technique and its implementation. Section 3 will present the validation results for the developed tool, and Section 4 will display the obtained results for two study-cases, namely, Bennu and Lutetia. Finally, Section 5 will sum up the main points of this research and will draw directions for future work.

2. From polyhedra to spherical harmonics

As the aforementioned shape models become more accurate, thanks to the images and observations taken on-board as the Spacecraft (S/C) gets closer to the small body, the computational burden of calculating the gravitational potential becomes heavier. When using shape models, polyhedron dynamics methods [12] are employed to compute the gravity potential at a certain coordinate, and given a set of vertices, edges, and faces for a constant density shape model. These methods, albeit very accurate even for irregularly-shaped bodies, require to run through all of the faces included in the shape model, which, for high-definition models, are in the order of hundreds of thousands, rendering them not fitting for on-board applications.

This is not compatible with the high degrees of autonomy a mission to an asteroid needs. Telecommanding every manoeuvre from Earth is not feasible due to the long time of travel of the signal (order of minutes) in comparison to the short reaction time needed in such delicate sequences (order of seconds). Alternatives exist, that reduce the computational effort of the orbital dynamics calculations, while maintaining high levels of accuracy, such as mascons or spherical harmonics. In [13], the authors perform an evaluation of how different mascons models perform in terms of both accuracy and computational effort. However, in this work, the analyses will be directed towards the spherical harmonics approach.

The main reference for this work is [1], where the authors developed an algorithm to compute the spherical harmonics coefficients of a given constant density polyhedron (which is not an unrealistic assumption for a wide set of asteroids), following a similar process to what was done in [14]. The key idea is to use recurrence relations for the integrands that appear when one computes the spherical harmonics coefficients for a given body. Such integrands usually involve Legendre functions and polynomials. The recurrence relations are presented in both their normalized and non-normalized form.

Finally the polyhedron is partitioned into a collection of simplices and the integration of the integrands is performed. These simplices are tetrahedra whose bases are the triangular faces of the polyhedron and the vertices are at the centre of the reference system used to define the coordinates of the points that form the shape model. A change of variable is used to ease the integration of the aforementioned simplices.

In terms of implementation, the main issues identified by [1] are to represent homogeneous polynomials in three variables and to operate with them without using symbolic manipulators. This is achieved by representing trinomials of degree n as arrays of length $(n+1)(n+2)/2$ whose elements are the coefficients of the trinomials ordered in such a way that each coefficients correspond to the right trinomial.

AstroHarm (AstroSim Harmonics) is a Python suite and module that takes the theory developed on [1] to a software materialisation. The capabilities that this module offers include: wavefront (.obj) files management, geometric assessment of shape models (reference radius, volume, centre of mass...), triplet management and operations, C and S matrices recursive computation, normalisation routines for coefficients, and file I/O.

3. Validation

In order to validate the coefficients obtained by *AstroHarm*, semi-analytical methods were used to obtain spherical harmonic coefficients from accurately known geometric bodies as a cube, a tetrahedron, and a double pyramid (octahedron). Using Wolfram Mathematica® [15] software and symbolic formulation, the semi-analytically-computed coefficients for various n and m † values were compared to the ones obtained from *AstroHarm*. The results for the cube can be observed in Figure 1.

Errors in the validation of the coefficients for the C and S matrices for the cube and the octahedron range from 1e-19 to 1e-17, which represents a totally negligible value, only due to floating-point precision errors. For the tetrahedron, the errors are also very small but significantly larger than the other cases, ranging from 1e-11 to 1e-10. This is due to the lower *sphericity* of the geometry, which, albeit being perfectly convex, leaves more empty volume when circumscribed by a sphere.

4. Results

Once these coefficients are obtained, a spherical harmonic model has to be implemented. Following the method shown in [16], the gravitational acceleration given by these coefficients can be computed and used as internal dynamics for a given propagator. To evaluate the accuracy of this spherical harmonics model, a ground truth trajectory is required. This ground truth is obtained from a polyhedron dynamics suite, developed by [17] and based on [18]. This choice is supported by the fact that this method is exact up to the surface of the given shape model, assuming constant density distribution.

† n and m refer to the order and degree of the spherical harmonics model, respectively.

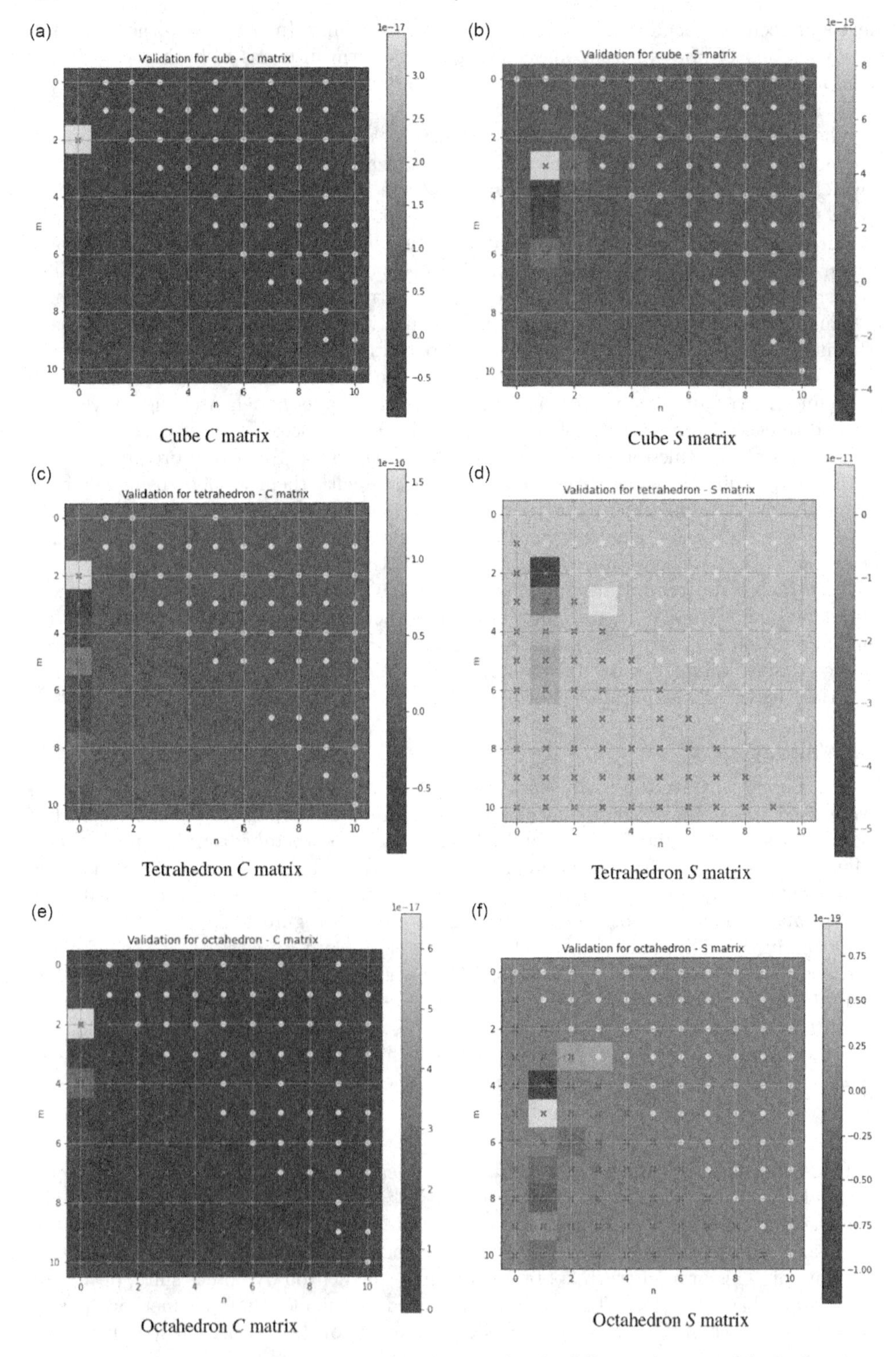

Figure 1. Validation heatmaps for C and S matrices for the different shape models. Red crosses point out cases were $m > n$, which should be disregarded due to its lack of conceptual meaning. Green dots represent values that analytically equal to zero.

(a)

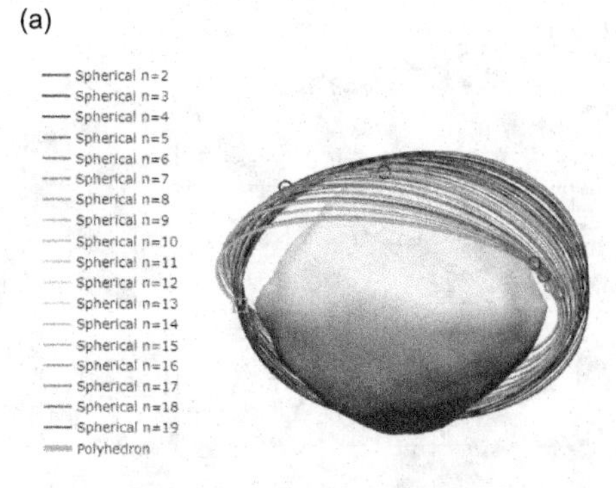

Shape model of asteroid Bennu used for simulation. Comparison of propagated trajectories using spherical harmonics with respect to the Polyhedron model (orange) for reference.

(b)

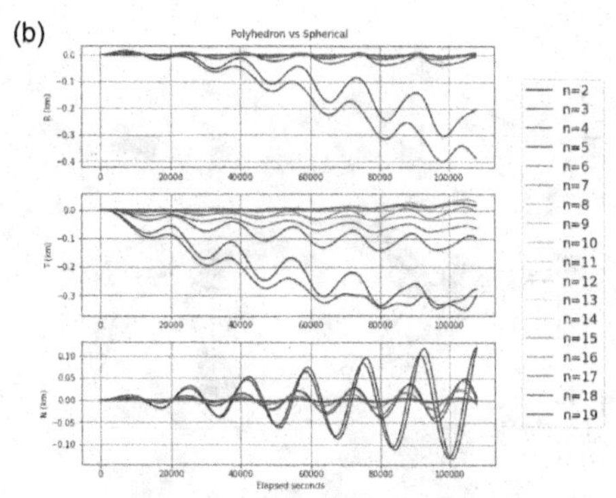

Radial, tangential, and normal component errors for different spherical harmonics models w.r.t. polyhedral model for Bennu.

Figure 2. Trajectories around Bennu for the different values of n after the 30 h propagation. $sma = 350$ m (1.20 Bennu radii), $ecc = 0.1$, $inc = 45°$, where sma, ecc, inc are the semi-major axis, eccentricity and inclination.

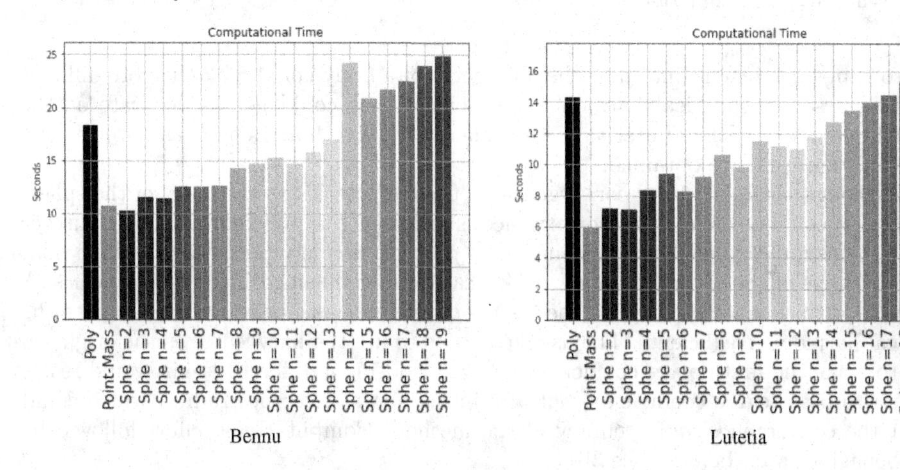

Figure 3. Computational effort for the different values of n after the 30 h propagation.

The two main drivers of this research are computational performance and results accuracy. Once the computation of the coefficients is validated, the next step is to take the experimentation to a real asteroid for which a shape model is given.

Starting with Bennu, the algorithm is run for different maximum orders for the coefficients going from 2 to 19, obtaining the results shown in Figure 2 in terms of trajectory difference w.r.t. the polyhedral model, which is taken as ground truth.

It can be clearly seen that, as the order of the coefficients increases, the error is reduced, arriving at values close to no error for $n > 10$ in this case. In terms of computational effort, Figure 3a shows that the computational effort for these levels of accuracy at $n > 10$ is reduced by around 17%, w.r.t. the baseline polyhedral model.

4.1. *Remarks*

The limitations of this algorithm were also explored within this project. In particular, two specific parameters were investigated: body shape and orbital distance. The former

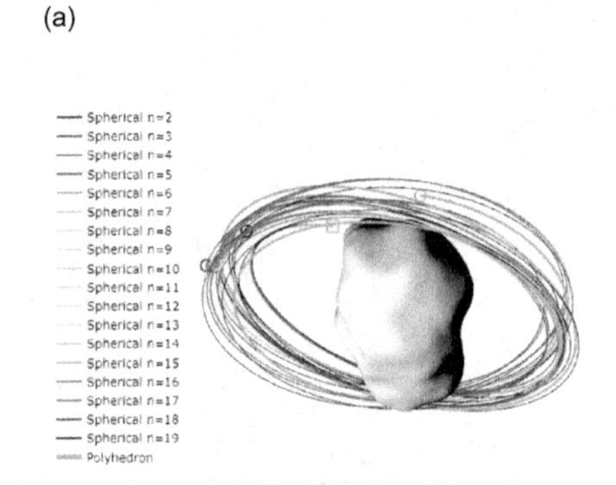

Shape model of asteroid Lutetia used for simulation. Comparison of propagated trajectories using spherical harmonics with respect to the Polyhedron model (orange) for reference.

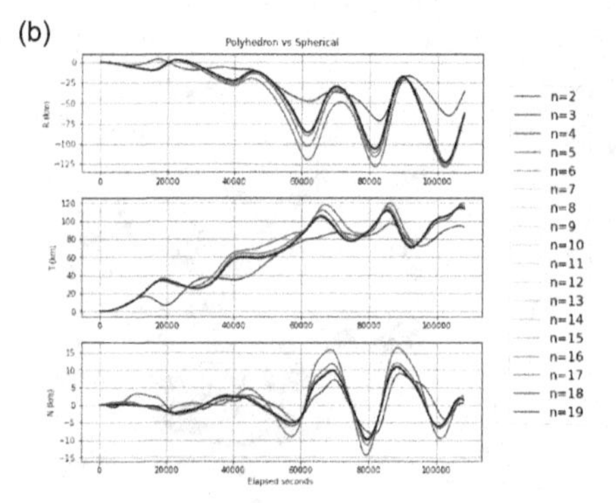

Radial, tangential, and normal component errors for different spherical harmonics models w.r.t. polyhedral model for Lutetia.

Figure 4. Trajectories around Lutetia for the different values of n after the 30 h propagation. $sma = 150$ km (1.55 Lutetia radii), $ecc = 0.1$, $inc = 45°$, where sma, ecc, inc are the semi-major axis, eccentricity and inclination.

tried to gain some insight into how the oblateness of a body could make it more difficult to obtain a set of spherical harmonics coefficients that could get as accurate as the polyhedron model. Using Lutetia, a less spherical body than Bennu, as subject, the results in Figure 4 were obtained.

Even though there is a clear convergence on the trajectories as the order of the spherical harmonics model rises, this convergence is far from the polyhedral model. Further investigations are required to determine if this problem is solely caused by the high non-spherical shape of the attracting body, or if it can be solved using different approaches. A possibility is to change the origin of the system of reference, which is used to compute the spherical harmonic coefficients. The usual choice is to use the barycenter as the origin, in order to make the first degree coefficient vanish. Choosing a different origin would result in non-zero first degree coefficients, but also in a different reference radius, which could benefit the convergence and accuracy of the method. Computational effort follows the same behaviour seen before, (see 3b).

When analysing the effect of the orbital distance (semi-major axis), the behaviour shown in Figure 5 was observed. When distance from the body grows larger, the significance of its irregular shape decays until it becomes no longer noticeable. This analysis serves to indicate up to which point these models can be used depending on the gravitational environment.

5. Conclusions and future work

A tool has been developed to obtain spherical harmonics coefficients from polyhedron shape models. The coefficients computed have been validated using semi-analytical methods. The further usage of these coefficients in an orbital propagator is assessed by comparing the integrated trajectories with the ones obtained using polyhedron dynamics.

The results show that, in terms of computational effort, the spherical harmonics implementation is far superior to the polyhedron dynamics implementation. Accuracy-wise, for more spherical bodies, the trajectories converge to the ground-truth.

However, this method finds it difficult to replicate the actual gravitational accelerations when orbiting too close to a highly non-spherical body. Future work will investigate on

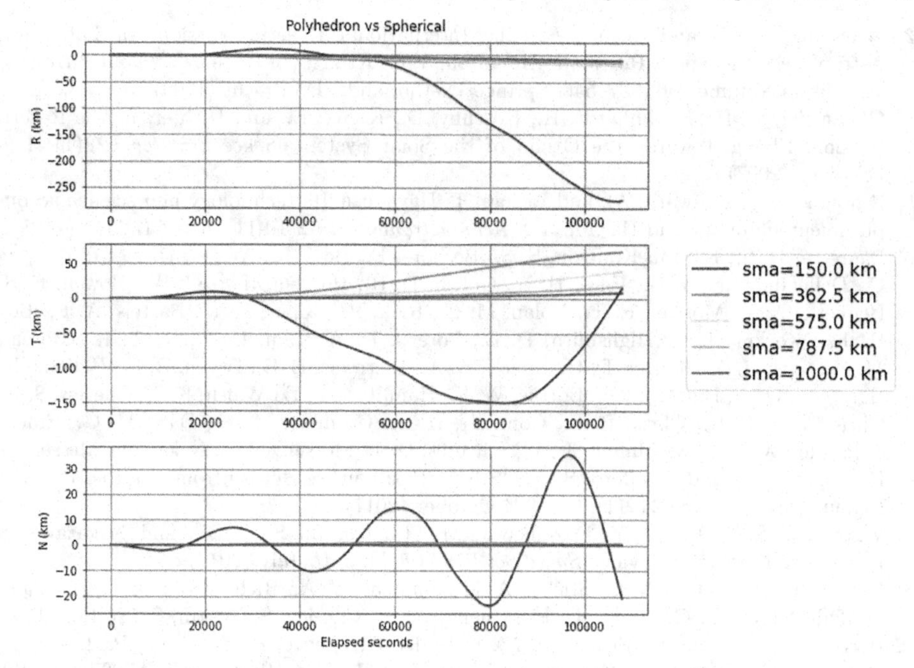

Figure 5. Radial, tangential, and normal component errors for different semi-major axis w.r.t. polyhedral model for Lutetia.

the influence of the choice of the origin of the reference frame for a certain body affects the accuracy of the propagated trajectories.

Pursuing the enhancement of this method could lead to the creation of a catalogue of spherical harmonic coefficients for bodies for which only shape models are available, given a certain constant density assumption. Such a catalogue could help future mission during their early stages by providing them with preliminary gravity potential models, which can be updated as more observations are gathered.

6. Acknowledgments

The authors would like to thank Cătălin Galeş for his suggestions. The authors would also like to acknowledge the funding received from the European Union's Horizon 2020 research and innovation programme under the Marie Skłodowska-Curie grant agreement No 813644.

References

[1] Werner, R. A. Spherical harmonic coefficients for the potential of a constant-density polyhedron. *Computers & Geosciences* **23**(10), 1071–1077 December (1997).

[2] Cheng, A. F., Santo, A. G., Heeres, K. J., Landshof, J. A., Farquhar, R. W., Gold, R. E., and Lee, S. C. Near-Earth Asteroid Rendezvous: Mission overview. *Journal of Geophysical Research: Planets* **102**(E10), 23695–23708 (1997).

[3] Schwehm, G. H. and Schulz, R. The International Rosetta Mission. In Laboratory Astrophysics and Space Research, Ehrenfreund, P., Krafft, C., Kochan, H., and Pirronello, V., editors, volume 236, 537–546. Springer Netherlands, Dordrecht (1999).

[4] Glassmeier, K.-H., Boehnhardt, H., Koschny, D., Kührt, E., and Richter, I. The Rosetta Mission: Flying Towards the Origin of the Solar System. *Space Sci Rev* **128**(1), 1–21 February (2007).

[5] Kawaguchi, J., Fujiwara, A., and Uesugi, T. Hayabusa–Its technology and science accomplishment summary and Hayabusa-2. *Acta Astronautica* **62**(10-11), 639–647 May (2008).

[6] Lauretta, D. S., Balram-Knutson, S. S., Beshore, E., Boynton, W. V., Drouet d'Aubigny, C., DellaGiustina, D. N., Enos, H. L., Golish, D. R., Hergenrother, C. W., Howell, E. S., Bennett, C. A., Morton, E. T., Nolan, M. C., Rizk, B., Roper, H. L., Bartels, A. E., Bos, B. J., Dworkin, J. P., Highsmith, D. E., Lorenz, D. A., Lim, L. F., Mink, R., Moreau, M. C., Nuth, J. A., Reuter, D. C., Simon, A. A., Bierhaus, E. B., Bryan, B. H., Ballouz, R., Barnouin, O. S., Binzel, R. P., Bottke, W. F., Hamilton, V. E., Walsh, K. J., Chesley, S. R., Christensen, P. R., Clark, B. E., Connolly, H. C., Crombie, M. K., Daly, M. G., Emery, J. P., McCoy, T. J., McMahon, J. W., Scheeres, D. J., Messenger, S., Nakamura-Messenger, K., Righter, K., and Sandford, S. A. OSIRIS-REx: Sample Return from Asteroid (101955) Bennu. *Space Sci Rev* **212**(1), 925–984 October (2017).

[7] Watanabe, S.-i., Tsuda, Y., Yoshikawa, M., Tanaka, S., Saiki, T., and Nakazawa, S. Hayabusa2 Mission Overview. *Space Sci Rev* **208**(1), 3–16 July (2017).

[8] Rivkin, A. S., Chabot, N. L., Stickle, A. M., Thomas, C. A., Richardson, D. C., Barnouin, O., Fahnestock, E. G., Ernst, C. M., Cheng, A. F., Chesley, S., Naidu, S., Statler, T. S., Barbee, B., Agrusa, H., Moskovitz, N., Daly, R. T., Pravec, P., Scheirich, P., Dotto, E., Corte, V. D., Michel, P., Küppers, M., Atchison, J., and Hirabayashi, M. The Double Asteroid Redirection Test (DART): Planetary Defense Investigations and Requirements. *Planet. Sci. J.* **2**(5), 173 August (2021).

[9] Michel, P., Küppers, M., and Carnelli, I. The Hera mission: European component of the ESA-NASA AIDA mission to a binary asteroid. **42**, B1.1–42–18 July (2018).

[10] Cheng, A. F., Michel, P., Jutzi, M., Rivkin, A. S., Stickle, A., Barnouin, O., Ernst, C., Atchison, J., Pravec, P., and Richardson, D. C. Asteroid Impact & Deflection Assessment mission: Kinetic impactor. *Planetary and Space Science* **121**, 27–35 February (2016).

[11] Oh, D. Y., Collins, S., Drain, T., Hart, W., Imken, T., Larson, K., Marsh, D., Muthulingam, D., Snyder, J. S., Trofimov, D., Elkins-Tanton, L. T., Johnson, I., Lord, P., and Pirkl, Z. Development of the Psyche mission for NASA's Discovery program. September (2019).

[12] Werner, R. and Scheeres, D. Exterior gravitation of a polyhedron derived and compared with harmonic and mascon gravitation representations of asteroid 4769 Castalia. *Celestial Mech Dyn Astr* **65**(3) (1997).

[13] Russell, R. and Wittick, P. *Mascon Models for Small Body Gravity Fields.* August (2017).

[14] Balmino, G. Gravitational potential harmonics from the shape of an homogeneous body. *Celestial Mech Dyn Astr* **60**(3), 331–364 November (1994).

[15] Inc, W. R. Mathematica, Version 12.3.1.

[16] Montenbruck, O., Gill, E., and Lutze, F. Satellite Orbits: Models, Methods, and Applications. *Appl. Mech. Rev.* **55**(2), B27 (2002).

[17] González, D. DaniGlez/polygrav, June (2020). original-date: 2020-06-18T18:16:52Z.

[18] Werner, R. A. The gravitational potential of a homogeneous polyhedron or don't cut corners. *Celestial Mech Dyn Astr* **59**(3), 253–278 July (1994).

Multi-scale (time and mass) dynamics of space objects
Proceedings IAU Symposium No. 364, 2022
A. Celletti, C. Galeş, C. Beaugé, A. Lemaitre, eds.
doi:10.1017/S1743921322000126

The semi-analytical motion theory of the third order in planetary masses for the Sun – Jupiter – Saturn – Uranus – Neptune's system

Alexander Perminov and Eduard Kuznetsov

Ural Federal University, Lenina Avenue, 51, Yekaterinburg, 620000, Russia

Abstract. The averaged four-planetary motion theory is constructed up to the third order in planetary masses. The equations of motion in averaged elements are numerically integrated for the Solar system's giant planets for different initial conditions. The comparison of obtained results with the direct numerical integration of Newtonian equations of motion shows an excellent agreement with them. It suggests that this motion theory is constructed correctly. So, we can use this theory to investigate the dynamical evolution of various extrasolar planetary systems with moderate orbital eccentricities and inclinations.

Keywords. celestial mechanics, methods: analytical, methods: numerical, planets and satellites: individual (Jupiter, Saturn, Uranus, Neptune)

The osculating Hamiltonian of the four-planetary problem is written in Jacobi coordinates. Then it is expanded into the Poisson series in the small parameter and orbital elements of the second Poincaré system. This algorithm is described in (Perminov, Kuznetsov (2015)).

The averaged Hamiltonian is constructed by the Hori–Deprit method. The implementation of the Hori–Deprit algorithm is considered in (Perminov, Kuznetsov (2016) and Perminov, Kuznetsov (2020)). The first-order terms of the averaged Hamiltonian are constructed up to 6th degree in eccentric and oblique Poincaré elements, the second-order and third-order terms are constructed up to 4th and 2nd degrees correspondingly.

The equations of motion in averaged elements are integrated by the Everhart method of 15th order (Everhart (1974)) with a time step of 1000 years to modeling of the orbital evolution of the Solar system's giant planets for different initial conditions (according to DE 432 ephemeris). Also, the same simulation is performed by Wisdom–Holman integrator with symplectic corrector of 11th order (Rein, Tamayo (2015)) and a time step of 4 days. The Time interval of the integration is 100 Myr for all cases. The limits of the change of osculating orbital eccentricities (e_{min}, e_{max}) and inclinations (I_{min}, I_{max}) in barycentric frame giving by semi-analytical (SA) motion theory and Wisdom–Holman (WH) methods are presented in Table 1 for four initial dates of the integration (at the moment $00^h00^m00^s$ UTC). The transition from averaged elements to osculating ones is carried out using the variables change functions of the first order. Table 2 presents the periods of the change of averaged orbital elements and the MEGNO indicator, which is computed in the process of the numerical simulation for each initial date.

The differences between periods of the change of orbital inclinations obtained by numerical methods and semi-analytical motion theory do not exceed 0.2% for all planets and all initial conditions. These discrepancies between periods of the change of orbital

Table 1. The range of osculating orbital eccentricities and inclinations.

Date	Theory	e_{min}	e_{max}	I_{min}, °	I_{max}, °	e_{min}	e_{max}	I_{min}, °	I_{max}, °
		Jupiter				Saturn			
31.01.2016	SA	0.025428	0.061603	1.093929	2.065336	0.012644	0.086018	0.557715	2.600120
	WH	0.025074	0.061959	1.093609	2.062945	0.009449	0.087273	0.561106	2.595098
29.02.2016	SA	0.025663	0.061743	1.093711	2.065397	0.012919	0.086132	0.559167	2.597615
	WH	0.025252	0.062026	1.095261	2.063808	0.009323	0.087498	0.561043	2.596079
31.01.2020	SA	0.026164	0.061377	1.089352	2.066233	0.012506	0.085095	0.559413	2.599554
	WH	0.025670	0.061914	1.094836	2.065492	0.009516	0.087377	0.560680	2.595056
29.02.2020	SA	0.025727	0.061232	1.094927	2.065816	0.012196	0.085157	0.563841	2.595928
	WH	0.025180	0.061789	1.094095	2.062955	0.009289	0.087229	0.565407	2.594549
		Uranus				Neptune			
31.01.2016	SA	0.006304	0.071602	0.391638	2.766937	0.003247	0.015126	0.775906	2.376785
	WH	0.003386	0.073773	0.378761	2.777150	0.002383	0.016046	0.773544	2.373315
29.02.2016	SA	0.006084	0.071265	0.425230	2.740155	0.003322	0.015133	0.780070	2.374410
	WH	0.003572	0.073201	0.427249	2.734788	0.002635	0.015978	0.783357	2.373475
31.01.2020	SA	0.005017	0.070079	0.446119	2.718907	0.003483	0.014972	0.769256	2.390563
	WH	0.002174	0.071974	0.444978	2.728521	0.002727	0.015824	0.777828	2.379745
29.02.2020	SA	0.005501	0.070569	0.414775	2.735945	0.003487	0.014914	0.776324	2.380880
	WH	0.002735	0.072512	0.430827	2.730216	0.002553	0.015786	0.779464	2.379941

Table 2. The periods of osculating orbital elements and the MEGNO indicator.

Date	Theory	Jupiter	Saturn	Uranus	Neptune	Jupiter	Saturn	Uranus	Neptune	MEGNO
		Periods of orbital eccentricities, years				Periods of orbital inclinations, years				
31.01.2016	SA	54 675	54 675	1 136 375	537 640, 364 967	49 213	49 213	432 905	1 886 811	2.28
	WH	54 735	54 735	1 136 375	534 765, 363 640	49 141	49 141	432 905	1 886 811	
29.02.2016	SA	53 764	53 764	1 136 375	537 640, 364 967	49 262	49 262	432 905	1 886 811	6.69
	WH	54 055	54 055	1 123 607	534 765, 362 323	49 189	49 189	432 905	1 886 811	
31.01.2020	SA	51 949	51 949	1 111 122	540 546, 363 640	49 432	49 432	432 905	1 886 811	103.63
	WH	51 841	51 841	1 098 912	537 640, 362 323	49 335	49 335	432 905	1 886 811	
29.02.2020	SA	54 113	54 113	1 123 607	540 546, 364 967	49 335	49 335	432 905	1 886 811	1.99
	WH	53 677	53 677	1 111 122	537 640, 362 323	49 237	49 237	432 905	1 886 811	

eccentricities of Jupiter and Saturn are in the range 0.1% – 0.8%. These differences do not exceed 1.1% and 0.7% correspondingly for Uranus and Neptune. The discrepancies for minimal and maximum values of the orbital eccentricities and inclinations do not exceed a few percent, except minimal orbital eccentricities of Saturn, Uranus, and Neptune.

The constructed semi-analytical motion theory can be used to study the orbital evolution and stability of extrasolar planetary systems with moderate orbital eccentricities and inclinations. The orbital elements of extrasolar systems are known from observations with highly uncertain, and some elements are not determined due to the specificity of the observation methods. We can vary unknown and known with errors orbital elements within allowable limits to determine the set of various initial conditions for modeling the orbital evolution. The limits of the change of orbital elements can be determined depending on the specific initial conditions. The assumption about the stability of observed planetary systems allows us to exclude the initial conditions leading to extreme values of the orbital eccentricities and inclinations that identify those under which these elements conserve small or moderate values over the whole modeling interval. Thus, it is possible to narrow the allowable range of unknown orbital elements and determine their most probable values in terms of stability.

The work was supported by the Ministry of Science and Higher Education of the Russian Federation via the State Assignment Project FEUZ-2020-0038.

References

Everhart, E. 1974, Celest. Mech. 10, 35–55
Perminov, A. S., Kuznetsov, E. D. 2015, Solar Syst. Res. 49(6), 430–441
Perminov, A. S., Kuznetsov, E. D. 2016, Solar Syst. Res. 50(6), 426–436
Perminov, A. S., Kuznetsov, E. D. 2020, Math. Comput. Sci. 14, 305–316
Rein, H., Tamayo, D. 2015, Mon. Not. R. Astron. Soc. 2015, 452(1), 376–388

Multi-scale (time and mass) dynamics of space objects
Proceedings IAU Symposium No. 364, 2022
A. Celletti, C. Galeş, C. Beaugé, A. Lemaitre, eds.
doi:10.1017/S1743921321001332

On the scattering and dynamical evolution of Oort cloud comets caused by a stellar fly-by

E. Pilat-Lohinger, S. Clees, M. Zimmermann and B. Loibnegger

Dept. of Astrophysics, University of Vienna, Türkenschanzstrasse 17, A-1180 Vienna, Austria
email: elke.pilat-lohinger@univie.ac.at

Abstract. Recent GAIA observations revealed that the K-type star Gliese 710 will cross the Oort cloud in a distance between approximately 4000 and 12000 au in about 1.3 Myrs. This occurrence motivated us to study the influence of a stellar encounter on comets in the outer region of the solar system. Even if the Oort cloud extends to 100000 au from the sun, we restrict our study to the region between 30 and 25000 au where 25 million objects are distributed randomly. Comets at larger distances are not taken into account as they hardly enter the observable region after a single stellar fly-by. An overview of all objects that are scattered towards the sun for the different fly-by distances at 4000, 8000 and 12000 au shows that only a handful of objects are moving towards the sun immediately after the stellar encounter.

However, a subsequent long-term study of all objects that are moved into highly eccentric motion by the stellar fly-by shows a significant increase of comets crossing Jupiter's orbit and entering into the observable region. In addition, our study shows the first comets crossing the orbit of Earth only about 2.5 Myrs after the stellar fly-by. Thus, the impact risk for the Earth increases only some million years after the stellar fly-by.

Keywords. comets: general, Oort Cloud, solar system: general

1. Introduction

Observations of the European spacecraft GAIA predict a fly-by of a K-type star named "Gliese 710" which will pass through the Oort cloud in about 1.3 Myrs. Gliese 710 is not the first stellar visitor of the solar system. About 70000 years ago the so-called "Scholz' star" also known as WISE J072003.20-084651.2 Mamajek et al. (2015) approached the sun within approximately 52000 au which is about $4 - 12$ times the predicted distance of Gliese 710. However, the closest approach distance is quite difficult to determine as it sensitively depends on the current position and velocity of the star. Thus, various minimum distances have been announced so far, e.g. GAIA data release 2 (DR2) based integrations revealed an enounter distance of about 13900 au Bailer-Jones et al. (2018); de la Fuente Marcos (2018) suggested a distance of 10700 au and figured out that the use of additional (prior) observational data would lead to a closest approach distance of 4300 au in 1.29 Myrs.

Therefore, we decided to study the influence of a stellar fly-by for three different closest approach distances: namely 4000, 8000, and 12000 au. For these three distances of the stellar fly-by we compare the number of comets that are scattered towards the sun.

Such comet showers have been studied in former times by means of impulse approximation or numerical simulations of the N-body problem using the restricted problem (see e.g. Dybczynski (2002); Rickman et al. (2008, 2012); Fouchard et al. (2011, 2013); Berski & Dybczynski (2016) where comets are considered as mass-less bodies that influence neither each other nor the planets and the star.

This investigation also used the restricted problem where the dynamical behaviour of 25 million comets have been studied in the graviational field of the outer solar system and the passing star. The mass-less objects were placed randomly between 30 and 25000 au. Larger distances of Oort cloud comets from the sun were not taken into account as we noticed in an earlier study that these objects remain outside 20000 au after a single stellar fly-by and can hardly enter into the observable region (< 5 au) of the solar system.

In the following sections, we first describe our numerical study and provide some details about the distribution of the initial cometary reservoir. Section 3 shows the influence of the stellar fly-by on the cometary motion and the results of a long-term study of all high-eccentric cometary orbits after the stellar passage are shown in section 4. These long-term study also indicates that comets could encounter our planet Earth.

2. Method and Numerical Setup

In our numerical study we solve the equations of motion using the Bulirsch-Stoer integration method Stoer et al. (2002) which is an extrapolation method with adaptive step size. Our code has been massively parallelized on graphical processing units (GPUs) Zimmermann (2021) to allow simulations of up to 10^4 interacting massive objects. In addition, a significantly larger number of mass-less bodies can be studied with this GPU program.

Our dynamical system investigates the interaction of six massive bodies, i.e. the sun, the planets of the outer solar system (from Jupiter through Neptune), and the passing star Gliese 710. The cometary reservoir consists of 25×10^6 mass-less objects without mutual interaction. Due to the capacity of our GPUs we splitted the cometary reservoir which extends from 30 au to 25000 au into three regions (**R1-R3**):

- **R1:** a **Disk** of 5×10^6 mass-less objects with semi-major axes (a_c) from 30 to 5000 au where all objects move in nearly circular orbits (comets' eccentricities $e_c < 0.0001$) and in the same plane (inclinations $i_c < 0.0001^o$).
- **R2:** a **Flared Disk** of 5×10^6 objects with a_c from 5000 to 10000 au and e_c close to zero (< 0.0001). The inclinations of the comets increase up to 45° with larger a_c.
- **R3:** a spherical **Cloud** of massless objects which expands from 10000 to 25000 au. The eccentricities are < 0.0001 and the inclinations vary between 0 and 180°.

In all regions the angles of argument of pericenter ω, ascending node Ω, and mean anomaly M are randomly chosen between 0 and 360°. Initial positions of the comets were determined using a Rayleigh distribution (where $F(x) = 1 - e^{-x^2/2\sigma^2}$ for $x \geq 0$ or $F(x) = 0$ for ($x < 0$). The value of σ defines the location of the maximum of $F(x)$).

Since R3 is significantly larger than R1 and R2, we divided the cloud region (up to a distance of 25000 au) into shells of 5000 au containing 5×10^6 comets each. Comets at larger distances† were not taken into account as they might not be scattered into the inner region of the solar system ($a < 15$ au) by a single stellar fly-by. Because test computations showed that all objects initially outside 25000 au were scattered to semi-major axes > 20000 au. Thus, these comets need eccentricities > 0.9993 to enter $a < 15$ au. However, such objects were not found in our computations.

3. Stellar Fly-by and Cometary Scattering

Perturbations of a passing star may lead to an influx of comets from the outer solar system into the observable region. During the passage the star is considered to move on a straight line with constant velocity neglecting the gravitational pull of the Galactic tides and the sun Rickman et al. (2005).

† Note that the Oort cloud extends to a distance of 100000 au from sun.

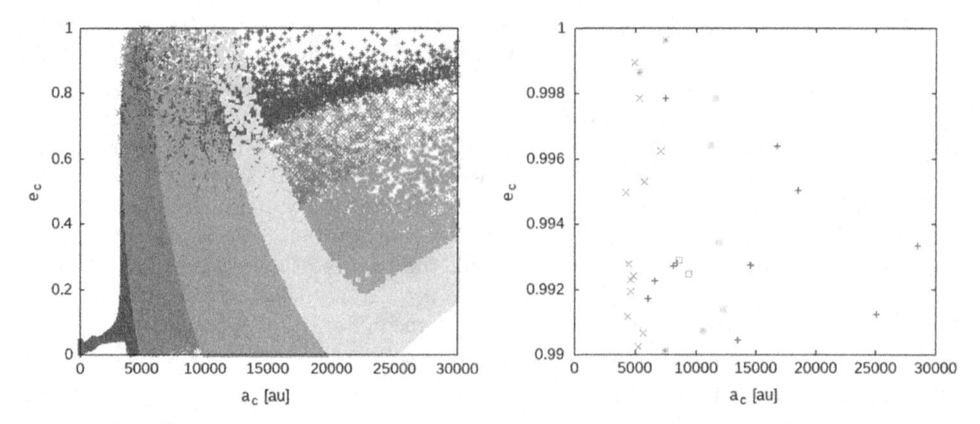

Figure 1. Scattered comets after the fly-by of Gliese 710 at 4000 au. Both panels show the scattering in the region between 30 au and 30000 au (x-axes). The upper panel shows the result for all eccentricities e_c from 0.0 to 1.0 (y-axis) and the lower panel is a zoom of the high eccentricity area between 0.99 and 1.0 (y-axis). Different colors indicate the different regions of the computations: **R1** purple, **R2** green, **R3** light blue, dark and light yellow. For details see the text.

The duration of the fly-by (i.e. the time when the star enters the Oort cloud at a distance of 100 000 au from the sun to the moment when it exits the cloud on the opposite side) depends on the encounter distance and the velocity of the passing star where the time to reach the minimum distance to the sun is given by:

$$t_d = \sqrt{\frac{b_{max}^2 - b_\star^2}{v_\star^2}}, \tag{3.1}$$

where b_{max}(= 110000 au) defines the maximum distance for the gravitational reach of the star; $b_\star$ is the impact parameter which defines the closest approach distance – in our study 4000, 8000, and 12000 au, respectively, and $v_\star$ in our study is 48 km/s or 0.02772 au/day which is the velocity of a typical K-type main sequence star according to Rickman et al. (2008).

Thus, the duration of the flyby is given by $2t_d$ which is about 20 000 years in our study.

Our simulations of stellar fly-bys indicate a large influx of objects that are close to the stellar trajectory while the region that is not crossed by the star remains quasi unperturbed. This can be seen in the results of R1 and R2 in Table 1 for a fly-by distance of 12000 au. Comets close to the stellar passage were scattered either towards the sun or are ejected from the system. This is shown in the upper panel of Fig. 1 which indicates two branches forming a "V-shape" in the (a_c, e_c)-plot of the comets after a simulation time of 20000 years. The V-shape of the evolved comets is visible for each region or shell that has been studied where (i) purple "+" label the disk objects of R1, (ii) green "x" the flared disk objects of R2, (iii) light blue stars indicate objects of the innermost shell of the cloud (from 10000 to 15000 au), (iv) dark yellow open squares mark objects of the second shell (from 15000 to 20000 au), and (v) light yellow squares label objects of the shell from 20000 to 25000 au. The branch of comets pointing towards the sun is steeper than the outward going branch. Thus comets of R2 (green objects) and R3 (light blue, dark and light yellow objects) that are scattered towards the sun have higher eccentricities than those moving outwards. Only for the disk objects of R1 we also observe a stronger increase in eccentricity for outward moving objects.

The lower panel of Fig. 1 shows objects of all studied regions that are scattered into high-eccentric orbits by the passing star. Some of these objects might enter the observable

Table 1. Overview of scattered comets.

Flyby distance [au]	Region	q < 100 au	q < 15 au	q < 5 au	e > 0.9	ejected comets
	R1	56612	0	0	1004	8351
	R2	24	2	0	231	1894
4000	**R3 - Shell 1**	7	2	1	57	645
	R3 - Shell 2	4	0	0	45	451
	R3 - Shell 3	4	0	0	31	355
	R1	56599	0	0	0	0
	R2	2	0	0	91	1068
8000	**R3 - Shell 1**	4	1	0	55	665
	R3 - Shell 2	2	0	0	44	443
	R3 - Shell 3	1	0	0	26	354
	R1	56599	0	0	0	0
	R2	0	0	0	0	0
12000	**R3 - Shell 1**	2	0	0	48	609
	R3 - Shell 2	5	0	0	443	445
	R3 - Shell 3	3	0	0	28	378

Notes:
$q = a_c(1 - e_c)$ is the pericenter distance.
The values of R1 in column 3 are that high since the disk objects are distributed between 30 and 5000 au.

Table 2. Comets approaching the inner solar system.

Flyby distance [au]	Region	q < 100 au	q < 15 au	q < 5 au
	R1	214	96	75
4000	**R2**	47	11	6
	R3	10	3	3
	R1	–	–	–
8000	**R2**	26	13	6
	R3	11	4	4
	R1	–	–	–
12000	**R2**	–	–	–
	R3	15	6	2

region ($q < 5$ au). According to Table 1 only one comet of the innermost shell of R3 is on an orbit towards this region after Gliese 710 has passed the solar system at 4000 au.

4. Long-term Evolution of Comets

With a long-term study of all comets with eccentricities > 0.9 (see Table 1) we checked the possibitlity that perturbed comets might enter the observable region at a later time after the stellar fly-by. Table 2 shows the number of comets entering a region of a certain distance from the sun (either at 100 or 15 or 5 au) during the long-term computation of 10 Myrs.

The result of the long-term computations of all high-eccentric comets ($e_c > 0.9$ of Table 1) shows clearly that a closer approach of Gliese 710 (to 4000 au from the sun) scatters a significantly higher number of comets into the observable region ($q < 5$ au). However, our long-term study shows that the number of such objects is quite low: 1368 / 216 / 519 objects for a considered stellar passages at 4000 / 8000 / 12000 au, respectively. Of course we expect a higher influx rate of comets when increasing the number of objects in the long-term computations. This study is still in progress and will be compared to previous studies of Fouchard et al. (2011). First results of 10^6 comets show a good agreement with the expectations resulting from the direct injection mechanism by Fouchard et al. (2011).

Even if our study included only 25×10^6 comets, we have found a couple of comets that approach or cross the orbit of Earth which is shown in Fig. 2 where all comets are plotted that enter the observable region within the simulation of 10 Myrs. In this plot a

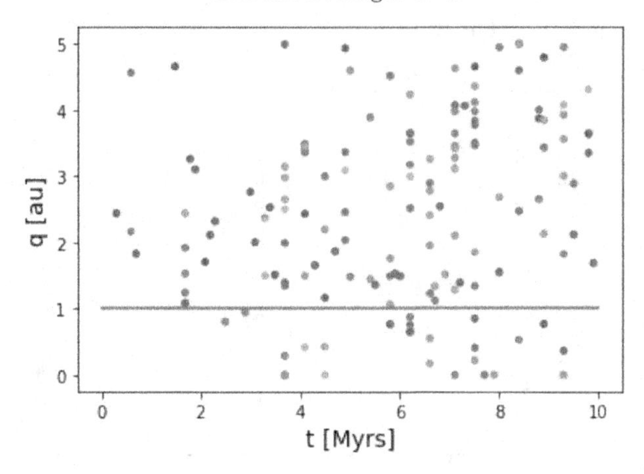

Figure 2. Perihelion distance of all comets entering in the observable region within 10 Myrs. Each color belongs to a certain comet.

certain color belongs to a certain object. Thus, we observe comets entering this region again and again. Moreover, Fig. 2 indicates that after ~ 2 Myrs the first object crosses the orbit of the Earth (i.e. the horizontal line), and the number of objects entering the region < 1 au increases visibly after 4 Myrs. Thus, a long time after the stellar fly-by the threat for planet Earth might increase.

5. Summary

Inspired by a recent observation of the European space mission GAIA, which indicates that the K-type star Gliese 710 will pass through the Oort cloud at a distance between 4000 and 12000 au from the sun, we studied the perturbations on comets moving in the outer region of the solar system and changes of their dynamical behaviour due to such stellar fly-bys .

When entering this cometary reservoir Gliese 710 perturbs the comets strongly by scattering them either towards the sun, or out of the solar system. The duration of the fly-by (to cross the whole Oort cloud) takes about 20000 years.

Our study of 25×10^6 comets randomly distributed between 30 and 25000 au shows that the strongest perturbations are close to the trajectory of Gliese 710. Moreover, cometary orbits with semi-major axes smaller than the closest approach distance of the star remain quasi unaffected.

Perturbations of the passing star create a “V-shape” of two branches with high-eccentricity orbits in the cometary reservoir. The branch pointing towards the sun is steeper and thus indicates higher eccentricities than the one directed outwards (see Fig. 1 upper panel). Only ~ 2100 of the 25×10^6 comets were scattered on high-eccentricity orbits with $e_c > 0.9$. Most of the inward moving comets showed a perihelion distance $q < 100$ au. A few comets crossed Uranus' orbit and only a single object was perturbed enough to enter the observable region ($q < 5$ au). We restricted this investigation to comets within 25000 au from the sun as a previous study by Clees (2021) indicated that a single perturbing event does not have enough impact to bring objects from distances > 20000 au into the observable region.

Our long-term study over 10 Myrs of the 2100 comets moving on high-eccentricity orbits after the stellar fly-by showed that the number of objects entering the observable region increases significantly with time, especially for the fly-by distance at 4000 au.

Moreover, there were no close encounters of comets with the Earth right after the stellar fly-by. The first comet reached the orbit of Earth about 2 Myrs after the passage of Gliese 710, and about 4 Myrs after the stellar encounter an increase of objects crossing the orbit of Earth was found. Thus, the stellar fly-by might increase the impact risk on our planet Earth but only long time after the stellar passage.

Acknowledgements

This research was funded in whole by the Austrian Science Fund (FWF) [P33351-N].

References

Bailer-Jones et al., 2018, A&A, 616, A37
Berski & Dybczynski, 2016, A&A, 595, L10
Clees S., 2021, Master thesis, University of Vienna
de la Fuente Marcos R.& de la Fuente Marcos, C., 2018, Research Notes of the American Astronomical Scociety, 2, 30
Dybczynski, P.A., 2002, A&A, 396, 283
Fouchard, M., Froeschle, C., Rickman, H., Valsecchi, G., 2011, Icarus, 214 , 334
Fouchard, M., Froeschle, C., Rickman, H., Valsecchi, G., 2011, Icarus, 222, 20
Mamajek, E.E., Barenfeld, S.A., Ivanov, V.D., et al., 2015, ApJ, 800, L17
Rickman, H., Fouchard, M., Valsecchi, G., Froeschle, C., 2005, EM&P, 97, 411
Rickman, H., Fouchard, M., Froeschlé, C., Valsecchi, G., 2008, CeMDA, 102,111
Rickman, H., Fouchard, M., Froeschlé, C., Valsecchi, G., 2008, P&SS, 73, 124
Stoer, J. & Bulirsch, R., 2002, Introduction to numerical analysis
Zimmermann, M., 2021, Master thesis, University of Vienna

Multi-scale (time and mass) dynamics of space objects
Proceedings IAU Symposium No. 364, 2022
A. Celletti, C. Galeş, C. Beaugé, A. Lemaitre, eds.
doi:10.1017/S1743921321001447

Planetary and lunar ephemeris EPM2021 and its significance for Solar system research

Elena Pitjeva[1], Dmitry Pavlov[2], Dan Aksim[1] and Margarita Kan[1]

[1]Insitute of Applied Astronomy RAS,
Kutuzova Embankment, 10, 191187 St. Petersburg, Russia
email: evp@iaaras.ru, danaksim@iaaras.ru, mo.kan@iaaras.ru

[2]St. Petersburg Electrotechnical University
ul. Professora Popova 5, 197376 St. Petersburg, Russia
email: dapavlov@etu.ru

Abstract. We present an updated public version of EPM (Ephemerides of Planets and the Moon). Since the last public version, EPM2017, many improvements were made in both the observational database and the mathematical model. Latest lunar laser ranging observations have been added, as well as radio ranges of Juno spacecraft and more recent ranges of Odyssey and Mars Reconnaissance Orbiter. EPM2021 uses a new improved way to calculate radio signal delays in solar plasma and has a major update in the method of determination of asteroid masses. Also, a delay-capable multistep numerical integrator was implemented for EPM in order to properly account for tide delay in the equations of the motion of the Moon. The improved processing accuracy has allowed to refine existing estimates of the mass of the Sun and its change rate, parameters of the Earth–Moon system, masses of the Main asteroid belt and the Kuiper belt; and also to raise important questions about existing numerical models of solar wind.

Keywords. celestial mechanics; astrometry; ephemerides; solar wind; minor planets, asteroids; Moon

---

1. Introduction

Numerical planetary and lunar ephemeris, being in constant development since 1960s in the United States (DE), since 1980s in Russia (EPM), and since 2000s in France (INPOP), have applications not only in space exploration, but also in realization of reference frames, fundamental astronomy, and physics. Each ephemeris theory is based on a mathematical model of the Solar system; the models are being improved as more (and more precise) planetary and lunar astronomical observations are made. The observational capabilities grow simultaneously with requirements for space missions and scientific experiments. Improvement of models may involve addition of bodies previously neglected or inclusion of physical effects previously unaccounted for. Recently, INPOP19a and DE440 were released (Fienga et al. 2019, Park et al. 2021). We present EPM2021, the latest public update of the EPM ephemeris. Compared to the last public release, EPM2017, EPM2021 has improvements in both the observational database and the mathematical model.

2. Observations

Several kinds of observations are processed to determine the parameters of the dynamical model of the ephemeris. The orbit of the Moon is determined from lunar laser ranging (LLR) observations. To determine the orbits of the inner planets, radio ranging observations of orbiters and landers are used, along with some Doppler observations of Martian landers. For the outer planets, there are fewer ranges obtained (and none at all for Pluto), thus optical observations of the outer planets and their satellites are used in the solution. Recent additions to the ground optical observations database came from Pulkovo Observatory in Russia (Ershova et al. 2016, Narizhnaya et al. 2018, 2019) and Yunnan Observatory in China (Wang et al. 2017, Xie et al. 2019). Also, some observations made at the Sheshan station of the Shanghai Observatory in 2003–2009 were added (Qiao et al. 2007, 2014). The next major release of data from the Gaia space telescope (Gaia DR3), planned for 2022, is expected to contain observations of planets and natural satellites, which will hopefully fill the current lack of data for the outer planets.

Important spacecraft ranging observations that were added to EPM2021 are: 4 normal points of the Jovian orbiter Juno obtained in 2016–2017, 780 normal points of Mars Reconnaissance Orbiter (MRO) obtained in 2014–2017, and 3226 observations of Mars Odyssey obtained too in 2014–2017. The MRO and Odyssey data were kindly provided by Dr. William Folkner of NASA JPL. The Juno data is published at the NASA website (`https://ssd.jpl.nasa.gov/?eph_data`), as are older observations of Odyssey and MRO, and other radio and optical observations used in the JPL DE ephemeris. However, the post-2017 observations of Odyssey, MRO, and Juno that were used for DE440 (Park et al. 2021) are not available.

In the lunar part, there was a major update in the LLR observations at the Apache Point Observatory in NM, USA: 1211 normal points were made available recently (`https://tmurphy.physics.ucsd.edu/apollo/norm_pts.html`), covering the timespan from late 2016 to late 2020. Côte d'Azur Observatory in Grasse, France regularly provides LLR data (Chabé et al. 2020) to the NASA CDDIS (`https://cddis.nasa.gov/archive/slr/data/npt_crd`) archive via the International Laser Ranging Service (ILRS). From September 2017 to July 2021, 3789 normal points were obtained with the 1024-nm infrared laser at Grasse. The "green" (532 nm) normal points seem to have ended in November 2020; 262 normal points produced since September 2017 have been added to the EPM database. 261 normal points (2017–present) were added from the Matera observatory in Italy, also via CDDIS. Finally, the Wettzell Observatory in Germany started to provide infrared laser ranges in 2018 (Eckl et al. 2019) and has since produced 101 normal points that can be found at `ftp://edc.dgfi.tum.de/pub/slr/data/npt_crd`.

3. Dynamical model

EPM's dynamical model of the Solar system includes all planets, Pluto, the Moon, the Sun, selected asteroids, the discrete uniform 180-point asteroid annulus (Pitjeva & Pitjev 2018a), selected Trans-Neptunian objects (TNOs), and the discrete uniform 160-point TNO ring (Pitjeva & Pitjev 2018b). Sixteen bodies (the Sun, the planets, Pluto, Ceres, Pallas, Vesta, Iris, Bamberga) obey Einstein–Infeld–Hoffmann equations of motion. Other bodies, for the sake of performance, are modeled as interacting with those 16 bodies with only Newtonian forces and not interacting with each other. Apart from point-mass interactions, the model includes additional accelerations from solar oblateness and Lense–Thirring effect. Earth also gets "point mass–figure" accelerations that come from the Sun, Venus, Mars, Jupiter, and the Moon. The Moon is modeled as an elastic body with a rotating liquid core (Pavlov et al. 2016).

After EPM2017, two point masses have been added that represent Jupiter trojans, placed at L_4 and L_5 Lagrange points of Jupiter's orbit. Also, the list of asteroids was revised. In EPM2017, 301 largest asteroids were present in the dynamical model as individual point masses. In EPM2021, the number of individual asteroids is 277. The source list of asteroids was compiled by merging (with removal of duplicates) the 343 asteroids of the DE430 model (Folkner et al. 2014) and the 287 asteroids from (Kuchynka et al. 2010).

The latter is believed to be the list of the most "non-ring-like-acting" asteroids, i.e. the ones whose cumulative effect on the inner planets cannot be modeled by a uniform ring. The resulting list contained 379 asteroids. Then, 102 asteroids whose masses were determined negative (though always within uncertainty) were excluded from the model.

4. Determination of masses of asteroids and TNOs

Of the 277 asteroids, only 17 masses are known with good accuracy because they are either binary asteroids (e.g. Kalliope) or had spacecraft orbiting them (e.g. Ceres). Masses of some other asteroids are estimated by deflections of other asteroids' orbits on approach (e.g. Iris), though estimates may differ across works. For the remaining majority of asteroids, only a weak estimate of the mass may be obtained from the estimate of the diameter based on the infrared observations of space telescopes IRAS and WISE, and the estimate of the density based on the taxonomical class (C/S/M). In EPM2017, masses of 30 asteroids were determined purely dynamically, by the perturbations that they inflict on orbits of the inner planets, while masses of other asteroids were determined as mean densities of the three taxonomic classes. In EPM2021, following the approach proposed in (Kuchynka & Folkner 2013), all the said estimates of masses were used as *a priori* estimates in the Tikhonov regularization scheme that extends the least-squares method. All 277 masses were then determined in the planetary solution along with other parameters. As said above, 379 masses were determined initially, then 102 masses that became negative (always within uncertainty) were excluded from the model. The gravitational effect of those 102 asteroids, as well as of all the others that were not selected in the first place, is approximated with the 180-point uniform annulus, whose mass is also determined.

Similar modification was made for determination of masses of 30 TNOs and the discrete 160-point TNO ring.

5. Reductions of spacecraft ranging observations

Two improvements were made to reduce systematic errors in the residuals of Martian orbiters MRO and Odyssey who provide Earth–Mars ranges (normal points) of sub-meter accuracy. One is accounting for measurement biases that come from miscalibrations on radio observatories. Following the decision from (Kuchynka et al. 2012), two sets of biases were determined for Deep Space Network (DSN) stations: one for Mars Global Surveyor (MGS) and Odyssey spacecraft, another for the MRO spacecraft.

The second improvement concerns the delay of the radio signal due to the free electrons in the solar plasma. In EPM, from 2004 to 2017 versions, the following model was used for the electron number density:

$$N_{\mathrm{e}} = \frac{A}{r^6} + \frac{B + \dot{B}t}{r^2},$$

where r is the distance to the center of the Sun. A was fixed to the value determined in DE200, and B with its linear drift $\dot{B}$ were determined from observations, per-planet, per-year. That makes more than 50 determined parameters in the planetary solution correlate with each other, with the biases, orbits, and the mass of the Sun. In EPM2019 (unreleased), A and $\dot{B}$ were set to zero, while B was determined per-conjunction from 2002 to 2018 plus a single B prior to 2002. That makes 10 parameters. In EPM2021, there is only one determined solar plasma parameter: C, the model being

$$N_{\mathrm{e}} = C\,\frac{N_1(t)}{r^2},$$

Table 1. Statistics of spacecraft ranging residuals. Ranges that pass closer to the Sun than 60 solar radii are excluded. The third column is the number of normal points that were formed from raw spacecraft ranging observations made during the specified timespan.

Spacecraft	Timespan	NPs	wrms
MGS	1999–2006	5590	91.5 cm
Odyssey	2002–2017	7988	56.3 cm
MRO	2006–2017	1924	60.2 cm
Mars Express	2005–2015	2888	2.46 m
Venus Express	2006–2013	1294	6.06 m
Cassini	2004–2014	161	17.2 m
MESSENGER	2011–2014	1141	60.9 cm
Juno	2016–2017	4	3.58 m

Table 2. Statistics of LLR residuals. Rejected observations are not counted. The third column is the number of normal points that were formed from raw LLR observations made during the specified timespan.

Station	Timespan	NPs	wrms, cm
McDonald, TX, USA	1969–1985	3554	21.4
Nauchny, Crimea, USSR	1982–1984	25	11.6
MLRS1, TX, USA	1983–1988	585	8.8
MLRS2, TX, USA	1988–2013	3280	3.6
Haleakala, HI, USA	1984–1990	747	5.2
Grasse, France (Ruby laser)	1984–1986	1109	16.8
Grasse, France (YAG)	1987–2005	8277	2.3
Grasse, France (MeO green)	2009–2020	2000	1.52
Grasse, France (infrared)	2015–2021	6179	1.15
Matera, Italy	2003–2021	358	3.43
Apache Point, NM, USA	2006–2020	3782	1.44
Wettzell, Germany	2018–2020	101	1.25
Total	**1969–2021**	**29997**	$\chi^2 = 1.358$

where the function $N_1(t)$ is equal to the smoothed *in situ* measurements of the electron density near Earth obtained from the NASA/GSFC's OMNI dataset (`https://spdf.gsfc.nasa.gov/pub/data/omni/low_res_omni`). The reference value of C is 1; *a priori* error estimate of this value was obtained from the formal errors provided in the OMNI dataset. Taking advantage of this estimate with the Tikhonov regularization has allowed to improve the accuracy of determination of the Sun's mass by more than 10%. The usage of time-varying electron density linked to *in situ* measurements has allowed to reduce the systematic errors in the ranging residuals.

6. Technical improvements

Versions of EPM since EPM2015 are made with the ERA-8 software (Pavlov, Skripnichenko 2015). Up to EPM2017, a single-step integrator was used (Avdyushev 2010). Another integrator was subsequently developed that allows to integrate differential equations that contain a time delay. Such a modification was needed to account for the tide delay in the equations of the motion of the Moon (Pavlov et al. 2016). The new integrator, called ABMD (Aksim & Pavlov 2020), is an extension of the multistep Adams–Bashforth–Moulton scheme and has allowed to improve the performance of numerical integration of orbits.

7. Residuals and determined parameters in the EPM2021 solution

Weighted root-mean-squares (wrms) of ranging residuals (one-way) of planetary orbiters are listed in Table 1. The wrms of the LLR residuals (one-way) are listed in Table 2. The values and uncertainties of selected parameters are given in Table 3.

Table 3. Selected parameters determined in EPM2021.

Parameter	Value and 3σ
$GM_\odot$	132712440043.17 ± 0.49 km^3/s^2
$J_{2\odot}$	$(2.252 \pm 0.024) \cdot 10^{-7}$
Asteroid belt mass	$(4.13 \pm 0.09) \cdot 10^{-4} M_\oplus$
Kuiper belt mass	$(1.74 \pm 0.42) \cdot 10^{-2} M_\oplus$
Earth–Moon GM	403503.23649 ± 0.00025 km^3/s^2
Lunar J_2	$(2.0321 \pm 0.0005) \cdot 10^{-4}$
Lunar tidal delay τ	(0.094 ± 0.002) days
Lunar $(C - A)/B$	$(631.022 \pm 0.001) \cdot 10^{-6}$
Lunar $(B - A)/C$	$(227.739 \pm 0.001) \cdot 10^{-6}$
Oblateness of lunar fluid core f_c	$(0.258 \pm 0.006) \cdot 10^{-3}$

8. Other results and future plans

Several scientific results have appeared during the development of EPM2021. Aside from the already mentioned determined masses of the asteroid belt and the Kuiper belt, they are:

- The Earth–Moon very-long-baseline Interferometry (VLBI) observations were modeled in order to estimate the astrometric outcome (Kurdubov et al. 2019).
- A two-delay model was proposed for the equations of lunar rotation (Pavlov 2019).
- A lunar reference frame was built with a decimeter accuracy; also, the potential of the modern LLR to determine the Earth orientation parameters was shown (Pavlov 2020).
- The change rate of the mass of the Sun has been estimated (Pitjeva et al. 2021).

Future plans include: using Gaia's observations of satellites of the outer planets (when observations are released); improving the models of the lunar core and lunar solid body tides; further tests of general relativity from planetary and lunar observations; application of the EPM model of the Solar system to pulsar timing.

References

Aksim, D., Pavlov, D. *Math Comp Sci* 14, 103

Avdyushev, V. 2010, *Computational technologies*, 15, 31 (in Russian)

Chabé, J., Courde, C., Torre, J.-M. 2020. *Earth and Space Science*, 7.

Eckl, J., Schreiber, U., Schüler, T. 2019. *Proc. SPIE 11027*, 1102708.

Ershova, A., Roshchina, E., Izmailov, I., 2016, *Planetary and Space Science*, 134.

Fienga, A., Deram, P., Viswanathan, V. et al. 2019, *N. Sci. et Tech. de l'Inst. de méc. cél.*, S109

Folkner, W., Williams, J., Boggs, D., Park, R., Kuchynka, P. 2014. *IPN Progress Report* 42-196.

Kuchynka, P., Laskar, J., Fienga, A., Manche, H. 2010. *A&A*, 514, A96.

Kuchynka, P., Folkner, W., Konopliv, A. 2012. *IPN Progress Report* 42-190.

Kuchynka, P., Folkner, W. 2013. *Icarus*, 222, 243.

Kurdubov, S., Pavlov, D., Mironova, M., Kaplev, S. 2019, *MNRAS*, 486, 815.

Narizhnaya, N., Khovrichev, M., Apetyan, A. et al. 2018, *Sol Syst Res* 52, 312.

Narizhnaya, N., Khovrichev, M., Bikulova, D. 2019, *Sol Syst Res* 53, 368.

Park, R., Folkner, W., Williams, J., Boggs, H. 2021, *AJ*, 161, 105

Pavlov, D., Williams, J.m Suvorkin, V. 2016, *Cel. Mech. Dyn. Astr*, 126, 61

Pavlov, D. 2019. In: Bizouard, C., Souchay, J. (eds), *Proceedings of the Journées 2019 "Systèmes de référence temps-espace*, p. 309.

Pavlov, D. 2020. *J. Geod.* 94, 5

Pavlov, D., Skripnichenko, V. 2015. In: Malkin, Z. and Capitaine, N. (eds), *Proceedings of the Journées 2014 "Systèmes de référence spatio-temporels"*, p. 243.

Pitjeva, E., & Pitjev, N. 2018a, *Astron. Lett.* 44, 554

Pitjeva, E., & Pitjev, N. 2018b, *Cel. Mech. Dyn. Astr*, 130, 57

Pitjeva, E., Pitjev, N. 2019, *Astron. Lett.* 45, 855

Pitjeva, E., Pitjev, N., Pavlov, D., Turygin, S. 2021, *A&A* 647, A141
Qiao, R., Yan, Y., Shen, K. et al. 2007. *MNRAS*, 376, 1707.
Qiao, R., Zhang, H., Dourneau, G. et al. 2014. *MNRAS*, 440, 3749.
Wang, N., Peng, Q., Peng, H. et al. 2017. *MNRAS*, 468, 1415.
Xie, H., Peng, Q., Wang, N. et al. 2019. *Planetary and Space Science*, 165, 110.

Multi-scale (time and mass) dynamics of space objects
Proceedings IAU Symposium No. 364, 2022
A. Celletti, C. Galeş, C. Beaugé, A. Lemaitre, eds.
doi:10.1017/S1743921321001320

Some of the most interesting cases of close asteroid pairs perturbed by resonance

A. Rosaev[1] and **Eva Plavalova**[2]

[1]Research and Educational Center "Nonlinear Dynamics",
Yaroslavl State University, Yaroslavl, Russia,
email: hegem@mail.ru

[2]Mathemailcal institute, Slovak academy of Science,
Bratislava, Slovakia
email: plavalova@mat.savba.sk, plavalova@komplet.sk

Abstract. We have randomly selected 20 close asteroid pairs (younger than 800 kyr) from known pairs, and by the application of backward numerical integration we have calculated their orbits. For the reason of speeding up the process of making the resonances visible, we have used a high value of Yarkowsky drift. The results of the calculation show that only two pairs appear to have a simple resonance with Earth and Jupiter while half of the tested pairs are visibly in the vicinity of three-body resonances.

We have found a 2-1J-1M resonance for the pair (56232) 1999 JM_{31} and (115978) 2003 WQ_{56}. Following our study of the pair (10123) Fideoja and (117306) 2004 VF_{21}, we discovered a different resonance than the 7-2J mean motion resonance previously published: we have proved that this pair is perturbed by 9-6J-4M three body resonance.

Keywords. asteroid pairs, orbital evolutions, resonance

---

1. Introduction

Asteroids tend to group into so-called families or into associations of objects sharing similar orbits. Most of them are the results of very old (about 1 Gyrs) collisions between asteroids (Spoto et al. 2015; Nesvorny et al. 2015). Since the beginning of the twentieth century, asteroid families and pairs have been the object of increasingly intensive studies.

The detection of several asteroid families and pairs with very recent formations (about 1.5 Myr or less) in the past decades has generated a new and exciting development. These discoveries are very important, because various collisional and dynamical processes have had little time to act on these families to alter their properties. Recent studies have shown that some asteroid families can also be the outcomes - in theist; however of a spin-up-induced fission of a critically rotating parent body (fission clusters, Jacobson and Scheeres 2011). Moreover, cases of subsequent breakups can take place in older families (Fatka et al. 2020).

It is important to note that the age of a young family can be determined by numerically integrating the orbits of its members backward in time and demonstrating that they converge to each other at some specific time in the past, however, the method of backward integration of orbits only works for families younger than a few million years.

Many cases of the resonance perturbations of young families and pairs are known. A prime example is the Datura family with its 9-16M resonance with Mars (Nesvorny et al. 2006). The chaotic orbits of the pair (49791) 1999 XF_{31} and (436459) 2011 CL_{97} may be explained by the 15-8M mean motion resonance with Mars (Pravec et al. 2018) and the pair (7343) Ockeghem and (154634) 2003 XX_{38} is in the 2-1J-1M three-body resonance (Duddy et al. 2012).

The main goal of this paper is to search for very young asteroid pairs (younger than 1 Myr) within the influence of resonances. We identify the types of resonances, their position and their chaotic zone.

2. Methods

To study the dynamic evolution of asteroid pairs, the equations of the motion of the systems were numerically integrated over 800 kyr using the N-body integrator Mercury (Chambers 1999) and the Everhart integration method (Everhart 1985).

Under the condition of finding two- or three- body mean motion resonances up to the order of 20, we tested 20 pairs of asteroids (Tab. 1), arbitrarily selected from Pravec et al. (2018). To calculate the nominal resonance positions, we used the values of the time-averaged planet semi-major axes: 1.52368 AU for Mars, 5.20259 AU for Jupiter and 9.5549 AU for Saturn. For the reason of speeding up the process of making the resonances visible, we set a very large value of Yarkovsky drift, applying non-gravitational force parameters for the asteroids $A_2 = 1 \cdot 10^{-13}$ and adapting numerical integrations with different planetary perturbations. The Yarkovsky effect is usually characterized by value da/dt; this parameter is directly proportional to value A_2 used in our integrations (see for example Farnocchia et al. 2013)

3. Results

Only one of the studied pairs, namely 21436 Chaoyichi and (334916) 2003 YK_{39} is close to the simple two-body mean motion resonance 11-3J with Jupiter and the pair (5026) Martes and 2005 WW_{113} is close to the simple two-body mean motion resonance 3-11E with Earth. However, for more than half of the studied pairs, we detect three-body resonances in distances less than d=0.0006 AU for at least one of the asteroids of the pair (Tab. 1). Asteroid-Jupiter-Saturn resonance was found in only five of the studied pairs (see Tab. 1) and all the other resonance cases include planets of the Earth group; asteroid-Earth (or Mars)-Jupiter. This fact highlights the role of Earth group planets on the dynamics of the inner asteroid belt.

The second half of the asteroid pair orbits we studied lie more distant from any resonance ($d > 0.0006$ AU) and have no notable perturbations. At this condition there are not any periodic variations of semi-major axis with large amplitudes and not any jumps. Therefore, this value of the distance of resonances may be considered as a rough boundary since which resonance perturbations are insignificant.

Such a high percentage of resonance-perturbed asteroid pairs allows us to suppose that the resonances play a significant role in the process of pair formation. In light of this relation, new theoretical researches on the process of the origin of asteroid pairs are required. Below we report about some of the most interesting cases we studied.

4. Two asteroid pairs in 2-1J-1M resonance

Pravec et al. (2018) estimated a lower limit of 382 kyr for the age of the pair (7343) Ockeghem and (154634) 2003 XX_{38}. Duddy et al. (2012) found that these two asteroids have very similar spectra to that of S class. They identified that this pair was orbiting in the 2-1J-1M three body resonance. The resonance argument is:

$$\varphi = 2\lambda - \lambda_{Jupiter} - \lambda_{Mars} \tag{4.1}$$

Here λ, $\lambda_{Jupiter}$, λ_{Mars} are the longitudes of the asteroid, Jupiter and Mars accordingly.

We conducted backward integrations on the orbit of this pair with different values of the Yarkovsky effect. We observe that in the case of medium values of the coefficient

Table 1. Selected close young asteroid pairs and their resonances.

Pair	Proper a [AU]	Resonance	a_r [AU]
(4765) Wasserburg (350716) 2001 XO_{105}	1.94542 1.94563	5-10J-1E	1.945479
(404118) 2013 AF_{40} (355258) 2007 LY_4	2.21744 2.21746	2+10J-7S	2.217453
(44620) 1999 RS_{43} (295745) 2008 UH_{98}	2.17644 2.17669	2+9J-4S	2.176198
(80218) 1999 VO_{123} (213471) 2002 ES_{90}	2.2185 2.21864	1+6J-6S	2.219288
(7343) Ockeghem (154634) 2003 XX_{38}	2.19254 2.19253	2-1J-1M	2.192728
(56232) 1999 JM_{31} (115978) 2003 WQ_{56}	2.19332 2.19328	2-1J-1M	2.192728
(26420) 1999 XL_{103} 2012 TS_{209}	2.19757 2.19749	6-10J-1E	2.196919
(2110) Moore-Sitterly (44612) 1999 RP_{27}	2.19804 2.19787	no	
(8306) Shoko 2011 SR_{158}	2.24159 2.24125	3-9J -4S	2.241658
(10123) Fideoja (117306) 2004 VF_{21}	2.26964 2.26962	9-6J-4M	2.269495
(17198) Gorjup (229056) 2004 FC_{126})	2.27969 2.27962	no	
(6369) 1983 UC (510132) 2010 UY_{57}	2.29324 2.29315	7-5J-3M	2.292687
(49791) 1999 XF_{31} (436459) 2011 CL_{97}	2.31665 2.31663	8-8J-3M	2.316370
(25021) Nischaykumar (453818) 2011 SJ_{109}	2.31788 2.31779	no	
(26416) 1999 XM_{84} (214954) 2007 WO_{58}	2.34257 2.34256	no	
(43008) 1999 UD_{31} (441549) 2008 TM_{68}	2.3481 2.34773	2+7J-1S	2.347592
(5026) Martes 2005 WW113	2.37752 2.37752	3-11E	2.377825
(46829)McMahon 2014 VR_4	2.39991 2.40048	no	
(42946) 1999 TU_{95} (165548) 2001 DO_{37}	2.56782 2.56761	no	
(4905) Hiromi (7813) Anderserikson	2.60102 2.60112	no	

Yarkovsky drift in semi-major axis $A_2 = 1 \cdot 10^{-14}$, asteroids stay within the neighbourhood of resonance in a stable orbit for at least 1 Myr. Conversely, when a larger value is used i.e. $A_2 = 1 \cdot 10^{-13}$, we observe a jump from one side of resonance to the other (Fig. 1). Both members of this pair are trapped in the considered resonance.

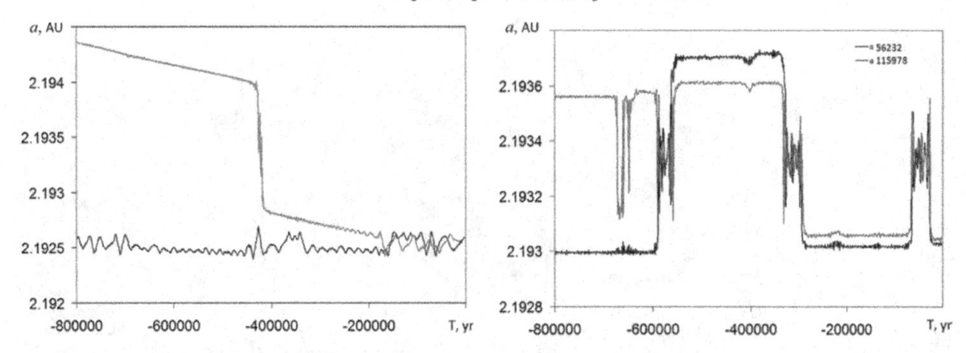

Figure 1. The left figure shows the evolution of the semi-major axes of the pair (7343) Ockeghem and (154634) 2003 XX_{38} - (Yarkovsky effect is set at $A_2 = \pm 10^{-13}$). The right figure shows the evolution of semi-major axes of pair 56232 (1999) JM_{31} and (115978) 2003 WQ_{56} (Yarkovsky effect is NOT accounted for).

We calculate that the observed centre of the chaotic resonance zone is a=2.19340 AU (Fig. 1, left figure). The nominal position of this resonance is 2.192728 AU.

We have found a second pair (56232) 1999 JM_{31} and (115978) 2003 WQ_{56} close to the same resonance (Fig. 1, right figure). Backward integration of their heliocentric orbits suggest that these two asteroids separated about 130 kyr ago (Pravec et al. 2018). The proper elements of both pairs are very similar and therefore the resonance 2-1J-1M has a significant effect on both pairs.

However, using backward integration with the Yarkovsky effect taken into account, we have found that the position of the centre of resonance for the second pair is different, at about a=2.19333 AU. Although the difference is small, it still requires an explanation.

In our previous paper Rosaev and Plavalova (2021) we have presented the approximation of orbital elements of some asteroids. Here we have applied this method to the considered pairs. The period of the (7343) Ockeghem perihelion precession is about 45.95 kyr. The detected period of short periodic eccentricity perturbation is about 46.06 kyr. The period of the 56232 (1999 JM_{31}) perihelion precession is about 45.49 kyr. The according period of short periodic eccentricity perturbation is about 45.40 kyr.

The long period in eccentricity is about 305.01 kyr for the (7343) Ockeghem - (154634) 2003 XX_{38} pair and 294.98 kyr for the (56232) 1999 JM31 - (115978) 2003 WQ56 pair.

The formal results of our eccentricity approximation for these two pairs using the method stated by Rosaev and Plavalova (2021) are:

$$e_{7343} = 0.15 + 0.054\cos(0.0206t + 2.3) - 0.025\cos(0.136t + 3.7) \tag{4.2}$$

$$e_{56232} = 0.15 + 0.049\cos(0.0213t + 0.2) - 0.024\cos(0.138t + 2.4) \tag{4.3}$$

We have highlighted the remarkable phase difference in eccentricity perturbations of these pairs (Fig. 2).

5. (10123) Fideoja and (117306) 2004 VF_{21} pair

Pravec et al. (2018) conducted backward numerical integrations of these two asteroids using a modest number of clones - that the pair orbits and revealing encounters about 1-2 Myr ago. They found that the primary (10123) Fideoja is a binary system while the satellite has a secondary-to-primary mean diameter ratio D1,s/D1,p =0.36±0.02 and an orbital period of 56.46 ±0.02 h. They note that the pair‘s orbits undergo irregular jumps over a 7-2J mean motion resonance with Jupiter.

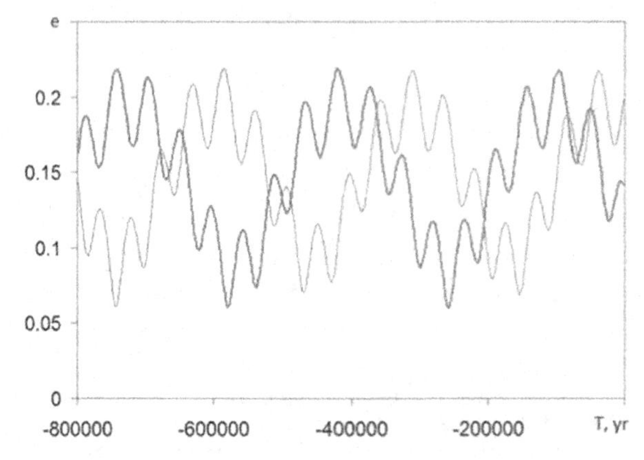

Figure 2. The nominal orbital evolution of eccentricity of (7343) Ockeghem (bold magenta line) and (56232) 1999 JM_{31} (thin gray line).

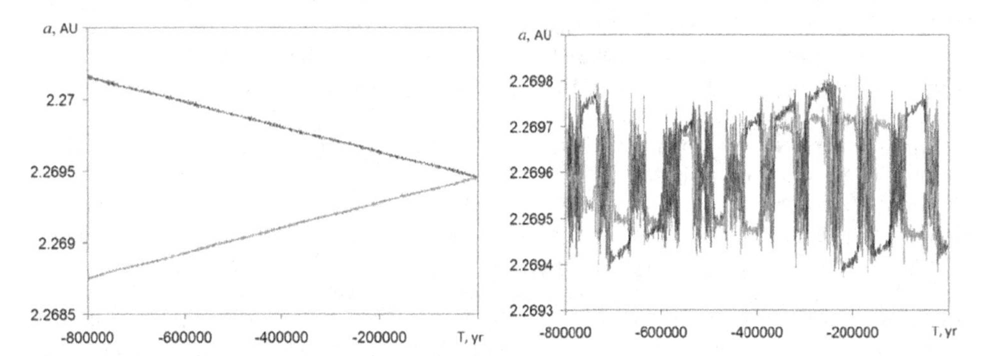

Figure 3. The left figure shows the evolution of the semi-major axis of pair Fideoja and (117306) 2004 VF_{21} for only Jupiter-Saturn-Earth perturbations. The right figure shows the evolution of the semi-major axis of the same pair for only Jupiter-Saturn-Mars perturbations.

We integrated the nominal orbits of this pair taking into account all large planetary perturbations. We obtained strong synchronous variations of the semi-major axis of both members as clear evidence of resonance perturbations. We determined the centre of the resonance related chaotic zone to be about 2.26960 ± 0.00001 AU. However, our study of the dynamics of this pair shows that they are not in 7-2J resonance, the nominal position of which would be about 2.2569 AU (i.e a difference of 0.011 AU).

In the search for resonance, we set a very large value of Yarkovsky drift ($A_2 = 1 \cdot 10^{-13}$) and applied numerical integrations with only one perturbing planet - Jupiter, or Earth, or Mars. These three calculations did not provide us with any resonance perturbations. We then repeated the integrations with perturbations from Jupiter-Saturn-Earth (Fig. 3 left) and Jupiter-Saturn-Mars (Fig. 3 right). As a result, we obtained clear evidence that the observed jumps in the semi-major axis of (10123) Fideoja and (117306) 2004 VF_{21} were caused by the three body resonance asteroid-Jupiter-Mars (9-6J-4M). Moreover, according to our numerical integration, the width of the chaotic zone related to the resonance is about 0.0003 AU. In this case the resonance argument has the form:

$$\varphi = 9\lambda - 6\lambda_{Jupiter} - 4\lambda_{Mars} \tag{5.1}$$

6. Conclusions

We have focused on possible resonances for 20 randomly selected close young asteroid pairs (younger than 800 kyr). In order to find possible resonances, we have applied backward numerical integration using a very high value of Yarkovsky drift (using the non-gravitational force parameter for asteroids $A_2 = 1 \cdot 10^{-13}$).

Our findings reveal that half of the tested pairs are affected by three-body resonances within a distance less than 0.0006 AU. We deduce from this that asteroid-Earth-Jupiter and asteroid-Mars-Jupiter resonances are very important in the dynamics of close asteroid pairs in the inner and middle asteroid belt.

Here we have considered three most interesting cases: the pair (10123) Fideoja and (117306) 2004 VF_{21}, the pair (56232) 1999 JM_{31} and (115978) 2003 WQ_{56} and the pair (7343) Ockeghem and (154634) 2003 XX_{38} in details. We conclude that resonance perturbations may play an important role in the dynamical evolution of very young asteroid families and close pairs. It is necessary to account for this fact in the models of their origin and dynamical evolution.

In closing, compact asteroid families and pairs near resonance provide a unique opportunity to study in detail the resonance perturbations and the dynamical interaction of minor bodies with resonances.

References

Chambers, J. E., 1999, MNRAS, 304, 793–799
Duddy S.R. et al. 2012, A&A 539, A36
Everhart, E., 1985. Astrophysics and Space Science Library, 115, 185
Farnocchia D.S., Chesley S.R., Vokrouhlicky D., Milani A., Spoto F., Bottke W.F., 2013, Icarus, 224, 1–13.
Fatka P., Pravec P. and Vokrouhlicky D., 2020, Icarus, 338, 113554 .
Jacobson S.A., Scheeres D.J. 2011, Icarus, 214, 161–178.
Knezevic, Z., Milani, A., 2003, Astronomy and Astrophysics, 403,(2003) 1165–1173.
Nesvorny, D., Vokrouhlicky, D., 2006, The Astronomical Journal 132, 1950–1958
Nesvorny, D., Broz, M., Carruba, V., Asteroids IV, (2015) 297.
Pravec P., and 28 colleagues, 2018, Icarus, 304, 110–126.
Rosaev A., Plavalova E. Galiazzo M., Res. Notes AAS Vol. 4,239, (2020)
Rosaev A., Plavalova E., 2021, PSS, 202, 105233
Spoto, F., Milani, A., Knezevic, Z., 2015, Icarus, 257, 275

Multi-scale (time and mass) dynamics of space objects
Proceedings IAU Symposium No. 364, 2022
A. Celletti, C. Galeş, C. Beaugé, A. Lemaitre, eds.
doi:10.1017/S1743921321001253

Characterization of the stability for trajectories exterior to Jupiter in the restricted three-body problem via closed-form perturbation theory

Mattia Rossi[1] **and Christos Efthymiopoulos**[2]

[1]Department of Mathematics "Tullio Levi-Civita", University of Padova,
Postal Code 35121, Via Trieste 63, Padova, Italy
email: mrossi@math.unipd.it

[2]Department of Mathematics "Tullio Levi-Civita", University of Padova,
Postal Code 35121, Via Trieste 63, Padova, Italy
email: cefthym@math.unipd.it

Abstract. We address the question of identifying the long-term (secular) stability regions in the semi-major axis-eccentricity projected phase space of the Sun-Jupiter planar circular restricted three-body problem in the domains i) below the curve of apsis equal to the planet's orbital radius (ensuring protection from collisions) and ii) above that curve. This last domain contains several Jupiter's crossing trajectories. We discuss the structure of the numerical stability map in the (a, e) plane in relation to manifold dynamics. We also present a closed-form perturbation theory for particles with non-crossing highly eccentric trajectories exterior to the planet's trajectory. Starting with a multipole expansion of the barycentric Hamiltonian, our method carries out a sequence of normalizations by Lie series in closed-form and without relegation. We discuss the applicability of the method as a criterion for estimating the boundary of the domain of regular motion.

Keywords. Asteroids, celestial mechanics, n-body simulations

---

1. Introduction

The secular (long-term) behaviour of the planetary problem, even in the restricted case, is a central question in the framework of the N-body problem. Several heuristic criteria, such as the orbital crossing or the Hill condition (see Ramos *et al.* (2015)) or AMD stability (Laskar & Petit (2017)), have been proposed to discriminate between stable and unstable orbits in phase space. However, there are numerical indications that such methods have some limits as regards their applicability both as a necessary and sufficient condition able to guarantee secular stability.

Here, we first briefly discuss the structure of regular and chaotic regions from a numerical point of view, using short- and long-period Fast Lyapunov Indicator (FLI) stability maps (Lega *et al.* (2016), Guzzo & Lega (2018)) in a very refined grid of initial conditions for the Sun-Jupiter planar circular system (pCR3BP hereafter). As in Todorović *et al.* (2020), we identify arch-like structures and the fractal boundaries discriminating between regular and chaotic orbits. For large values of the semi-major axis and correspondingly increasing eccentricities, a wide set of regular orbits emerges in the FLI diagram. These are clearly protected from collisions and we call them the "lower stability region" (low part of the (a, e) plane).

The main purpose of our semi-analytical work consists of formulating a normalization scheme via closed-form theory capable to deal with considerably high eccentricities and capture topological details of the boundary motion of the lower stability region.

Closed-form perturbation theory provides a framework for series calculations in perturbed Keplerian problems without expansions in powers of the bodies' orbital eccentricities. This is mainly motivated by the necessity to construct secular models for sufficiently eccentric orbits. A main obstruction for the application of closed-form theory in the restricted three-body problem stems from the difficulty to solve the homological equation explicitly when the kernel contains addenda beyond the Keplerian ones. An effective procedure to overcome this issue has been proposed by Deprit *et al.* (2001), called relegation algorithm, which, however, comes with intrinsic poor convergence properties: convergence occurs only in the limit when one of the frequencies is dominant. Such hypothesis cannot be adopted in our case. Hence, in our work we propose a normalization algorithm avoiding relegation supported by numerical verifications, like the accurate reproduction of the orbital elements' variations and detection of mean motion resonances.

The method presented below applies to particles with trajectory completely external to the trajectory of Jupiter. For an analogous method in the case of internal trajectories, see, instead, Cavallari & Efthymiopoulos (this volume of the proceedings).

2. FLI stability map of the Sun-Jupiter pCR3BP

The pCR3BP is defined by the planar motion of a body $\mathcal{P}$ of negligible mass in the gravitation field of two massive bodies $\mathcal{P}_0$ (the primary) and $\mathcal{P}_1$ (the secondary), which perform a circular trajectory around the common barycenter. Let $\vec{R}(t)$ be the barycentric radius vector of the particle and $\vec{r}_1(t)$ the relative radius vector of $\mathcal{P}_1$ with respect to $\mathcal{P}_0$.

The starting Hamiltonian of the model written in barycentric coordinates (i.e. Jacobi variables when $||\vec{R}|| > ||\vec{r}_1||$) reads

$$\mathcal{H}(\vec{R}, M_1, \vec{P}, J_1) = \frac{||\vec{P}||^2}{2} - \frac{\mathcal{G}m_0}{||\vec{R} + \mu\vec{r}_1(M_1)||} - \frac{\mathcal{G}m_1}{||\vec{R} - (1-\mu)\vec{r}_1(M_1)||} + n_1 J_1 \, , \tag{2.1}$$

where $\vec{R} = (X, Y)$, $\vec{P} = (P_X, P_Y) \in T^*(\mathbb{R}^2 \setminus \{-\mu\vec{r}_1, (1-\mu)\vec{r}_1\})$ is the position-momentum couple of $\mathcal{P}$, $\mathcal{G}$ is the gravitational constant, $M_1 = n_1 t$ is the mean anomaly of $\mathcal{P}_1$ (n_1 is the mean motion of the $\mathcal{P}_0, \mathcal{P}_1$ system), J_1 is a dummy action variable canonically conjugate to the angle M_1 and

$$\mu = \frac{m_1}{m_0 + m_1} \in (0, 1/2)$$

is the mass parameter. For a circular trajectory of the primary we have

$$\vec{r}_1(M_1) = ||\vec{r}_1|| \, (\cos M_1, \sin M_1) \ . \tag{2.2}$$

Figure 1 shows the short-term and long-term FLI stability maps in the semi-major axis-eccentricity (a, e) plane when $\mathcal{P}_0$ is the Sun and $\mathcal{P}_1$ is Jupiter ($\mu = 9.5364 \cdot 10^{-4}$) for particle trajectories computed numerically in the above model. The initial conditions are such that the particle starts orbiting from its pericenter positioned on the X axis.
The top diagram shows how regions of regular orbits permeate the whole phase space, even above the line of pericenter crossing. It is worth noticing that the line of pericenter overestimates the boundary of the lower stability region. This boundary has a fractal shape whose form becomes clearer increasing the integration time, as displayed in bottom panel. Also, mean motion resonances are depicted as spikes penetrating the regular regions of the stability map.

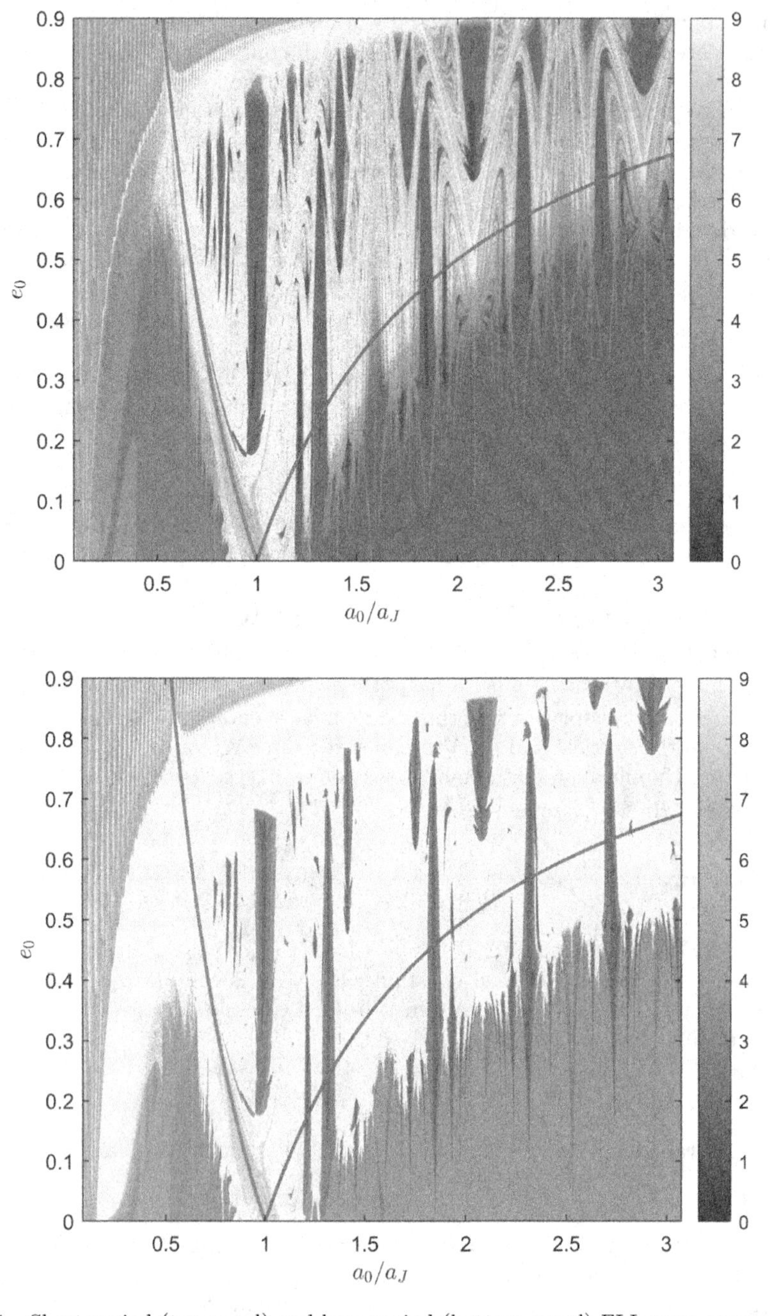

Figure 1. Short-period (top panel) and long-period (bottom panel) FLI maps computed over a grid of 300×900 initial data, where $a_J = \|\vec{r}_1\|$. Integration times are respectively 50 and 1000 Jupiter's orbital periods. The two curves represent the lines of constant apocenter and pericenter of the particle's trajectory equal to a_J.

We also observe in the same plot intricate structures created by the manifolds of the unstable orbits of various mean motion resonances (the "arches of chaos", see Todorović *et al.* (2020)).

3. Closed-form method for trajectories exterior to the trajectory of Jupiter

Assuming $\|\vec{r}_1\|/\|\vec{R}\| \ll 1$, we introduce a book-keeping symbol σ, with numerical value equal to 1, that keeps trace of the order of magnitude of the eccentricity e and of the small mass ratio μ at the same time via the powers σ^1, σ^ν. Expanding (2.1) up to $\mathcal{O}(\mu^{k_\mu}, (\|\vec{r}_1\|/\|\vec{R}\|)^{k_{\rm mp}})$, for $k_\mu, k_{\rm mp} \in \mathbb{N} \setminus \{0\}$ with $k_\mu > 1$, we pass to Delaunay elements (ℓ, g, L, G), defined by

$$\begin{aligned} L &= \sqrt{\mathcal{G} m_0 a}\,, & \ell &= M\,, \\ G &= L\sqrt{1-e^2}\,, & g &= \omega\,, \end{aligned} \tag{3.1}$$

where a, M, ω stand for the semi-major axis, the mean anomaly and the argument of pericenter respectively. We then write $L = L_* + \delta L$, $L_* = \sqrt{\mathcal{G} m_0 a_*}$ and $n_* = \sqrt{\mathcal{G} m_0} a_*^{-3/2}$ for constant reference values a_*, e_* dependent on initial conditions.

Given the above definitions, we have the following.

Proposition. *There exists a canonical transformation conjugating* (2.1) *to the secular normal form with respect to the fast angles* f, M_1 *of the system provided by*

$$\mathscr{H} = \mathscr{H}_0 + \mathscr{R}\,, \tag{3.2}$$

with

$$\mathscr{H}_0 = n_* \delta L - \frac{3n_*}{2L_*}\delta L^2 + n_1 J_1 + \sum_{j=\nu}^{\nu k_\mu - 1} c_j(\delta L, e; \mu)\sigma^j + \mathcal{O}(\delta L^3)\,, \tag{3.3}$$

$$\begin{aligned} \mathscr{R} = &\sum_{\substack{s\in\mathbb{Z}^3 \\ s=(s_1,s_2,s_3)}} d_{\nu k_\mu, s}(\delta L, e; \mu)\cos(s_1 f + s_2 g + s_3 M_1)\sigma^{\nu k_\mu} \\ &+ \mathcal{O}\left(\sigma^{\nu k_\mu + 1}; \left(\frac{\|\vec{r}_1\|}{\|\vec{R}\|}\right)^{k_{mp}+1}, \delta L^3\right)\,, \end{aligned} \tag{3.4}$$

where f *denotes the particle's true anomaly and*

$$\nu = \left\lceil \frac{\log_{10}\mu}{\log_{10} e_*} \right\rceil\,, \tag{3.5}$$

for $c_j, d_{\nu k_\mu, s} \in \mathbb{R}$.

The details of the proof of the above proposition will be presented elsewhere (Rossi & Efthymiopoulos, in preparation). Briefly, expanding the Hamiltonian in power series of the quantity $\delta L = L - L_*$, we obtain

$$\begin{aligned} \mathcal{H} &= -\frac{\mathcal{G}^2 m_0^2}{2L_*^2}\sum_{l=1}^{\infty} l\left(-\frac{\delta L}{L_*}\right)^{l-1} + n_1 J_1 + \mu \sum_{l=0}^{\infty}\frac{1}{l!}\left.\frac{\partial^l \mathcal{H}_1}{\partial L^l}\right|_{L=L_*}\delta L^l \\ &= n_*\delta L + n_1 J_1 + \mu\left(\mathcal{H}_1|_{\delta L=0,\ \mu=0} + \left.\frac{\partial \mathcal{H}_1}{\partial \delta L}\right|_{\delta L=0,\ \mu=0}\delta L\right) + \mathcal{O}(\mu^2, \delta L^2)\,, \end{aligned} \tag{3.6}$$

where we drop the constant $-\mathcal{G}^2 m_0^2/(2L_*^2)$, and $n_* = \mathcal{G}^2 m_0^2/L_*^3$, with the function $\mathcal{H}_1$ given by

$$\mathcal{H}_1 = -\frac{\mathcal{G} m_0}{\|\vec{R}\|} + \mathcal{O}\left(\left(\frac{\|\vec{r}_1\|}{\|\vec{R}\|}\right)^2, \mu^2\right). \tag{3.7}$$

Let us rewrite (3.6) in Fourier expansion taking advantage of the periodicity of the angles and making the book-keeping symbol explicit:

$$\mathcal{H} = n_* \delta L + n_1 J_1 + \sum_{s \in \mathbb{Z}^3} q_s(\delta L, e; \mu) \cos(s_1 f + s_2 g + s_3 M_1) \sigma_s \tag{3.8}$$

where $\sigma_s \in \{\sigma^\nu, \sigma^{\nu+1}, \ldots\}$ and, by D'Alembert rules, only cosines and real coefficients q_s appear.
Setting $\mathscr{Z}_0 = n_* \delta L + n_1 J_1$ and $\mathscr{R}_\nu^{(0)} = \mathcal{O}(\sigma^\nu)$ the remaining summation in (3.8), we define the Lie series operator by

$$\exp\left(\mathcal{L}_\chi\right) = \sum_{n \geq 0} \frac{1}{n!} \mathcal{L}_\chi^n = \mathbb{I} + \mathcal{L}_\chi + \frac{1}{2} \mathcal{L}_\chi \circ \mathcal{L}_\chi + \ldots, \tag{3.9}$$

where $\mathcal{L}_\chi \cdot = \{\cdot, \chi\}$ is the Poisson bracket operator.
Applying (3.9) to (3.8) we get the first transformed Hamiltonian

$$\mathscr{H}^{(1)} = \mathscr{Z}_0 + \mathscr{R}_\nu^{(0)} + \{\mathscr{Z}_0, \chi_\nu^{(1)}\} + \{\mathscr{R}_\nu^{(0)}, \chi_\nu^{(1)}\} + \frac{1}{2}\{\{\mathcal{H}, \chi_\nu^{(1)}\}, \chi_\nu^{(1)}\} + \ldots \tag{3.10}$$

with respect to the generating function $\chi_\nu^{(1)}$, found out as solution of the homological equation

$$\{\mathscr{Z}_0, \chi_\nu^{(1)}\} + \mathscr{R}_\nu^{(0)} = \mathcal{O}(\sigma^{\nu+1}) \tag{3.11}$$

which cancels out σ^ν-terms.

By means of an appropriate rearrangement of the Poisson structure using the chain rule, one can show that

$$\chi_\nu^{(1)} = \sum_{\substack{s \in \mathbb{Z}^3 \\ (s_1, s_3) \neq (0,0)}} \frac{q_{s_\nu}}{s_1 n_* + s_3 n_1} \sin(s_1 f + s_2 g + s_3 M_1) \sigma^\nu, \tag{3.12}$$

with coefficients $q_{s_\nu} = \mathcal{O}(\sigma^\nu)$.

The procedure can be repeated at successive normalization steps.

4. Numerical tests

We consider the example reported in Figure 2. The left panel shows in logarithmic scale the quantity:

$$\mathscr{E}^{(j)} = \sum_{l=\nu+j}^{\nu k_\mu} \sum_{s \in \mathbb{Z}^3} |d_{l,s}^{(j)}| \geq \|\mathscr{R}_{\nu+j}^{(j)}\|_\infty, \quad j = 1, \ldots, \nu(k_\mu - 1), \tag{4.1}$$

where $\|\cdot\|_\infty$ is the sup norm, j is the number of normalization steps and $\mathscr{R}_{\nu+j}^{(j)} = \mathcal{O}(\sigma^{\nu+j})$ is the normal form remainder of the j-th order.

The plot gives an estimate of the error of the semi-analytical method at the j-th step. The right panel shows a remarkably good agreement of the evolution of the semi-major axis $a(t)$ between a direct integration of the Cartesian equations of motion and a semi-analytic integration using the normal form part of (3.3).

Figure 3, finally, shows in shade scale the size of the remainder as a function of (a, e). We observe that the semi-analytical method allows to define with good precision the

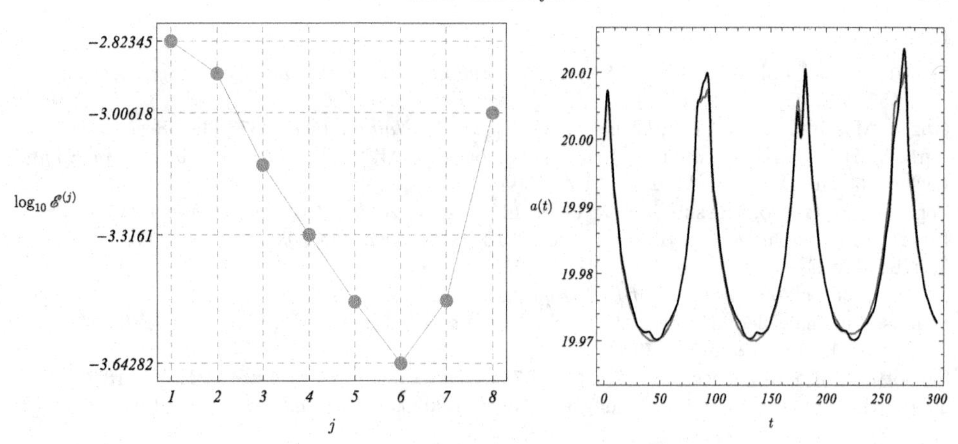

Figure 2. (Left) Evolution of the size of the remainder of the normal form construction of Section 3 as a function of the normalization order j, in an example with $a_* = 20$, $e_* = 0.4$ ($\nu = 8$), $k_\mu = 2$, $k_{\rm mp} = 3$ and order of δL expansion equal to 1. (Right) Comparison of the semi-analytic (dark curve) with the numerical propagation (light curve) of a trajectory.

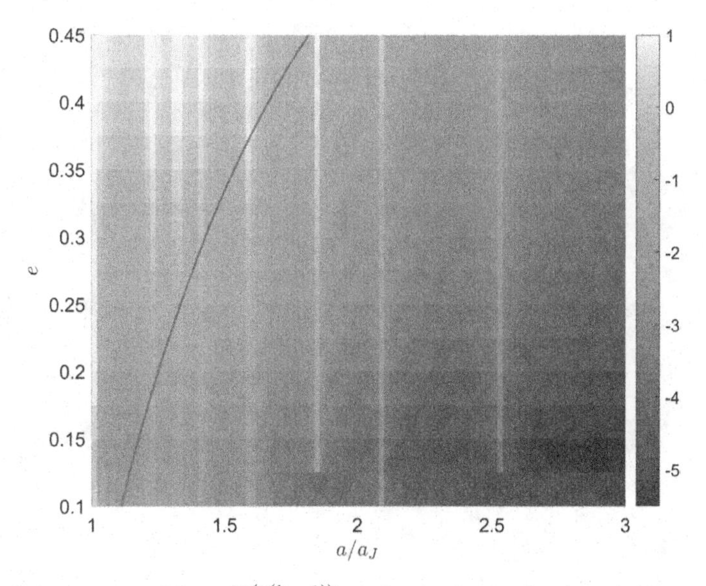

Figure 3. Computation of $\log_{10}(\mathscr{E}^{(\nu(k_\mu-1))})$ over a 100×15 (a, e) grid. For every $e = e_*$ a different normalization is derived and then evaluated for each $a = a_*$. The traced curve is again the constant pericenter line.

lower stability region in Figure 1: the error increases for trajectories close to Jupiter's orbit. The vertical strips correspond to mean motion resonances, as illustrated also in the FLI maps.

Acknowledgements

C.E. was partially supported by the MIUR-PRIN 20178CJA2B New Frontiers of Celestial Mechanics: Theory and Applications.

References

Deprit, André, Palacián, Jesúus & Deprit, Etienne 2001, *Cel. Mech. and Dyn. Astron.*, 79, 157–182

Guzzo, Massimiliano & Lega, Elena 2018, *Physica D: Nonlin. Phen.*, 373, 38–58

Lara, Martin, San-Juan, Juan F. & López-Ochoa, Luis M. 2013, *Mathem. Problems in Engin.*

Laskar, Jacques & Petit, AC 2017, *A&A*, 605, A72

Lega, Elena, Guzzo, Massimiliano & Froeschlé, Claude 2016, *Chaos Detec. and Predict.*, 35–54

Libert, Anne-Sophie & Sansottera, Marco 2013, *Cel. Mech. and Dyn. Astron.*, 117, 149–168

Métris, G & Exertier, Pierre 1995, *A&A*, 294, 278–286

Palacián, Jesús 2002, *Journal of Diff. Equat.*, 180, 471–519

Ramos, Ximena Soledad, Correa-Otto, Jorge Alfredo & Beauge, Cristian 2015, *Cel. Mech. and Dyn. Astron.*, 123, 453–479

Sansottera, Marco & Ceccaroni, Marta 2017, *Cel. Mech. and Dyn. Astron.*, 127, 1–18

Todorović, Nataša, Wu, Di & Rosengren, Aaron J 2020, *Science advances*, 6, eabd1313

Multi-scale (time and mass) dynamics of space objects
Proceedings IAU Symposium No. 364, 2022
A. Celletti, C. Galeş, C. Beaugé, A. Lemaitre, eds.
doi:10.1017/S1743921322000047

Astrometry and photometry of asteroids from the UkrVO database of astroplates

I.B. Vavilova[1], S.V. Shatokhina[1], L.K. Pakuliak[1], O.M. Yizhakevych[1], I. Eglitis[2], V.M. Andruk[1] and Yu.I. Protsyuk[3]

[1]Main Astronomical Observatory of the NAS of Ukraine,
27 Akademik Zabolotny str., Kyiv, 03143, Ukraine
email: irivav@mao.kiev.ua

[2]Institute of Astronomy, University of Latvia, Raina blvd. 19, Riga, LV-1586, Latvia

[3]Mykolaiv Astronomical Observatory, MES of Ukraine,
1 Observatorna str., Mykolaiv, 54000, Ukraine

Abstract. We present the developed methods of digitization, image processing, reduction, and scientific data mining with the latest reference catalogs, which allowed us to obtain a good positional and photometric accuracy in B-band of 6,500 asteroids down to 17.5^m from the Ukrainian Virtual Observatory database of astroplates. The archive includes FON-Kyiv, FON-Kitab, FON-Dushanbe sky surveys (1981–1996) and astroplates of the Baldone and Tautenburg observatories. For some of asteroids, observations are either completely absent or not enough over the certain time interval to the moments of their official discoveries (about 300 such objects were found). Positional observations during these time scales are highly useful for a more detailed study of the dynamics and orbital parameters of asteroids as well as the obtained photometric parameters are very complementary with present-day data for studying changes in brightness and light curves.

Keywords. methods: data analysis; techniques: image processing, photometric; asteroids

---

1. Introduction

The value of old photographic archives as a source of "new" scientific data does not diminish over the years. The digitizing of observational archives all over the observatories possessing the glass collections allows successful involving these data into modern-day science, including for small Solar System bodies research (Ruphy et al. (1994), Lupishko 1997, Mahabal et al. (2002); Ivezic et al. (2002); Davis et al. (2004); Goodman (2012); Pakuliak et al. (2014); DiCarlo (2018); Lehtinen et al. (2018); Villarroel et al. (2020); Vavilova et al. (2020); Khovritchev et al. (2021)). The Ukrainian Virtual Observatory (UkrVO, http://ukr-vo.org) database covers data of about 40,000 astroplates exposed in 1898–2018, from which 15,000 are digitized (Vavilova (2016); Vavilova et al. (2017)). The most of them is related to the stellar FON project (Northern Sky Photographic Survey, 1981–1996). But these images have produced a large number of faint asteroids down to 17.5^m. For some of them, observations are either completely absent or not enough over a certain time interval to the moments of their official discoveries.

Allowing to enlarge observational timescales, these data are helpful to improve the orbital elements of asteroids (Veres et al. (2015); Savanevych et al. (2015); Savanevych et al. (2015); Savanevych et al. (2018); Eggl et al. (2020)) and to validate simulations of collisions in the dynamical and kinematic studies. For example, Chernetenko

(2019) estimated the possible change in the velocity of the (596) Sheila active mass-losing asteroid. Fuentes-Munoz and Scheeres (2021) developed semi-analytical long-term propagation model of near-Earth asteroids (NEA) to calculate their close flyby probabilities. Such validation permits us to characterize the nature of objects whose orbits are potentially hazardous in the future, or which may have had a period of close passages in the past to the Earth and other planets (Farnocchia and Chodas (2021); Forgács-Dajka et al. (2021)). In this context we also note work by Pravec et al. (2012), who used albedo estimates from WISE thermal observations for revealing size dependencies of surface properties on absolute magnitude of 583 asteroids observed in 1978-2011. Solano et al. (2014) accomplished a project by the Spanish Virtual Observatory to improve the orbits of 551 NEA using old astronomical archives and 938,046 images from the SDSS DR8 to confirm or discard the presence of NEA images from previous observing epochs.

The aim of our paper is to demonstrate reliability of our approach in digitization, image processing, and reduction of the UkrVO astroplates, which allowed us to obtain a good positional and photometric accuracy in B-band of 6,500 asteroids down to 17.5^m.

2. Image processing and plate reduction scheme

All astroplates were digitized on commercial scanners Microtek and Epson with scanning mode of 1200 dpi and the gray scale of 16-bit color. The original software for image processing and further data mining was developed in LINUX MIDAS-ROMAFOT environment (Andruk (2018)). In general, the scheme of the image processing and the further reduction includes the following main steps (see, in detail, Andruk et al. (2019); Protsyuk et al. (2019); Pakuliak & Andruk (2020)):

- Conversion of files from 16-bit tiff format into 8-bit format fits, while after the preliminary tests 16-bit resolution turned out to be excessive.
- Photometric equalization of a digitized star field without applying the auxiliary flat-field images.
- Calculation of the rectangular coordinates and the photometric instrumental values for all objects registered on the astroplate.
- Separation of registered objects by exposure (case of the multiply exposed plate).
- Identification of stars of the reference catalog (Tycho-2 or GAIA DR2) by their rectangular and equatorial coordinates.
- Astrometric reduction of all objects into the reference frame at the epoch of the plate exposure with the assessment of the accuracy. The rectangular coordinates obtained for objects, corrected for systematic errors of scanners, were transformed into a system of tangential coordinates by full polynomials of 6 or 3 orders (depending on the telescope) taking into account coma and the magnitude equation.
- Photometric reduction of instrumental stellar magnitudes to a system of photoelectric standards with the assessment of the accuracy.
- Final analysis of objects to reject fictitious and erroneous images.

Of all the objects recorded on the scan, each asteroid was identified by ephemeris coordinates and magnitude at the time of observation of the JPL online service (http://ssd.jpl.nasa.gov). The diameters of the images and the maximum intensity of the central pixel of the image on the scan were also taken into account.

3. Asteroids in the FON-Kyiv sky survey

Catalog of 2292 astrometric positions and B-magnitudes of asteroids and comets with magnitudes from 16^m to 8^m (Shatokhina et al. (2019)) have been compiled from digitized photographic observations of FON-Kyiv obtained in 1985–1991. Coordinates of objects

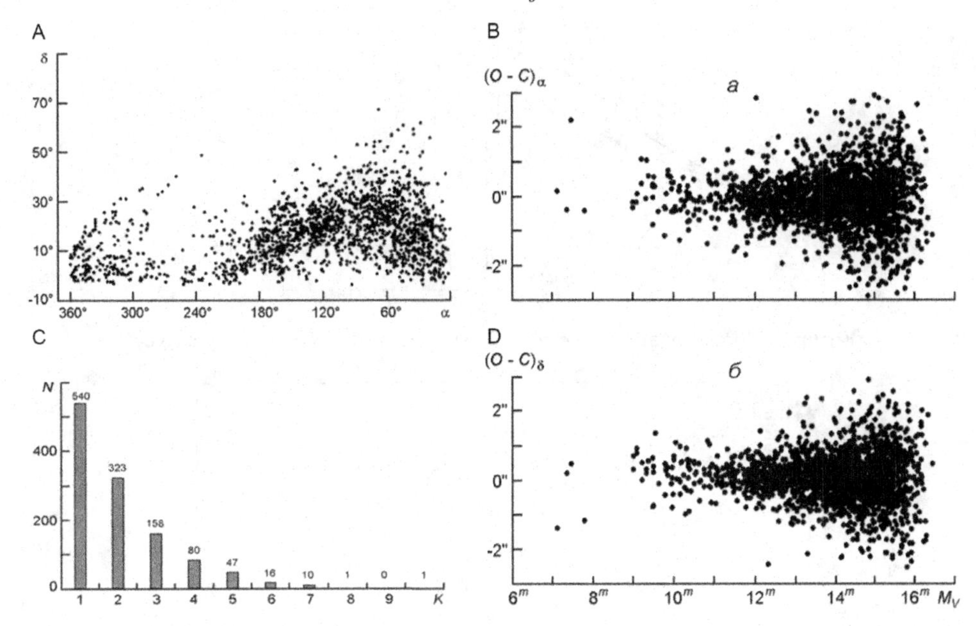

Figure 1. Parameters of array of registered asteroids from the FON-Kyiv astroplate archive.

were reduced to the Tycho-2 reference system and B-values to the system of photoelectric standards. The results of software testing and application for the determination of positions and magnitudes are given by Eglitis et al. (2016); Yizhakevych et al. (2017) and a number of related publications. Most of the registered asteroids are in the declination zone below +30 degrees. A little sample is in the high-declination zone up to +70 degrees.

Fig. 1, panel A, shows the distribution of asteroids across the celestial sphere. Fig. 1, panel B, shows the distribution of asteroids by the multiplicity of positions of each of the asteroids. All the asteroid positions were compared with JPL DE431 (http://ssd.jpl.nasa.gov/horizons) ephemeris. The obtained coordinates differences (O-C) between the observed and computed asteroid positions are given in Fig. 1, panels C and D. It was found that for 54 asteroids the observational moment precedes the moment of the asteroid discovery. Among them there are 4 asteroids with their chronologically the earliest observations in the world according the MPC data.

4. Asteroids in the FON-Kitab sky survey

FON-Kitab photographic observations were carried out in 1981–1993 with the Zeiss Double Wide Angle Astrograph (D/F = 40/300, 69"/mm) at the Kitab observatory (Uzbekistan). The application of the above described scheme (Chapter 2) resulted in a few catalogs of coordinates and B-magnitudes for more than 13 million stars and galaxies from the FON-Kitab part. Based on the FON-Kitab digitized astroplates, the catalog of 2728 topocentric positions and B-magnitudes of asteroids and comets have been compiled. Asteroid coordinates and B-magnitudes were obtained in the Tycho-2 reference catalog system and the system of photoelectric standards, respectively.

The catalog's contents is given by Shatokhina et al. (2018). The observations were conducted with both telescope tubes, so two plates with the same center were exposed simultaneously. As a result, 1412 asteroid positions are paired, and the remaining 1316 positions are single. Based on the paired positions, the rms errors of the equatorial coordinates and B-magnitudes of asteroids were determined. Their values averaged over

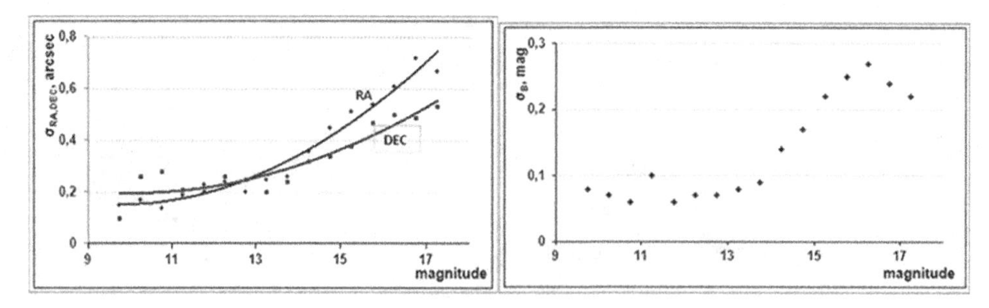

Figure 2. Positional r.m.s. errors of asteroids in FON-Kitab sky survey.

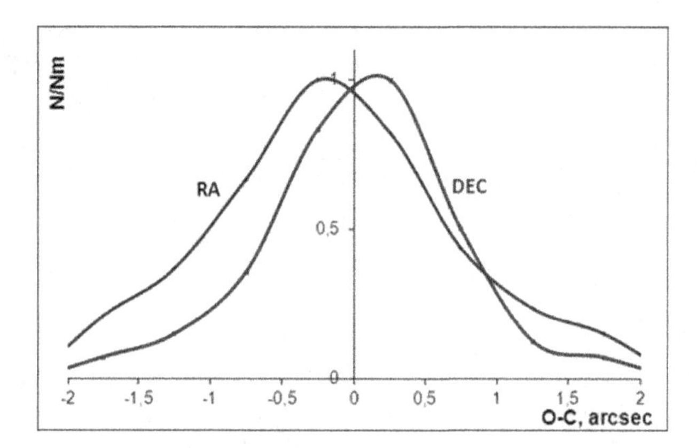

Figure 3. (O-C) differences for all asteroid positions in FON-Kitab sky survey.

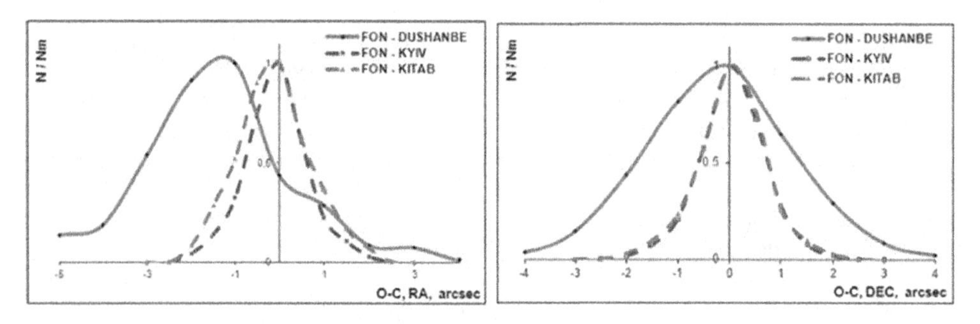

Figure 4. (O-C) differences on both coordinates for 302 asteroid positions from the FON-Dushanbe sky survey.

the magnitude intervals are given in Fig. 2. The histogram of O-C differences for all positions of the asteroids constructed from the results of comparison with the ephemeris is shown in Fig. 3. A good agreement can be noted between the results obtained from FON-Kyiv and FON-Kitab surveys.

5. Asteroids in the FON-Dushanbe sky survey

The Dushanbe part of the FON project is represented by about 1570 astroplates obtained in 1985–1992 with a Zeiss-400 astrograph at the Hissar Astronomical Observatory of the Institute of Astrophysics of the NAS of Tajikistan (Shatokhina et al. (2020)). Image processing and plate reduction scheme (Section 2) were the same as for

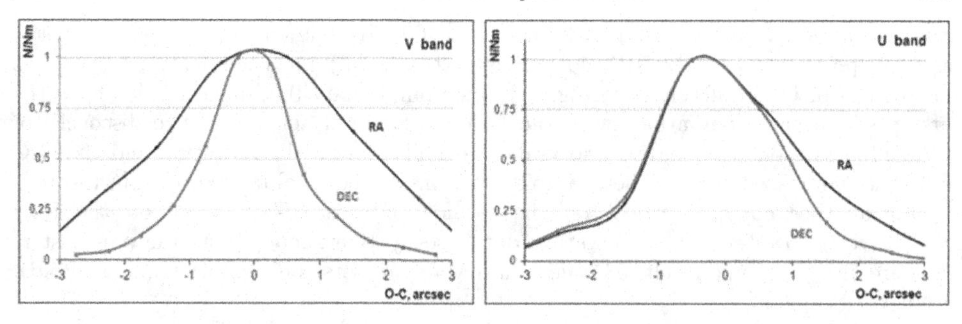

Figure 5. (O-C) differences from comparison of asteroid positions from the Baldone astroplate archive with the JPL ephemeris.

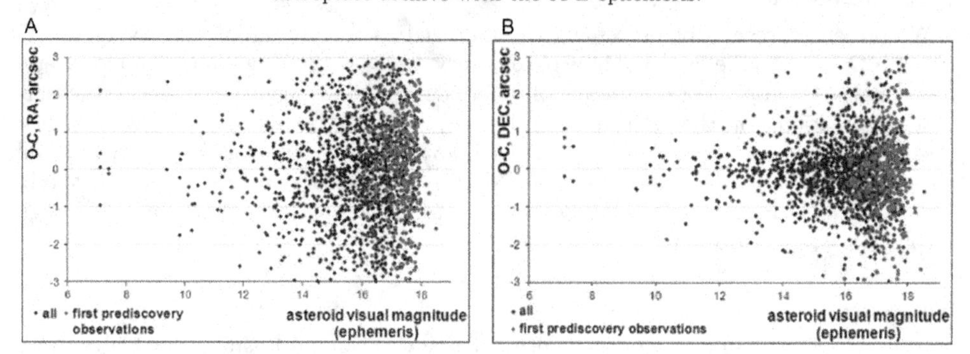

Figure 6. (O-C) differences in RA (A) and in DEC (B) for all and first prediscovery asteroid observations from the Baldone astroplate archive.

the FON-Kyiv and FON-Kitab surveys. This permits to get a star catalog with millions of objects and a preliminary list for the upcoming catalog of 300 positions of asteroids and comets with visual magnitudes from 7^m to 16.5^m.

The O-C differences in both coordinates for all asteroids are presented in Fig. 4 in comparison to similar data for FON-Kyiv and FON-Kitab. A systematic O-C shift is noticeable in the RA coordinate for all asteroid positions from the two FON-Dushanbe zones. The further analysis using data of orbital velocities of asteroids near the observational moments shows a clear correlation of the O-C with the value of their orbital velocities. The last could be the result of systematic underestimation of time in positions of asteroids. In the future, this systematic component should be clarified and excluded.

6. Asteroids in the Baldone astroplate archive

The Baldone Observatory of the Institute of Astronomy of the Latvian University owns a photographic collection of 22,633 astroplates obtained with a 1.2 m Schmidt telescope in 1966–2005. Among them there are about 780 plates in the U photometric band and 4600 film negatives in the V band, which are close to the Johnson's photometric system. In 2013–2018, the Baldone Observatory performed a digitization of these astroplates by the scheme described in Section 2. At present, 281 plates and 2167 film negatives exposed in U and V photometric ranges have been used for asteroid searching.

Based on the scan processing results, a catalog of 1848 topocentric positions and magnitudes of asteroids and comets was compiled. It includes 1678 and 170 positions and magnitudes from observations in the V and U bands, respectively, as well as 7 positions and magnitudes of comets. Astrometric and photometric reduction of digitized astroplates is performed in the Tycho-2 reference system and photoelectric standards, respectively (Eglitis et al. (2016), Eglitis et al. (2019). The comparison results of the asteroid positions with the JPL ephemeris (https://ssd.jpl.nasa.gov) are presented as histograms of O-C differences in Fig. 5. It was found that 490 faint asteroids have the first prediscovery observations with the Baldone Schmidt telescope. The discovery of these objects took place only 20-40 years later. Their individual location and the O-C difference compared to the other asteroid positions in the catalog are shown in Fig. 6.

Our another current project on the catalog of asteroids is based on the Karl Schwarzschild Tautenburg Observatory database by Boerngen (1991): the most asteroids are objects of main belt, 53 objects are Mars crossers, 110 double/triple asteroids, 1 NEA, 19 comets.

7. Conclusions

We presented results of processing the digitized astroplates from FON-Kyiv, FON-Kitab, FON-Dushanbe, and Baldone observatory archives allowing us to compile several positional and photometric catalogs of asteroids. Some of them could occur the earliest observations of the objects long before their official discovery. The obtained (O-C) differences evident about good accuracy, so these catalogs are highly useful for dynamical and kinematic research. All the published catalogs are displayed in VizieR as well as current databases of digitized astroplates are available through http://ukr-vo.org/digarchives/index.php?b1&1.

References

Andruk, V. (Ed.), Results of Processing of Digitized Astronomical Photographic Plates. Riga: LAMBERT Academic Publish

Andruk, V., Eglitis, I., Protsyuk, Yu. et al. 2019, *Odessa Astron. Publ.*, 32, 181

Boerngen F. 1991, *AN*, 312, 65

Chernetenko, Yu. A. 2019, *Astrophys. Bull.*, 74, 203

Davis A., Barkume K., Springob C. et al. 2004, *JAVSO*, 32, 117

Di Carlo M., Vasile M., Dunlop J. 2018, *AdSpR*, 62, 2026

Eggl, S. et al. 2020, *Icarus*, 339, 113596

Eglitis, I., Eglite, M., Shatokhina, S.V. et al. 2016, *Odessa Astron. Publ.* , 29, 123

Eglitis, I., Yizhakevych, O., Shatokhina, S.V. et al. 2019, *Odessa Astron. Publ.* , 32, 189

Farnocchia, D. and Chodas, P.W. 2021, *Res. Notes of the American Astron. Society*, 5, 257

Forgács-Dajka E., Sándor Zs., Sztakovics, J. 2021, *arXiv:2110.11745*

Fuentes-Munoz, O. and Scheeres, D. 2021, *AAS/Div. of Dynamical Astron. Meeting*, 53, 106.05

Goodman A. A. 2012, *AN*, 333, 505

Ivanov G., Pakuliak L., Shatokhina S. et al. 2013, *IzPul*, 220, 501

Ivezic Z., Juric M., Lupton R. H. et al. 2002, *SPIE*, 4836, 98

Khovritchev M. Y., Robert V., Narizhnaya N. V. et al. 2021, A&A, 645, A76

Lehtinen K., Prusti T., de Bruijne J. et al. 2018, A&A, 616, A185

Lupishko D. F., Vasil'Ev S. V. 1997, *KPCB*, 13, 12

Mahabal A., Djorgovski S. G., Gal R. et al. DPOSS Team, 2002, *adaa.conf*, 281

Pakuliak L., Shlyapnikov A., Rosenbush A., Gorbunov M. 2014, ASInC, 11, 103

Pakuliak, L. & Andruk, V. 2020, In *Knowledge Discovery in Big Data from Astronomy and Earth Observation*, p. 325. doi: 10.1016/B978-0-12-819154-5.00029-1

Pravec, P. et al. 2012, textitIcarus, 221, 365

Protsyuk, Yu. I., Andruk, V.N., & Relke H. 2019, in: P. Skala(eds.), *Astroplate 2016*, p. 47

Ruphy S., Epchtein N., Bec-Borsenberger A. 1994, *Ap&SS*, 217, 97

Savanevych V. E., Briukhovetskyi A. B., Ivashchenko Y. N. et al. 2015, *KPCB*, 31, 302.

Savanevych V. E., Briukhovetskyi O. B., Sokovikova N. S. et al. 2015, MNRAS, 451, 3287

Savanevych V. E., Khlamov S. V., Vavilova I. B. et al. 2018, A&A, 609, A54

Shatokhina, S. V.; Relke, H.; Yuldoshev, Q. et al. *Odessa Astron. Publ.*, 31, 235

Shatokhina, S.V., Kazantseva, L.V., Izhakevich, E.M. et al. 2019, *2019yCatp0030034O1S*
Shatokhina, S.V. , Relke, H., Mullo-Abdolov, A. Sh. et al. 2020, *Odessa Astron. Publ.* , 33, 154
Solano, E. et al. 2014, *AN*, 335, 2, 142
Vavilova I. B., Pakuliak L. K., Protsyuk Y. I. et al. 2012, *BaltA*, 21, 356
Vavilova I. B. 2016, Odessa Astron. Publ, 29, 109
Vavilova, I.B., Yatskiv, Y.S., Pakuliak, L.K. et al. 2017, *IAUS, 325*, 361.
Vavilova, I., Pakuliak, L., Babyk, I. et al. 2020. In *Knowledge Discovery in Big Data from Astronomy and Earth Observation*, p. 57. doi:10.1016/B978-0-12-819154-5.00015-1
Vereš P., Jedicke R., Fitzsimmons A. et al. 2015, *Icarus*, 261, 34
Villarroel, B., Soodla, J., Comerón, S. et al. 2020, *AJ* 159, 8
Yuldoshev, Q., Protsyuk, Y., Relke, H. et al. 2019, *AN*, 340, 6, 494
Yizhakevych, O. M., Andruk, V. M., & Pakuliak, L. K. 2017, *KPCB*, 33, 3, 142

Multi-scale (time and mass) dynamics of space objects
Proceedings IAU Symposium No. 364, 2022
A. Celletti, C. Galeş, C. Beaugé, A. Lemaitre, eds.
doi:10.1017/S174392132100123X

Families of periodic orbits around asteroids: From shape symmetry to asymmetry

G. Voyatzis, D. Karydis and K. Tsiganis

Section of Astrophysics, Astronomy and Mechanics, Dept. of Physics,
Aristotle University of Thessaloniki,
GR 54124, Thessaloniki, Greece
email: voyatzis@auth.gr, dkarydis@auth.gr, tsiganis@auth.gr

Abstract. In Karydis *et al.* (2021) we have introduced the method of shape continuation in order to obtain periodic orbits in the complex gravitational field of an irregularly-shaped asteroid starting from a symmetric simple model. What's more, we map the families of periodic orbits of the simple model to families of the real asteroid model. The introduction of asymmetries in a gravitational potential may significantly affect the dynamical properties of the families. In this paper, we discuss the effect of the asymmetries in the neighborhood of vertically critical orbits, where, in the symmetric model, bifurcations of 3D periodic orbit families occur. When asymmetries are introduced, we demonstrate that two possible continuation schemes can take place in general. Numerical simulations, using an ellipsoid and a mascon model of 433-Eros, verify the existence of these schemes.

Keywords. Asteroids, Orbital mechanics, Periodic orbits

1. Introduction

Many space missions to small NEA have taken place recently or are planned in the coming years. Close proximity operations around such small bodies, which have irregular shape in general, demand sufficient knowledge of their gravitational field and their dynamics. In orbital mechanics, periodic orbits play an important role in understanding the dynamics and have been studied widely in celestial mechanics and especially in the three body problem. In addition, they can find direct applications in astrodynamics as parking orbits for a spacecraft or the unstable ones may be used for computing landing or escape paths (Scheeres (2012)). In such complex gravitational fields, which can be sufficiently modeled e.g. by polyhedrals or mascons (see Scheeres (2012)), the computation of periodic orbits is a challenge. The grid search method introduced by Yu & Baoyin (2012) has been proved very efficient and applied for various asteroids (e.g. Jiang *et al.* (2018)).

In Karydis *et al.* (2021), which will be referred in the following as 'Paper I', we approach the potential of an irregular body by starting from the symmetric potential of a simplified model (an ellipsoid), where the families of periodic orbits can be easily computed and show particular structures and types. Then, asymmetric terms are gradually introduced in the potential and periodic orbits are continued along this procedure, which is called *shape continuation* and ends when the 'real' potential of the target asteroid is adequately approximated. In this way, we assign families of the simplified model to families of the 'real' model and we can study the effect of the symmetric perturbations in the characteristic curves of the families and their stability. In the present study, we use a theoretical analysis and numerical simulations in order to show how families are affected by asymmetric forces when they are close to vertically critical orbits, where planar and 3D orbit families intersect in the symmetric model.

2. Description of the orbital mechanics

We consider the motion of a mass-less body in the gravitational field of an irregularly shaped asteroid which rotates with angular velocity ω. If the center of mass of the asteroid is considered as the origin point of a reference frame which rotates with the asteroid (i.e. it is a body-fixed frame), and $\mathbf{r}=(x, y, z)$ is the position vector of the mass-less body, its motion is described by the Hamiltonian

$$H(\mathbf{r},\mathbf{p})=\frac{1}{2}\mathbf{p}^2-\mathbf{p}(\omega\times\mathbf{r})+U(\mathbf{r}), \tag{2.2}$$

where the generalized momenta are given by $\mathbf{p}=\dot{\mathbf{r}}+\omega\times\mathbf{r}$ and U is the gravitational potential of the asteroid. If ω is constant, which is the case considered in this study, then H is also constant (H being the Jacobi integral or, simply, the energy).

Let $\mathbf{X}=(x, y, z, \dot{x}, \dot{y}, \dot{z})$ denote a phase space point and $\mathbf{X}=\mathbf{X}(t;\mathbf{X}_0)$ a trajectory with initial conditions $\mathbf{X}_0$. The system is autonomous and the condition $\mathbf{X}(T;\mathbf{X}_0)=\mathbf{X}_0$ implies a periodic orbit of period T. Supposing that the orbit intersects a Poincaré section, say $x=0$ with $\dot{x}>0$ and energy h, the orbits can be defined explicitly by a point in the $4D$ space of section, called Π_4, which is defined by vector $\mathbf{Y}=(y, z, \dot{y}, \dot{z})$. Thus, the periodicity conditions are reduced to

$$\mathbf{Y}(t^*;X_0)=\mathbf{Y}_0, \tag{2.3}$$

where t^* is the time of the mth intersection of the orbit, with a section that satisfies (2.3) for the first time. In this case, t^* and period T coincide and m denotes the multiplicity of the section.

In general, in space Π_4, periodic orbits are isolated and analytically continued with respect to h, forming mono-parametric families (Meyer *et al.* (2009), Scheeres (2012)). In computations, we may consider a continuation with respect to any variable but it is more convenient to continue the families by using an extrapolation procedure and considering as parameter the length s of the characteristic curve of the family in Π_4 (see Paper I). In this way, the numerical continuation is still successful at energy extrema that may exist along the family.

Let ξ denote a variation vector that satisfies the system of linear variational equations of system (2.2), namely

$$\dot{\xi}=\mathbf{A}(t)\xi \quad\Rightarrow\quad \xi=\mathbf{\Phi}(t)\xi(0). \tag{2.4}$$

Matrix $\mathbf{A}$ is computed along a periodic orbit and thus, it is also periodic. $\Phi(t)$ is the fundamental matrix of solutions and the constant matrix $\mathbf{M}=\mathbf{\Phi}(T)$ is the monodromy matrix, which is symplectic. Therefore, two eigenvalues are equal to unit and the rest four form reciprocal pairs. If we remove the rows and columns that correspond to the variables which define the Poincaré section (e.g. x and $\dot{x}$) from $\mathbf{M}$, then we obtain the reduced monodromy matrix $\mathbf{M}'$ of size 4×4 and the unit eigevalues are removed. The periodic orbit is stable if the two reciprocal pairs of eigenvalues of $\mathbf{M}'$ lie on the complex unit circle. In computations, we use the Broucke's stability indicies b_1 and b_2, which are computed from the elements of $\mathbf{M}'$ and their stability implies that they are real and $|b_i|<2$ (Broucke (1969)).

When $\mathbf{M}'$ is computed for a planar orbit, then it is decomposed in two 2×2 sub-matrices, $\mathbf{M}_h$ and $\mathbf{M}_v$ that refer to horizontal stability (index b_1) and vertical stability (index b_2), respectively. If $b_2=2$ then, the planar orbit is called *vertically critical orbit*

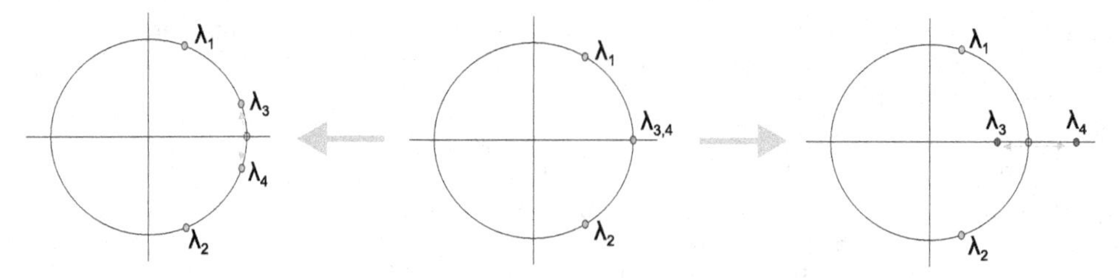

Figure 1. Distribution of eigenvalues for a v.c.o. of the symmetric model (center) and their displacement after introduction of asymmetry (*scheme I* in the left panel and *scheme II* in the right panel).

(v.c.o.) and signifies a bifurcation for another family of 3D periodic orbits (Hénon (1973)). We note that b_2 may also take the value of two when the planar orbit needs m times to complete a period (multiplicity). Then, if T is the period of the v.c.o., the 3D bifurcating orbit close to the v.c.o. will be of period mT.

3. Continuation near a v.c.o. : from a symmetric to an asymmetric model

Suppose that U_{ast} is a potential model of the asteroid provided by a 'real' model (e.g. by mascons or a polyhedral model). Let us define a mono-parametric set of potentials

$$U(\varepsilon)=U_0+\varepsilon U_1, \quad 0\leqslant\varepsilon\leqslant\varepsilon_0, \tag{3.1}$$

where U_0 is the symmetric potential of the ellipsoid that approximates the potential of the asteroid and U_1 includes the asymmetric part of the potential such that $U(\varepsilon_0)=U_{ast}$ with e_0 being sufficiently small.

Let F_p be a symmetric planar family of periodic orbits with a potential of U_0 and O a v.c.o. of F_p. We suppose that in the neighborhood of O the planar orbits of F_p are of the same horizontal stability type. In the present study, we consider that they are stable so, the eigenvalues of M_h are of the form $\lambda_1,\ \lambda_2{=}e^{\pm i\phi}$, $\phi\in(\delta,\pi-\delta)$, $\delta>0$. The eigenvalues of M_v are critical for O, i.e. $\lambda_{3,4}=1$ when the appropriate multiplicity m is taken into account. The distribution of λ_i on the unit circle is shown in the middle panel of Fig. 1. Suppose we perform an analytic continuation of the v.c.o. O with respect to parameter ε. As ε increases smoothly towards value ε_0, the eigenvalues $\lambda_{1,2}$ should move smoothly on the unit circle due to the analyticity (see Meyer *et al.* (2009)) and if δ is sufficiently large, the eigenvalues do not reach the critical values ±1 as $\varepsilon\to\varepsilon_0$. On the other hand, the critical eigenvalues $\lambda_{3,4}$, as ε varies, may move either on the unit circle or on the real axis. These cases are called *scheme I* and *scheme II*, respectively, and are presented in Fig. 1. Which one of the two schemes will take place, depends on the term U_1, which represents the asymmetric part of the asteroid's potential.

Applying analytic continuation to all orbits of F_p in the neighborhood of O,with respect to ε, we obtain the set of families $F(\varepsilon)$, with $F(0)=F_p$. All orbits of $F(\varepsilon)$ with $\varepsilon\neq0$ are spatial and asymmetric and family $F_{ast}=F(\varepsilon_0)$ is the family of orbits of the real asteroid originating from the planar family of the ellipsoid. The initial orbit $O\in F_p$ is mapped to the orbit $O'\in F_{ast}$. When *scheme I* takes place, F_{ast} should consist of stable orbits at least near O'. Instead, in *scheme II* the orbit O' is unstable and there should exist a continuous segment on F_{ast} near O' consisting of unstable orbits.

Let us consider the family, $F_{3D}(0)$, of three dimensional orbits that bifurcates from O. Similarly to the planar family, analytic continuation with respect to ε can be also applied providing the set of families $F_{3D}(\varepsilon)$. All orbits should be asymmetric for $\varepsilon\neq0$ and family $F_{3D}(\varepsilon_0)$ is the asteroid's family of periodic orbits associated to the family $F_{3D}(0)$ of the

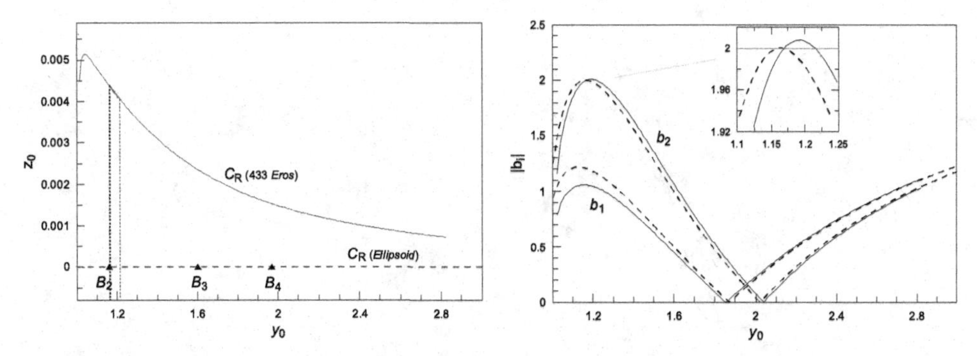

Figure 2. (left) The characteristic curve of the circular family C_R for the ellipsoid (dashed curve) and 433-Eros (solid curve) projected on the plane $y_0 - z_0$. The points B_m indicate the y_0-position of the v.c.o. with the subscript m being the multiplicity. The red segment indicates the part of the family with unstable orbits. (right) The variation of the stability indices b_1 and b_2 along the C_R-family of ellipsoid and Eros.

symmetric ellipsoid model. When *scheme I* takes place, the families F and F_{3D}, which for $\varepsilon = 0$ intersect at O, should be detached for $\varepsilon > 0$ since no bifurcation point exists on $F(\varepsilon)$ (whole family near O is stable). However, in *scheme II* the edges of the unstable segment formed on $F(\varepsilon_0)$ may be bifurcation points for the family $F_{3D}(\varepsilon_0)$. The above assumptions are verified by the numerical computations presented in the next section.

4. Numerical computations : The asteroid 433-Eros

In Paper I, we used the symmetric ellipsoid model (with normalized maximum semi-axis, $a = 1$, and angular velocity, $\omega = 1$) to initially approximate the potential of asteroid 433-Eros. Then, we applied shape-continuation to identify families of periodic orbits for the 'real' gravitational potential of 433-Eros, implemented with a sufficient number of mascons (Soldini *et al.* (2020)). In the ellipsoid model, we consider the family of planar ($z = 0$) circular retrograde orbits, C_R, which is fully stable. The family is also vertically stable but there are v.c.o. for higher period multiplicities ($m = 2, 3, 4, ..$). Their y_0-position (where y_0 is the approximate radius of the orbit) is shown in the left panel of Fig. 2. The right panel shows the stability indicies b_i along the family (dashed curves). The C_R is continued when asymmetric terms are added in the potential in order to simulate the potential of the asteroid. The computed family for 433-Eros consists of orbits which are no longer planar and symmetric but are almost circular. The family is presented in Fig. 2 with solid curves. The major part of C_R of Eros consists of stable orbits and this is also the case close to the radius of the v.c.o. B_3 and B_4. Therefore, such a situation implies *scheme I* for the 3D orbits emanating in the symmetric model from these v.c.o.. However, it is evident that the introduced asymmetries caused an unstable segment close to B_2 and this implies *scheme II.* It should be noted that this instability has been also mentioned in Ni *et al.* (2016) who used a polyhedral model for 433-Eros.

Scheme I is shown by considering the 3D family L_{24} of the ellipsoid, which bifurcates from the v.c.o. B_4. The family near B_4 is stable but becomes unstable when it becomes significantly inclined as shown in the left panel of Fig. 3. For the asymmetric potential of 433-Eros the family is represented by the characteristic curve in the right panel of Fig. 3. We can see that in the asymmetric asteroid case, family L_{24} does not intersect the planar family C_R and the two families are now separated. The stability type of orbits is not affected by the asymmetry for orbits close to the plane $z = 0$. However, a break of family L_{24} arises because of the irregular shape of Eros. After this break, the family continues with the unstable segment L'_{24}. Such family breaks are discussed also in Paper

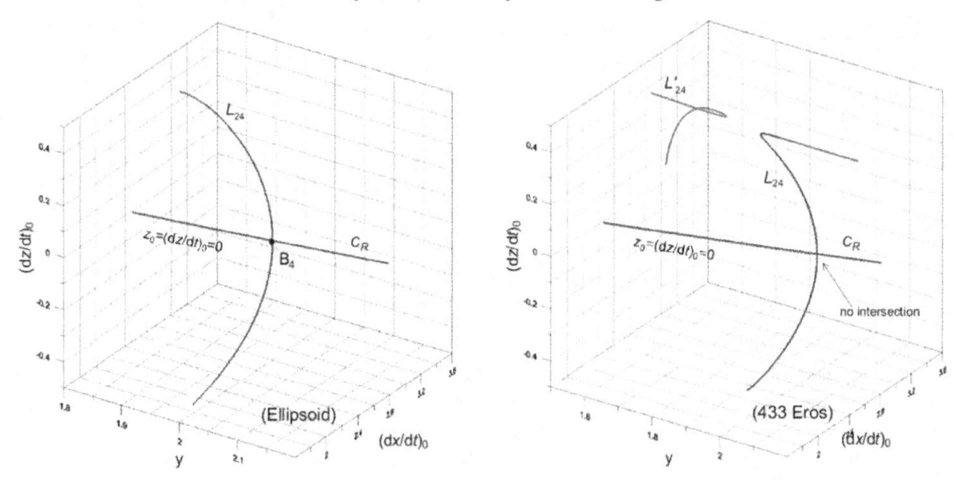

Figure 3. (left) The characteristic curves of the planar family C_R and the 3D family L_{24} of the ellipsoid. Blue (red) color indicates stability (instability). B_4 is the v.c.o. where the two families intersect. (right) The characteristic curves for the corresponding families of 433-Eros potential. The transition from the ellipsoid (left) to the mascon model of 433-Eros (right) indicates that *scheme I* takes place.

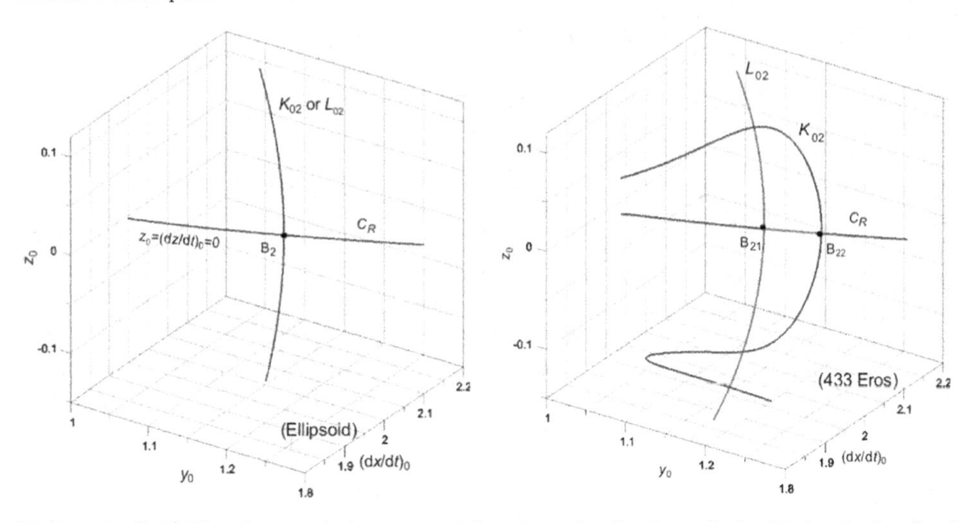

Figure 4. (left) The characteristic curves of the planar family C_R and the 3D family L_{02} (and its equivalent K_{02}) of the ellipsoid. Blue (red) color indicates stability (instability). B_2 is the vco where the two families intersect. (right) The characteristic curves for the corresponding families of 433-Eros potential. L_{02} and K_{02} are families of different orbits. The transition from the ellipsoid (left) to the mascon model of 433-Eros (right) indicates that *scheme II* takes place.

I. In the same paper, where family L_{13} is studied, *scheme I* also holds true, with a change of stability at $z \approx 0$.

Scheme II holds true for the case of v.c.o. B_2 of the ellipsoid from which the 3D families L_{02} and K_{02} originate (see Paper I). The two families are equivalent because they consist of the same doubly symmetric periodic orbits but their characteristic curves are presented in different spaces of initial conditions. In the left panel of Fig. 4, we present the initial conditions of the orbits in K_{02} family. As we have already mentioned, the C_R family of 433-Eros shows an unstable segment at B_2, defined by the points B_{21} and B_{22}. These points should be bifurcation points of other families. By computing the families K_{02} and L_{02} in the asymmetric potential of 433-Eros (see right panel of Fig. 4) we obtain that i) the two families are separated and they now consist of different asymmetric periodic orbits ii) the families pass from the points B_{21} and B_{22} and, therefore, the continuation *scheme II* is valid here. K_{02} consists of unstable orbits and L_{02} of stable ones (at least in the neighborhood of the bifurcation points). However, we cannot claim that the appearance of a stable and an unstable family is a general property for *scheme II.*

Acknowledgments

The authors acknowledge funding support from the European Unions Horizon 2020 research and innovation program under grant agreement No. 870377 (project NEO-MAPP).

References

Broucke, R. 1969 *Stability of periodic orbits in the elliptic, restricted three-body problem* AIAA Journal, 7:1003-1009

Hénon, M. 1973 A&A, 28, 415

Jiang, Y., Schmidt, J., Li, H., Liu, X. & Yang, Y. 2018 *Astrodynamics*, 2, 69

Karydis, D., Voyatzis, G., & Tsiganis, K. 2021 *Advances in Space Research* (to appear)

Meyer, K.R., Hal, G.R., & Offin, D. 2009 *Introduction to Hamiltonian dynamical systems and N-body problem* (Springer)

Ni, Y., Jiang, Y., & Baoyin, H. 2016 *ASS*, 361, 170

Scheeres, D.J. 2012 *Orbital motion in strongly perturbed environments* (Springer, Berlin Heidelberg)

Scheeres, D.J. 2012 *Acta Astronautica*, 72, 1

Soldini, S., Takanao, S., Ikeda, H., Wada, K, Yuichi, T., Hirata, N, & and Hirata, N. 2020 *Planetary and Space Science* 180, 104740

Yu, Y., & Baoyin, H. 2012 *MNRAS*, 427, 872

Multi-scale (time and mass) dynamics of space objects
Proceedings IAU Symposium No. 364, 2022
A. Celletti, C. Galeş, C. Beaugé, A. Lemaitre, eds.
doi:10.1017/S1743921321001289

Oscillations around tidal pseudo-synchronous solutions for circumbinary planets

F. A. Zoppetti[1,2], H. Folonier[3], A. M. Leiva[1] and C. Beaugé[1,2]

[1]Observatorio Astronómico de Córdoba, Universidad Nacional de Córdoba, Laprida 854, Córdoba X5000BGR, Argentina
email: federico.zopetti@unc.edu.ar

[2]CONICET, Instituto de Astronomía Teórica y Experimental, Laprida 854, Córdoba X5000BGR, Argentina

[3]Instituto de Astronomia Geofísica e Ciências Atmosféricas, Universidade de São Paulo, 05508-090, Brazil

Abstract. Tidal evolution of low-eccentric circumbinary planets is expected to drive the rotational evolution toward a pseudo-synchronous solution. In this work, we present a study of the oscillation amplitudes around this state by considering that the two central stars exert creep tides on the planet. These amplitudes are computed by direct numerical integrations of the creep equations and also by means of the calculation of the coefficients of the periodic terms in this stationary solution. As in the two-body-problem, the planetary spin and lag-angle are observed to have maximum oscillation amplitudes for stiff bodies and almost null oscillation for the gaseous regime, while the opposite behaviour is observed in the equatorial and polar flattenings. Our analytical approximation shows to be very accurate and specially necessary for very-low eccentric planets. However, the magnitudes of the oscillation amplitudes around the pseudo-synchronous solution in the circumbinary problem appears to be very small respect to the mean value. Thus, considering these oscillation in the computation of the tidal energy dissipation may not have a substantial contribution in the results, at least compared to the case in which only the mean values are taken into account.

Keywords. planet-star interactions – methods: analytical

1. Introduction

One of the main characteristics shared by most circumbinary (CB) planets discovered by the *Kepler* mission is their location very close to the binary system. At such distances, tidal torques are expected to play an important role on the rotational evolution of the planet, with the peculiarity that in this context they are exerted by two central bodies of comparable mass, instead of just one.

In Zoppetti *et al.* (2019), we studied the rotational evolution of a CB planet due to the tidal interaction of the central binary, using the classical *Constant time-lag* model (Mignard 1979). Interestingly, we found that the typical stationary state of a CB planet is sub-synchronous respect to its mean motion, and this is exclusively due to the presence of the secondary star.

More recently, in Zoppetti *et al.* (2021) (hereafter Z2021) we investigate the effects of the binary tides on the CB planet with a more realistic formalism such as the creep tide model (Ferraz-Mello 2013). This formalism considers the bodies as Maxwellian fluids without the elastic component, and one of its main advantages is that the tidal-lags are not quantities *ad hoc* included in the theory, but are calculated from solving the set of differential equations derived from the Newtonian creep equation. Moreover, the theory does not need to assume weak friction, so this lag can be large and the model can be applied to bodies with arbitrary viscosities.

With this more general model, in Z2021 we could also find the sub-synchronous stationary solution for gaseous bodies, previously reported in Zoppetti *et al.* (2019). Furthermore, we provided a set of high-order analytical expressions for the mean values of the rotational stationary state around the 1:1 spin-orbit resonance. This last configuration, was shown to be like the most probable for low eccentric systems with viscosities in the range estimated for the planets of the Solar System (Ferraz-Mello 2013).

Although the mean values for the rotational stationary state obtained in Z2021 reproduce very well the behavior of the numerical simulations, we did not study in that article the oscillation of the real solution around these values. According to the 2-body-problem experience (e.g. Folonier et al. 2018), the oscillation amplitudes of the spin, for example, can become important for stiff bodies located very close to the perturber. In addition to the consequences to the potential habitability of these worlds, considering these amplitudes in the models is essential when estimating the dissipation of energy due to tides on the body and, also, the timescales and magnitudes of orbital evolution.

This article is organized as follows. In Section 2, we present the creep tide model applied to the case of the rotational evolution of a CB planet. In Section 3, we explain the method we used to obtain an analytical solution of the oscillation amplitudes around the pseudo-synchronous solution. Section 4 shows the results obtained for the case of Kepler-38 system. Finally, we discuss our main results and its implication in Section 5

2. The creep tide model for the rotational evolution of a CB planet

Let us consider an extended CB planet m_2 with radius $\mathscr{R}_2$, perturbed by the tidal interaction of a central binary with components m_0 and m_1. Additionally, let us consider the planar problem where the spin vector of the CB planet is perpendicular to the orbital plane.

As a consequence of the gravitational interaction exerted by the central stars, the CB planet undergoes a tidal deformation. In addition to this, in this work we also take into account the rotational flattening on m_2 due to its own spin, and assume that the resulting deformation due to these effects is small enough that a model developed up to the first order of the flattenings represents an accurate description of the problem.

In the creep tide theory (Ferraz-Mello 2013), the real shape and orientation of the planet is computed by means of the distance ζ_2 (measured from the center of mass of the body) of an arbitrary point on its surface with co-latitude ϑ and longitude φ. Its explicit form is given by

$$\zeta_2(\vartheta,\varphi) = \mathscr{R}_2\left[1 + \mathscr{E}_2^z\left(\frac{1}{3} - \cos^2\vartheta\right) + \frac{\mathscr{E}_2^\rho}{2}\sin^2\vartheta\cos\left(2\varphi - 2(\varphi^{eq} + \delta_2)\right)\right], \tag{1}$$

where $\mathscr{E}_2^r$ and $\mathscr{E}_2^z$ are the equatorial and tidal flattenings, δ_2 is the lag angle and φ^{eq} is the position of a fictitious body that exerts a tide on the planet equivalent to that exerted by the two central stars (see Section 2.1 of Z2021)

The parameters that characterize the shape ($\mathscr{E}_2^\rho$ and $\mathscr{E}_2^z$) and orientation (δ_2) of the planet, can be calculated by solving the creep tide equation

$$\dot{\zeta}_2 + \gamma_2\zeta_2 = \gamma_2\rho_2, \tag{2}$$

where γ_2 is the relaxation factor of the planet, which is inversely proportional to its viscosity and assumed as constant in this work, while ρ_2 is the surface equation of the equilibrium figure (see Equation (3) of Z2021). The explicit set of differential equations for the shape and orientation of the planet is

$$\begin{aligned}\dot{\delta_2} &= \Omega_2 - \varphi^{eq} - \frac{\gamma_2 \varepsilon_2^\rho}{2\mathscr{E}_2^\rho}\sin(2\delta_2) \\ \dot{\mathscr{E}}_2^\rho &= \gamma_2(\varepsilon_2^\rho \cos(2\delta_2) - \mathscr{E}_2^\rho) \\ \dot{\mathscr{E}}_2^z &= \gamma_2(\varepsilon_2^z - \mathscr{E}_2^z),\end{aligned} \tag{3}$$

where ε_2^ρ and ε_2^z are the equivalent equatorial and polar flattenings of the equilibrium figure (for its explicit expressions, see equations (2) and (4) of Z2021).

For a given planetary spin Ω_2, the real shape and orientation of the body can be calculated by solving the differential equation system (3). Then, the variation in the planetary rotation is derived from the reaction torques that the extended body feels and the conservation of total angular momentum. Neglecting the term corresponding to the variation of the polar moment of inertia, its explicit form is

$$\dot{\Omega}_2 = -\frac{2\mathscr{G}m_2}{5\mathscr{R}_2^3}\varepsilon_2^\rho \mathscr{E}_2^\rho \sin(2\delta_2). \tag{4}$$

The set of differential equations (3) and (4) describe the rotational evolution of the CB planet due to the creep tides of the central binary.

In Z2021, we consider planetary relaxation factors in the range estimated for the Solar System planets, and found that the most probable rotational stationary state is the pseudo-synchronous solution, at least for low eccentric CB systems. For this reason, the oscillation amplitudes around this particular solution will be the target of this work.

On the other hand, in the framework of the *Constant-time-lag* model, in Zoppetti *et al.* (2019) and Zoppetti et al. (2020) we observed that the characteristic timescales of the rotational evolution of *Kepler* CB planets are typically much shorter than the timescales of the orbital evolution. For this reason, in this work we also take advantage of this adiabatic nature of the problem and solve the set of differential equations given by (3) and (4), assuming fixed values for all the orbital elements except the mean anomalies.

3. Analytical resolution method for the pseudo-synchronous stationary state

We adopt here the same procedure adopted in Z2021. We then choose a Jacobi reference frame and propose a particular stationary solution inspired in the functional dependence of the elliptical expansions of ε_2^ρ, ε_2^z and φ^{eq} of the form

$$\begin{aligned}\Omega_2 &= \sum_{\vec{\ell}} \{\Omega_2\}_{\vec{\ell}} \cos(l_1 M_1 + l_2 M_2 + l_3 \varpi_1 + l_4 \varpi_2 - \Phi_{\vec{\ell},\Omega_2}) \\ \mathscr{E}_2^\rho &= \sum_{\vec{\ell}} \{\mathscr{E}_2^\rho\}_{\vec{\ell}} \cos(l_1 M_1 + l_2 M_2 + l_3 \varpi_1 + l_4 \varpi_2 - \Phi_{\vec{\ell},\mathscr{E}_2^\rho}) \\ \delta_2 &= \sum_{\vec{\ell}} \{\delta_2\}_{\vec{\ell}} \cos(l_1 M_1 + l_2 M_2 + l_3 \varpi_1 + l_4 \varpi_2 - \Phi_{\vec{\ell},\delta_2}) \\ \mathscr{E}_2^z &= \sum_{\vec{\ell}} \{\mathscr{E}_2^z\}_{\vec{\ell}} \cos(l_1 M_1 + l_2 M_2 + l_3 \varpi_1 + l_4 \varpi_2 - \Phi_{\vec{\ell},\mathscr{E}_2^z}),\end{aligned} \tag{5}$$

where M_1 and ϖ_1 are the mean anomaly and pericentre longitude of the secondary star, while M_2 and ϖ_2 are those corresponding to the planet. We note that each of the rotational evolution variables $w = \Omega_2, \mathscr{E}_2^\rho, \delta_2$ or $\mathscr{E}_2^z$, the amplitudes $\{w\}_{\vec{\ell}}$ and the constant phases $\Phi_{\vec{\ell},w}$ depend on the subscripts $\vec{\ell} = (l_1, l_2, l_3, l_4)$.

Table 1. Initial conditions of our Kepler-38 like system (Orosz et al. 2012). Orbital elements are given in a Jacobi reference frame.

body	m_0	m_1	m_2
mass	$0.95\, m_\odot$	$0.25\, m_\odot$	$10\, m_\oplus$
radius	$0.84\, \mathscr{R}_\odot$	$0.27\, \mathscr{R}_\odot$	$4.35\, \mathscr{R}_\oplus$
a_i [AU]		0.15	0.45
e_i		0.15	$0.0 - 0.1$

The particular solution given in (5) and its derivatives are then introduced into the system (3) and (4). Having expanded the forced terms (ε_2^ρ, ε_2^z and φ^{eq}) in power series of semimajor axis ratio $\alpha = a_1/a_2$ and the eccentricities e_1 and e_2, the coefficients can be explicitly calculated by equating the terms with the same trigonometric argument and neglecting terms of higher order.

The terms in system (5) with $l_1 = l_2 = 0$ correspond to the mean values of the rotational stationary solution and were explicitly given in Equation (16)-(19) of Z2021.

In this work, we focus on the amplitudes of the periodic terms. The calculation of these coefficients is even more cumbersome than in the case of the mean values. For this reason, we computed them up to 3rd order in α and only up to 1st-order in the eccentricities. However, as we will see in Section(4), we are able to predict very well the oscillation amplitudes of low-eccentric CB planets close to their host binary. On the other hand, due to space limitations, the coefficients can not be explicitly shown in this manuscript but can be provided by contacting any of the authors.

4. Results

We apply the results of our model to a Kepler-38-like CB planet (Orosz et al. 2012). This system has been the test case of most of our recent investigations about the tides on CB planets. The chosen nominal values for system parameters and initial orbital elements are detailed in Table 1. The value of m_2 was estimated from a semi-empirical mass–radius fit (Mills & Mazeh 2017). Note that two different values of the planetary eccentricity e_2 are considered in this work: $e_2 = 0.0$ and $e_2 = 0.1$

In Figure 1, we consider a Kepler-38-like CB planet and show the oscillation amplitudes, predicted by our creep tide model, for the rotational evolution around the pseudo-synchronous solution, as a function of the normalized relaxation factor γ_2/n_2. The wide black lines correspond to the amplitudes obtained by numerically integrating the system (3) and (4), once the planet reaches the stationary solution, while the dashed black curves correspond to the mean values obtained in this integration. The red curves represent the results of our analytical solution up to 3rd-order in α, obtained by means of the method described in Section 3, while the results of our 0th-order analytical solution is represented by green curves. Note that this last case is equivalent to the 2-body-problem in which the planet only feels the tides of one central body with mass equal to the sum of both stellar masses.

We first note from Figure 1 that, independently of the planetary eccentricity, the oscillation around the pseudo-synchronous planetary spin shows the same behavior than the one around the mean lag-angle: it is maximum for $\gamma_2 << n_2$ (i.e. typically stiff bodies) and decays to zero for $\gamma_2 >> n_2$ (i.e. gaseous bodies). On the other hand, the completely opposite behavior is observed for the flattenings $\mathscr{E}_2^\rho$ and $\mathscr{E}_2^z$. The magnitudes of the oscillation amplitudes do depend (proportionally) on the planetary eccentricity, although in our case they are much smaller than the ones obtained for the 2-body-problem Saturn-Enceladus in Folonier et al. (2018).

Regarding to the accuracy of our analytical solution, we note from Figure 1 that our model up to 3rd-order in α fits very well the behaviour of the numerical integrations,

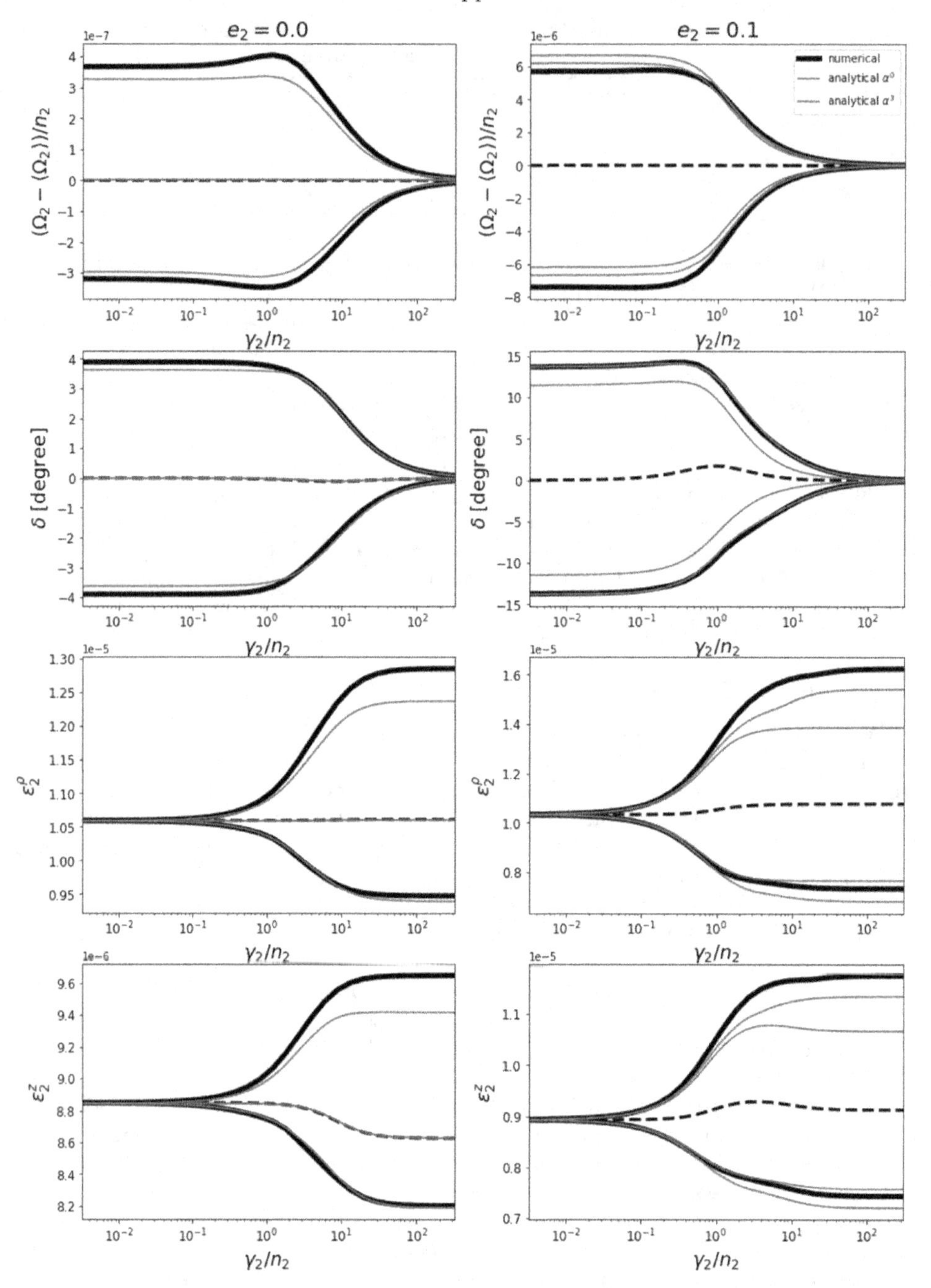

Figure 1. Oscillation amplitudes of the spin Ω_2 (first row), the lag-angle δ_2 (second row), the equatorial flattening ε_2^ρ (third row) and polar flattening ε_2^ρ (bottom row), of a Kepler-38-like planet around the pseudo-synchronous solution. Different colors indicate different methods for obtaining the amplitudes: numerical in black, analytical up to 0th-order in green and analytical up to 3th-order in red. The dashed black lines represent the numerical mean solution. Different columns represent different planetary eccentricities: $e_2 = 0.0$ (left column) and $e_2 = 0.1$ (right column).

for any arbitrary planetary viscosity in low eccentric orbits. The accuracy of the model is even more remarkable (for example, see the second row panels) when we take into account that in CB environments, tides are non-negligible for planets very close to the binary and high α-order expansions are required for the analytical models (Figure 1 is built for $\alpha = 0.33$).

On the other hand, the comparison between the red and the green curves for the panels located on the left columns, shows that the 3BP approximation is specially necessary to compute the oscillation amplitudes of very-low-eccentric orbits, where the 2BP approach predicts null oscillations (see Equations (53) of Folonier et al. (2018)) and the numerical solution is far from fulfilling it. However, the 0th-order solution seems to be quite acceptable for moderate eccentricities.

5. Discussion

In this work, we study the oscillation amplitudes that exhibits the rotational evolution of a CB planet around the pseudo-synchronous tidal solution. We employ the creep tide model (Ferraz-Mello 2013), which is equivalent to considering bodies as Maxwellian fluids without the elastic component (e.g. Ferraz-Mello 2015).

We compute the oscillation amplitudes in two ways: by a direct numerical integration of the differential equation system of the rotational evolution, and by analytically calculating the coefficients of the periodic terms of the stationary pseudo-synchronous state. As discussed in Z2021, having analytical expressions is a very cumbersome task. However, it allows to carry out semi-analytical studies of the orbital evolution for very long periods of time, by introducing *ad-hoc* the solutions found for the rotational evolution, and avoiding having to solve our entire multi-timescales system. The analytical approach presented in this work fits very well the numerical predictions of low eccentric systems, even with high semimajor axis ratio, so it can be an important contribution in this problem.

For CB planets, the behaviour observed in the oscillation amplitudes is analogous to the one observed in the 2-body-problem: maximum oscillation amplitudes for the spin and lag-angles of stiff planets and almost zero amplitudes in the gaseous regime, while the completely opposite behaviour is observed for the flattenings. However, the magnitudes of the oscillation observed for our Kepler-38-system (whose parameters and orbital elements are typical within the planets observed by the *Kepler* mission) are very small, specially when we compare with the ones obtained for the Saturn-Enceladus system (Folonier et al. 2018), this last being a much tighter system. The presence of an inner instability limit for CB orbits (e.g. Holman & Wiegert 1999), prohibits to have planets much closer to the binary than Kepler-38. Thus, the expected oscillation amplitudes for the parameters that characterise the rotational evolution of CB planets are expected to be typically very small.

We mention two important consequences of this result. On one hand, the very low oscillation amplitudes of the spin of CB planets that have reached the pseudo-synchronous solution should be taken into account in habitability studies of these planets. On the other hand, from a technical point of view, in this type of environment, building models that consider only the mean behavior and neglect the oscillation amplitudes seems to be very accurate.

References

Ferraz-Mello, S. 2013, Celestial Mechanics and Dynamical Astronomy, 116, 109.

Ferraz-Mello, S. 2015, A&A, 579, A97.

Folonier, H. A., Ferraz-Mello, S., & Andrade-Ines, E. 2018, Celestial Mechanics and Dynamical Astronomy, 130, 78.

Holman, M. J. & Wiegert, P. A. 1999, AJ, 117, 621.
Mignard, F. 1979, Moon and Planets, 20, 301.
Mills, S. M. & Mazeh, T. 2017, ApJL, 839, L8.
Orosz, J. A., Welsh, W. F., Carter, J. A., et al. 2012, ApJ, 758, 87.
Zoppetti, F. A., Beaugé, C., Leiva, A. M., et al. 2019, A&A, 627, A109.
Zoppetti, F. A., Leiva, A. M., & Beaugé, C. 2020, A&A, 634, A12.
Zoppetti, F. A., Folonier, H., Leiva, A. M., et al. 2021, A&A, 651, A49.

Multi-scale (time and mass) dynamics of space objects
Proceedings IAU Symposium No. 364, 2022
A. Celletti, C. Galeş, C. Beaugé, A. Lemaitre, eds.
doi:10.1017/S1743921322000035

Apsidal alignment in migrating dust - Crescent features caused by eccentric planets

Maximilian Sommer[1], Petr Pokorný[2], Hajime Yano[3,4] and Ralf Srama[1]

[1]University of Stuttgart, Institute of Space Systems, Germany
email: sommer@irs.uni-stuttgart.de

[2]NASA Goddard Spaceflight Center, Greenbelt, USA

[3]JAXA Institute of Space and Astronautical Science, Sagamihara, Japan

[4]Graduate University for Advanced Studies (SOKENDAI), Japan

Abstract. Circumstellar discs are known to exist in great variety, from gas-rich discs around the youngest stars to evolved debris discs such as the solar system's zodiacal cloud. Through gravitational interaction, exoplanets embedded in these discs can generate density variations, imposing potentially observable structural features on the disc such as rings or gaps. Here we report on a mirrored double crescent pattern arising in simulations of discs harbouring a small, moderately eccentric planet - such as Mars. We show that the structure is a result of a directed apsidal precession occurring in particles that migrate the planet's orbital region under Poynting-Robertson drag. We further analyze the strength of this effect with respect to planet and particle parameters.

Keywords. Circumstellar matter, meteoroids, celestial mechanics

1. Overview

The zodiacal cloud, a circumsolar disc of dust, has long been known to pervade the space inhabited by the planets. Constantly replenished by comets and asteroids, dust grains from 10s to 100s of microns in size slowly migrate inward under Poynting-Robertson drag, until being destroyed in the vicinity of the Sun. As the dust moves inward, its spatial distribution is shaped by the gravitational interaction with the planets, whose orbital regions it traverses. The emerging structures, known to exists not only in our own solar system but also around distant stars, can act as signatures of planets and have been subject of numerical as well as observational studies (e.g. Jackson and Zook (1989); Wyatt et al. (1999); Stark and Kuchner (2008)). A recent numerical study primarily concerned with resonant structures arising in the zodiacal cloud hinted at a further feature forming mechanism associated with the planet Mars, which embodied as a mirror-inverted double crescent pattern, locked to the apseline of the planet (Sommer et al. 2020).

In this follow-up, we revisit the structure's formation principle and analyse the strength of this effect with respect to planet, as well as particle parameters. Figure 1 shows the simulated surface density of a disc made up of particles of one discrete size harbouring a Mars-like planet, posing an example of the aformentioned structure. The effect of the planet on the disc manifests as a crescent with increased particle density, roughly spanning the inbound—that is from apocentre to pericentre—half of the planet orbit, and a mirrored but density-wise inverted crescent, spanning the outbound half of the

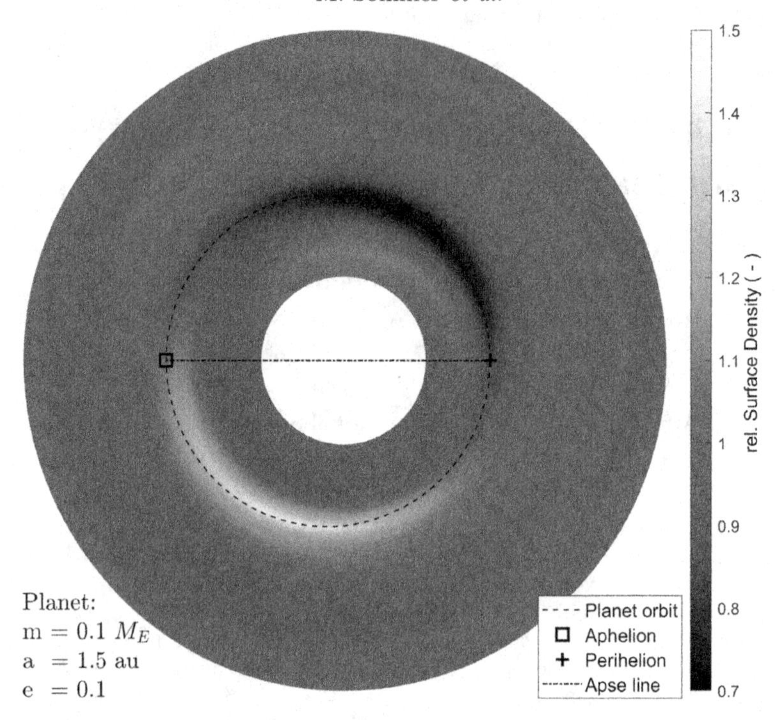

Figure 1. Exemplary surface density arising in a system with a Mars-like planet. The density distribution is normalized to the one produced by a simulation without planet but otherwise unchanged parameters. The orbital elements of the initial particle population are uniformely distributed in: 2.1 au $\leqslant q \leqslant$ 2.4 au, $0 \leqslant e \leqslant 0.5$, and $0° \leqslant i \leqslant 20°$. ($q$: pericentre distance, e: eccentricity, i: inclination). Ratio of forces acting on particles resulting from central star radiation pressure and central star gravity used here is: $\beta = F_{r\star}/F_{g\star} = 0.002$.

planet orbit. This can be explained by the evolution of longitude of pericentres, ϖ, of particles as they migrate the planet's orbital region, driven by Poynting-Robertson drag. As displayed in Fig. 2 (a), upon droping inside the planet orbit, the orbit of a typical particle starts to experience an apsidal drift that is maintained until it revolves entirely within the planet orbit. Effectively, the orbit dacay is accelerated in the region that the pericentre rotates towards and halted in the region that the apocentre rotates towards, apparent in the varying gap width between the *before* and *after* orbits (indicated in blue color in Fig. 2 (a)). Due to the asymmetric decay, the cumulated particle dwell time is lengthened in the region of halted decay, and shortened in the other. Thus, the opposed depletion-enhancement zones are consequential, if all particles experience an apsidal drift with a preferred final apse line orientation relative to that of the planet. This is evident in Fig. 2 (b), showing the relative frequency distribution of ϖ for all particles making up the synthetic disc over bins of their semi-major axes. When still distant from the planet, at semi-major axes larger than 2 au, particle pericentres start off distributed uniformely. However, once particle's semi-major axes decrease below the planet's aphelion distance, an alignment of their apselines occurs, with ϖ reaching a relative frequency of a factor of 3 over that of the uniform distribution, at a longitude around 60° to 80° ahead of the planet pericentre. This aggregation of particle pericentres along the outbound half of the planet orbit (corresponding to positive longitudes in Fig. 2 (b)) confirms the asymmetric orbit decay within the whole particle population, thus generating the opposed depletion-enhacement structure.

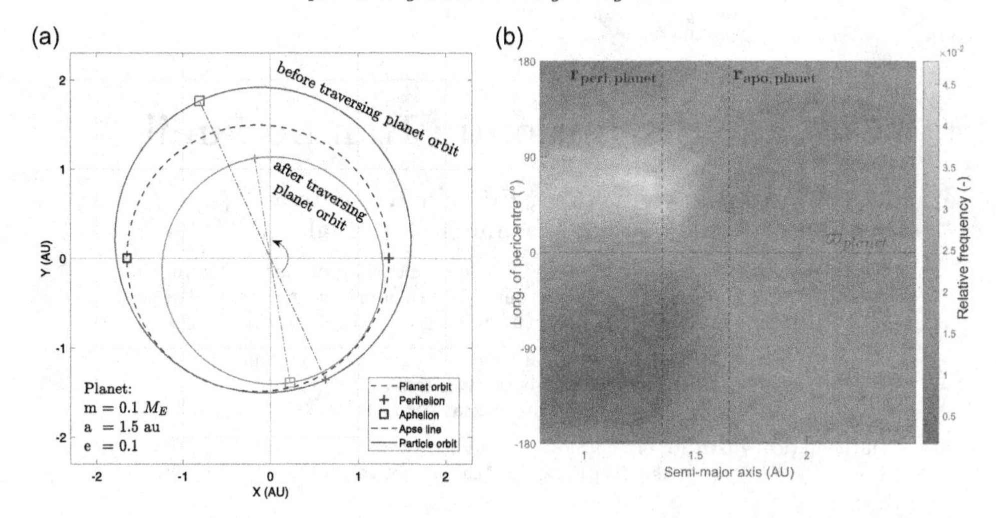

Figure 2. (a) Apsidal precession of an exemplary particle ($\beta = 0.002$, $i = 7°$, no close enctounter with the planet occurred) as it traverses the orbital region of a Mars-like planet under PR-drag. The particle's orbit is shown before and after crossing the planet orbit. (b) Evolution of the distribution of ϖ in a migrating particle population (same as that used in Fig. 1). Relative frequency of ϖ is recorded at semi-major axis bins of 0.035 au and is normailized for each bin.

2. Conclusions

We have run simulations with a multitude of planet and particle population parameters (with mass of the central star fixed at 1 solar mass). We conclude that the effect favours particles less influenced by radiation ($\beta < 0.01$), as well as moderate particle eccentricities ($e < 0.4$) and inclinations ($i < 20°$). The formation of these crescents most strongly occurs in the presence of low-mass (0.1 to 0.3 Earth-masses) planets of moderate eccentricity ($0.1 \leqslant e \leqslant 0.2$) and with a semi-major axis of 1.5 au to 5 au. For heavier planets and/or planets further out, the crescents disappear in favour of enhacements caused by resonant trapping or distant secular perturbation (akin to Wyatt et al. (1999)), as the planets' capacity to influence particles outside their orbital region increases. The fact that these features are readily produced by sub-Earth-mass planets, which are incapable of clearing a gap in the vicinity of their orbit or produce a meaningful resonant enhancement is especially noteworthy. In light of ongoing advances in observational astronomy and our increasing capability to resolve structures in exozodiacal clouds, these findings may become relevant in tracking down exoplanets in a planet-mass regime hardly accessible through other methods.

References

Jackson, A. & Zook, H. A. 1989, A solar system dust ring with the earth as its shepherd. *Nature*, 337(6208), pp. 629–631.

Sommer, M., Yano, H., & Srama, R. 2020, Effects of neighbouring planets on the formation of resonant dust rings in the inner solar system. *Astronomy & Astrophysics*, 635, A10.

Stark, C. C. & Kuchner, M. J. 2008, The detectability of exo-earths and super-earths via resonant signatures in exozodiacal clouds. *The Astrophysical Journal*, 686(1), 637.

Wyatt, M., Dermott, S., Telesco, C., Fisher, R., Grogan, K., Holmes, E., & Pina, R. 1999, How observations of circumstellar disk asymmetries can reveal hidden planets: Pericenter glow and its application to the hr 4796 disk. *The Astrophysical Journal*, 527(2), 918.

Multi-scale (time and mass) dynamics of space objects
Proceedings IAU Symposium No. 364, 2022
A. Celletti, C. Galeş, C. Beaugé, A. Lemaitre, eds.
doi:10.1017/S1743921322000746

Cascade disruption in Rampo family

Mariia Vasileva[1], **Eduard Kuznetsov**[1], **Alexey Rosaev**[2] **and Eva Plávalová**[3]

[1]Department of Astronomy, Geodesy, Ecology and Environmental Monitoring, Ural Federal University, Lenina Avenue, 51, Yekaterinburg, 620000, Russia
emails: vasilyeva.maria@urfu.ru, eduard.kuznetsov@urfu.ru

[2]Research and Educational Center "Nonlinear Dynamics", Yaroslavl State University, Yaroslavl, Russia
email: hegem@mail.ru

[3]Mathematical Institute of the Slovak Academy of Sciences, Bratislava, Slovakia
email: plavalova@mat.savba.sk

Abstract. We have found three new members of the Rampo asteroids family: 2009 HD_{95}, 2010 VO_{19}, 2013 JF_{69}. We estimated the Yarkovsky semimajor axis drift rate. Based on the simulation results, estimates of the asteroid pairs' age included in the family are obtained. In the scenario of the cascade disruption of the parent body of the asteroid (10321) Rampo, one can note the concentration of estimates of the pairs' age to values of 900, 750, 500, and 250 kyr.

Keywords. Celestial mechanics, methods: numerical, minor planets, asteroids.

Pravec and Vokrouhlický (2009) discovered the Rampo family with only three members since that number of members belonging to this cluster has increased up to 7 (Pravec et al. 2018). Kuznetsov and Vasileva (2019) discovered six new members.

Search for new members of the Rampo family was carried out by calculating the Kholshevnikov metrics ρ_2 and ρ_5 (Kholshevnikov et al. 2016). The selection criterion was the simultaneous fulfillment of two conditions: $\rho_2 < 0.008$ au$^{1/2}$ and $\rho_5 < 0.002$ au$^{1/2}$. As a result, three new members of the (10321) Rampo family were found: 2009 HD_{95}, 2010 VO_{19}, 2013 JF_{69}.

The dynamic evolution of Rampo family asteroids was simulated numerically using the Orbit9 program of the OrbFit complex for 1 Myr. Perturbations from major planets and the dwarf planet Pluto, the Sun oblateness, relativistic effects, and the Yarkovsky effect's influence were considered. For each asteroid, based on nominal orbital elements, five evolution scenarios were considered for different values of the semimajor axis drift rate $\dot{a}$ corresponding to the different orientation of the asteroid rotation axis relative to its orbit plane: $\dot{a} = 0$ at $\varphi = 90°$ or $270°$; $\dot{a} = \pm|\dot{a}|_{max}$ at $\varphi = 0°$ or $180°$, respectively; $\dot{a} = \pm 0.5 \times |\dot{a}|_{max}$ at $\varphi = 60°$ or $240°$, respectively. The maximum absolute values of the semimajor axis drift $|\dot{a}|_{max}$ caused by the Yarkovsky effect were estimated by normalization using the asteroid's parameters (101955) Bennu (Del Vigna et al. 2018).

We assume the scenario of cascade disruptions of the parent body of asteroid (10321) Rampo. In this case, four groups of orbits can be distinguished in the Rampo family. Asteroids whose orbits approach the orbit of the asteroid (10321) Rampo about 900 kyr ago – (451686) 2013 BR_{67}, 2009 HD_{95}, 2009 SR_{371}, 2013 JF_{69}, 2015 TM_{372} and 2017 UH_{21}; about 750 kyr ago – 2006 UA_{169}, 2010 VO_{19}, 2013 RL_{101}, 2013 VC_{30},

2014 HS_9, 2015 HT_{91} and 2016 TE_{87}; about 500 kyr ago – (294272) UM_{101}; about 250 kyr ago – 2015 TA_{367}.

We can conclude that there is a very low possibility that the Rampo family was formed in a single breakup event. This conclusion agrees with a novel idea about the cascade breakup of some young asteroid families (Fatka et al. 2020).

At the stage of current knowledge about the physical and dynamic properties of members of this family, without accurate estimates of the semimajor axis drift rates or non-gravitational parameter A_2, it is impossible to reconstruct the true picture of the formation of the family.

The next necessary step in studying the Rampo asteroids family will be to obtain information about the surface's thermophysical properties, the shape, and the parameter of the axial rotation of asteroids and refine the orbital parameters. These data will help determine the Yarkovsky semimajor axis drift rate and simulate realistic scenarios of the evolution of asteroids within limits of parameter determination errors.

Rampo family asteroids have very small but nonzero eccentricities. For this reason, the perihelion argument sometimes has change precession to regression motion. This fact complicates the condition of the orbital elements convergence and required future studying.

Acknowledgements

The work was supported by the Ministry of Science and Higher Education of the Russian Federation via the State Assignment Projects FEUZ-2020-0030 (EDK) and FEUZ-2020-0038 (MAV).

References

Pravec, P. & Vokrouhlický, D. 2009, *Icarus*, 204, 580–588

Pravec, P., Fatka, P., Vokrouhlický, D., Scheeres, D. J., Kušnirák, P., Hornoch, K., Galád, A., Vraštil, J., Pray, D. P., Krugly, Yu. N., Gaftonyuk, N. M., Inasaridze, R. Ya., Ayvazian, V. R., Kvaratskhelia, O. I., Zhuzhunadze, V. T., Husárik, M., Cooney, W. R., Gross, J., Terrell, D., Világi, J., Kornoš, L., Gajdoš, Š., Burkhonov, O., Ehgamberdiev, Sh. A., Donchev, Z., Borisov, G., Bonev T., Rumyantsev V. V., & Molotov, I. E. 2018 *Icarus*, 304, 110–126

Kuznetsov, E. D. & Vasileva, M. A. 2019, *Meteoritics and Planetary Science*, 54, S2

Kholshevnikov, K. V., Kokhirova, G. I., Babadzhanov, P. B., & Khamroev, U. H. 2016, *Mon. Not. R. Astron. Soc.*, 462, 2275

Del Vigna, A., Faggioli, L., Milani, A., Spoto, F., Farnocchia, D., & Carry, B. 2018, *Astronomy and Astrophysics*, 617, A61

Fatka, P., Pravec, P., & Vokrouhlický, D., 2020, *Icarus*, 338, 113554

Multi-scale (time and mass) dynamics of space objects
Proceedings IAU Symposium No. 364, 2022
A. Celletti, C. Galeş, C. Beaugé, A. Lemaitre, eds.
doi:10.1017/S1743921322000023

An algorithm for automatic identification of asymmetric transits in the TESS database

M. Vasylenko[1,2], Ya. Pavlenko[1], D. Dobrycheva[1], I. Kulyk[1], O. Shubina[1,3] and P. Korsun[1]

[1]Main Astronomical Observatory of the NAS of Ukraine,
27 Akademika Zabolotny Str., Kyiv, 03143, Ukraine

[2]Institute of Physics of the National Academy of Sciences of Ukraine,
46 avenue Nauka, Kyiv, 03028, Ukraine

[3]Astronomical Observatory of Taras Shevchenko National University of Kyiv,
3 Observatorna Str., Kyiv, 04053, Ukraine

Abstract. Currently, the Transiting Exoplanet Survey Satellite (TESS) searches for Earth-size planets around nearby dwarf stars. To identify specific weak variations in the light curves of stars, sophisticated data processing methods and analysis of the light curve shapes should be developed and applied. We report some preliminary results of our project to find and identify minima in the light curves of stars collected by TESS and stored in the MAST (Mikulski Archive for Space Telescopes) database. We developed Python code to process the short-cadence (2-min) TESS PDCSAP (Pre-search Data Conditioning Simple Aperture Photometry) light curves. Our code allows us to create test samples to apply machine learning methods to classify minima in the light curves taking into account their morphological signatures. Our approach will be used to find and analyze some sporadic events in the observed light curves originating from transits of comet-like bodies.

Keywords. TESS, MAST, exoplanets, exocomets.

---

1. Introduction

During the last ten years, more than 4,300 exoplanets orbiting their parent stars have been discovered, mostly by the space missions Kepler (Borucki (2010)) and TESS (Ricker (2015)). However, we still have a little information about the populations of extrasolar subplanetary bodies in these systems, such as planetesimals, asteroids, and comets. Meanwhile, the modern theories of the planetary system formation assume a large population of planetesimals, which play an important role in the dynamics and physical evolution of the planetary system (see Lagrange (2020) and references herein). Currently, only a few solid detections of exocomets have been reported by Rappaport (2018), and Zieba (2019) for star KIC3542116 and β Pictoris system based on analysis of the Kepler and TESS data bases, respectively.

Our project is to contribute to studying cometary activity in extrasolar systems based on the TESS high-precision observations. The obtained and pre-proceeded light curves are available in the MAST archive (Ricker (2015)). Our work aims to analyze the light curves of stars to find asymmetric transits caused by the passage of comet-like bodies across the stars' disks. Machine learning (ML) methods can be used to optimize this process.

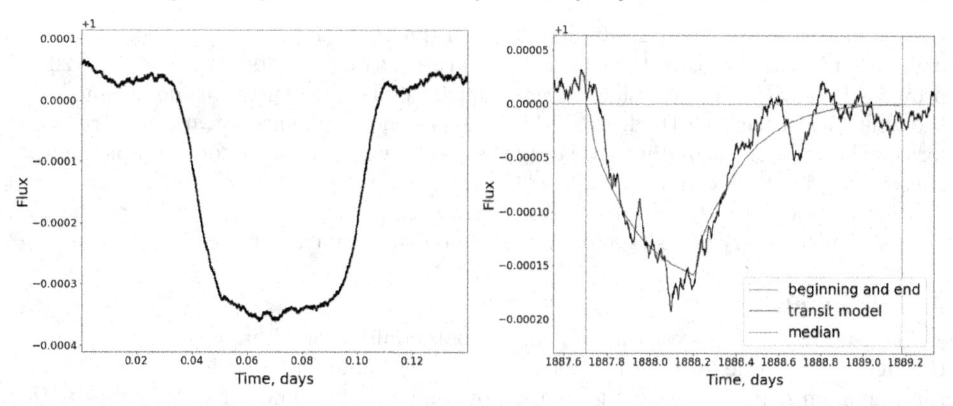

Figure 1. *Left:* The profile of the "55 Cnc e" planet transit, processed by our program "*Lc_cuter*". *Right:* The simulation of comet Hale-Bopp transit encapsulated into the observed light curve is marked by black solid line. The start and the end moments of the comet transit are marked by the red vertical lines; the exocomet transit model is marked by the blue curve; the star flux median calculated apart from the transit interval is marked by the green line.

2. Data analysis and results

More than 200,000 stars were selected as primary TESS mission targets and included in the TESS Input Catalogue to produce time sequences of the brightness measurements at every 2 min during the long enough observation sets in 27 days. To analyse so large amount of the data novel approaches should be used, such as ML techniques, which have been already successfully applied for searching for exoplanet transits (Shallue (2018), Malik (2020)). In order to find and identify the asymmetric decrease in the star brightness caused likely by a comet-like body and separate them from exoplanet transits we apply the classical ML Random Forest method, for which we constructed the classifier using the Python module Scikit-learn (Pedregosa (2011)). The two different samples of data were used to train the ML model: a) the light curve profiles caused by the already identified exoplanet transits and b) the simulated brightness profiles due to exocomet transits. The latter is calculated using the Monte-Carlo approach to model the exocomet dusty tail taking into account orbital characteristics of the exocomet and physical properties of particles that populated its dusty atmosphere at the given distance from the star (Korsun (2010)). We developed a program code "*Lc_cuter*", based on the Python package "lightkurve2.0" (Borucki (2010)) to process the short-cadence (2-min) TESS PDCSAP light curves and create the sample of exoplanet transits. The example of the brightness profiles from the two different data samples is presented in Fig. 1.

To generate the sample for the ML we chose about 6000 PDCSAP light curves which have no signs of transits. Then we selected ~50% of the light curves and artificially put planetary transits with random periods larger than 2 days. If the period is less than the observation sector base line, it means that the planet transit occurs more than once. There may also be a case when no planetary transit can be seen in the light curve for the certain sector. The remaining light curves were populated with the simulated cometary transits. The important step of the analysis is to evaluate the noise level which restrict the ability to detect a transit. As an indicator of the noise we used a simplified proxy algorithm to calculate CDPP (Combined Differential Photometric Precision) metric based on the Lightkurve 2.0 package. The package implements the Savitzky-Golay filter to remove frequency signals. To calculate the CDPP we applied a window of 1515 points for the Savitzky-Golay filter and transit duration of 6.5 hours. For the trained samples we used the initial light curves with the CDPP less than 30 ppm (part per million). We focused the ML on two options of classifying signals in the light curves: the presence or absence cometary transits. For this, the time series feature extraction ("tsfreash", Christ (2018)) provides features for the classifier. The significance of the features was determined with the Scikit-learn package (Pedregosa (2011)). The most significant features were tested for correlation with each other. To verify the results we divided the total sample into two subsamples, i.e. 80% training and the 20% test one. We found that the Random Forest method allows us to separate the light curves with accuracy of 97% (98% for the light curves with cometary transits and 95% with no cometary transits).

3. Brief conclusions

The ML Random Forest method allows potentially identify asymmetric minima in the light curves due to the transits of comet-like bodies with an accuracy of 97% if the minimum depth is larger than 0.02% of the star's flux and for light curves with a noise level corresponding to CDPP < 30. Since the detection of any signal becomes more difficult with increasing in the noise level, our result is the first step towards the ability to detect shallower comet transits in the light curves of poorer quality, using more advanced data processing methods, a wider set of features for the classifier, and more powerful deep neural networks.

Acknowledgements

This study was performed in the frames of the government funding program for institutions of the National Academy of Sciences of Ukraine (NASU) and supported by the National Research Foundation of Ukraine (№ 2020.02/0228). All data presented in this paper were obtained from the Mikulski Archive for Space Telescopes (MAST). STScI is operated by the Association of Universities for Research in Astronomy, Inc., under NASA contract NAS5-26555. Support for MAST for non-HST data is provided by the NASA Office of Space Science via grant NNX09AF08G and by other grants and contracts. This paper includes data collected by the TESS mission. We thank the referee for their helpful comments on an earlier version of this contribution.

References

Barentsen, G., Hedges, C. L., *et. al* 2019, *American Astronomical Society Meeting Abstracts*, 233, 109.08
Borucki, W. J., Koch, D., Basri, G., *et. al* 2010, *Science*, 327, 977
Christ, M., Braun, N., Neuffer, J., and Kempa-Liehr, A. W. 2018, *Neurocomputing*, 307, 72
Korsun, P. P., Kulyk, I. V., Ivanova, O. V., *et. al* 2019, *Icarus*, 210, 916
Lagrange, A. M., Rubini, P., Nowak, M., *et. al* 2020, *A&A*, 642, A18
Malik, A., Moster, B. P. and Obermeier, C., 2020, *arXiv e-prints*, arXiv:2011.14135
Pedregosa, F., Varoquaux G., Gramfort A., *et al* 2011, , 12, 2825
Rappaport, S., Vanderburg, A., Jacobs, T.,*et. al* 2018, *MNRAS*, 474, 1453
Ricker, G. R., Winn, J. N., Vanderspek, R., *et. al* 2015, *Journal of Astronomical Telescopes, Instruments, and Systems*, 1, id. 014003
Shallue, C. J. and Vanderburg, A., 2018, *AJ*, 155 (2), 94
Zieba, S., Zwintz, K., Kenworthy, M. A., *et. al* 2019, *A&A*, 625, L13

Author index